ECLECTIC EDUCATIONAL SERIES

RAY'S™

NEW HIGHER

ARITHMETIC

BY

JOSEPH RAY, M.D.

Originally published by
Van Antwerp, Bragg & Co.

This edition published by

Copyright
1880
By Van Antwerp, Bragg, & Co.

This edition copyright © 1985 by Mott Media ®
1130 Fenway Circle, Fenton, MIchigan 48430
www.mottmedia.com.

ISBN-10: 0-88062-063-3
ISBN-13: 978-0-88062-063-5
Printed by Dickinson Press, Inc., Grand Rapids, Michigan, USA
Batch #4350000; July 2014

PRESENT PUBLISHER'S PREFACE

We are honored and happy to bring to you the classic Arithmetics by Joseph Ray. In the 1800s these popular books sold more than any other arithmetics in America, in fact over 120,000,000 copies. Now with this reprinting, they are once again available for America's students.

Ray's Arithmetics are organized in an orderly manner around the discipline of arithmetic itself. They present principles and follow up each one with examples which include difficult problems to challenge the best students. Students who do not master a concept the first time can return to it later, work the more difficult problems, and master the concept. Thus in these compact volumes is a complete arithmetic course to study in school, to help in preparing for ACT and SAT tests, and to use for reference throughout a lifetime.

In order to capture the spirit of the original Ray's, we have refrained from revising the problems and prices. Only a few words have been changed, as we felt it wise. Thus students will have to rely on their arithmetic ability to solve the problems. Also the charm of a former era lives on in this reprinting. Flour and salt are sold by the barrel, kegs may contain tar, and postage stamps cost 3¢ each. Through this content, students learn social history of the 1800s in a unique, hands-on manner at the same time they are mastering arithmetic.

The series consists of four books ranging from Primary Arithmetic to Higher Arithmetic, as well as answer keys to accompany them. We have added a teacher's guide to help today's busy teachers and parents.

We wish to express our appreciation to the staff of the Special Collections Library at Miami University, Oxford, Ohio, for its cooperation in allowing us to use copies of their original Ray's Arithmetics.

George M. Mott, Founder
Mott Media

PREFACE

Ray's Higher Arithmetic was published nearly twenty-five years ago. Since its publication it has had a more extensive circulation than any other similar treatise issued in this country. To adapt it more perfectly to the wants of the present and future, it has been carefully revised.

It has been the aim of the revision to make Ray's New Higher Arithmetic thoroughly practical, useful, and teachable. To this end the greatest care has been given to securing concise definitions and explanations, and, at the same time, the systematic and thorough presentation of each subject. The pupil is taught to think for himself correctly, and to attain his results by the shortest and best methods. Special attention is given to modern business transactions, and all obsolete matter has been discarded.

Almost every chapter of the book has been entirely rewritten, without materially changing the general plan of the former edition, although much new, and some original matter has been introduced. Many of the original exercises are retained.

Particular attention is called to the rational treatment of the Arithmetical Signs, to the prominence given to the Metric System, and to the comprehensive, yet practical, presentation of Percentage and its various Applications. The method of combining the algebraic and geometric processes in explaining square and cube root will commend itself to teachers. The chapter on Mensuration is unusually full and varied, and contains a vast amount of useful information.

The Topical Outlines for Review will prove invaluable to both teachers and pupils in aiding them to analyze and to classify their arithmetical knowledge and to put it together so as to gain a comprehensive view of it as a whole.

Principles and Formulas are copiously interspersed as summaries, to enable pupils to work intelligently.

The work, owing to its practical character, logical exactness, and condensation of matter, will be found peculiarly adapted to the wants of classes in High Schools, Academies, Normal Schools, Commercial Schools and Colleges, as well as to private students.

The publishers take this opportunity of expressing their obligations to J. M. Greenwood, A.M., Superintendent of Public Schools, Kansas City, Mo., who had the work of revision in charge, and also to Rev. Dr. U. Jesse Knisely, of Newcomerstown, Ohio, for his valuable assistance in revising the final proof-sheets.

Cincinnati, July, 1880.

CONTENTS.

RAY'S

HIGHER ARITHMETIC

I. INTRODUCTION.

Article 1. A **definition** is a concise description of any object of thought, and must be of such a nature as to distinguish the object described from all other objects.

2. Quantity is any thing which can be increased or diminished; it embraces number and magnitude. Number answers the question, "How many?" Magnitude, "How much?"

3. Science is knowledge properly classified.

4. The primary truths of a science are called **Principles.**

5. Art is the practical application of a principle or the principles of science.

6. Mathematics is the science of quantity.

7. The elementary branches of mathematics are Arithmetic, Algebra, and Geometry.

8. Arithmetic is the introductory branch of the science of numbers. Arithmetic as a science is composed of definitions, principles, and processes of calculation; as an **art**, it teaches how to apply numbers to theoretical and practical purposes.

9. A **Proposition** is the statement of a principle, or of something proposed to be done.

10. Propositions are of two kinds, demonstrable and indemonstrable.

Demonstrable propositions can be proved by the aid of reason. *Indemonstrable propositions* can not be made simpler by any attempt at proof.

11. An **Axiom** is a self-evident truth.

12. A **Theorem** is a truth to be proved.

13. A **Problem** is a question proposed for solution.

14. Axioms, theorems, and problems are propositions.

15. A process of reasoning, proving the truth of a proposition, is called a **Demonstration.**

16. A **Solution** of a problem is an expressed statement showing how the result is obtained.

17. The term **Operation,** as used in this book, is applied to illustrations of solutions.

18. A **Rule** is a general direction for solving all problems of a particular kind.

19. A **Formula** is the expression of a general rule or principle in algebraic language; that is, by symbols.

20. A **Unit** is one thing, or one. *One thing* is a concrete unit; *one* is an abstract unit.

21. **Number** is the expression of a definite quantity. Numbers are either abstract or concrete. An *abstract number* is one in which the kind of unit is not named; a *concrete number* is one in which the kind of unit is named. Concrete numbers are also called *Denominate Numbers.*

22. Numbers are also divided into Integral, Fractional, and Mixed.

An *Integral number*, or *Integer*, is a whole number; a *Fractional number* is an expression for one or more of the equal parts of a divided whole; a *Mixed number* is an Integer and Fraction united.

23. A **Sign** is a character used to show a relation among numbers, or that an operation is to be performed.

24. The signs most used in Arithmetic are

$$+ \quad - \quad \times \quad \div \quad \surd$$

$$= \quad : \quad :: \quad (\,) \quad \overline{\qquad} \quad .\,.\,.\,. \quad \therefore$$

25. The *sign of Addition* is [+], and is called *plus.* The numbers between which it is placed are to be added. Thus, $3 + 5$ equals 8.

Plus is described as a perpendicular cross, in which the bisecting lines are equal.

26. The *sign of Subtraction* is [—], and is called *minus.* When placed between two numbers, the one that follows it is to be taken from the one that precedes it. Thus, $7 - 4$ equals 3.

Minus is described as a short horizontal line.

Plus and *Minus* are Latin words. *Plus* means *more; minus* means *less.*

Michael Steifel, a German mathematician, first introduced + and — in a work published in 1544.

27. The *sign of Multiplication* is [×], and is read *multiplied by*, or *times.* Thus, 4×5 is to be read, 4 *multiplied by* 5, or 4 *times* 5.

The sign is described as an oblique cross.

William Oughtred, an Englishman, born in 1573, first introduced the sign of multiplication.

28. The *sign of Division* is [÷], and is read *divided by.* When placed between two numbers, the one on the left is to be divided by the one on the right. Thus, $20 \div 4$ equals 5.

The sign is described as a short horizontal line and two dots; one dot directly above the middle of the line, and the other just beneath the middle of it.

Dr. John Pell, an English analyst, born in 1610, introduced the sign of division.

29. The *Radical sign*, [$\sqrt{\ }$], indicates that some root is to be found. Thus, $\sqrt{36}$ indicates that the *square root of* 36 is required; $\sqrt[3]{125}$, that the *cube root* of 125 is to be found; and $\sqrt[4]{625}$ indicates that the *fourth root* of 625 is to be extracted.

The root to be found is shown by the small figure placed between the branches of the Radical sign. The figure is called the *index*.

30. The signs, $+$, $-$, $\times$, $\div$, $\sqrt{\ }$, are symbols of operation.

31. The *sign of Equality* is [=], two short horizontal parallel lines, and is read *equals* or *is equal to*, and signifies that the quantities between which it is placed are equal. Thus, $3 + 5 = 9 - 1$. This is called an *equation*, because the quantity $3 + 5$ is equal to $9 - 1$.

32. Ratio is the relation which one number bears to another of the same kind. The *sign of Ratio* is [:]. Ratio is expressed thus, $6 : 3 = \frac{6}{3} = 2$, and is read, the ratio of 6 to $3 = 2$, or is 2.

The sign of ratio may be described as the sign of division with the line omitted. It has the same force as the sign of division and is used in place of it by the French.

33. Proportion is an equality of ratios. The *sign of Proportion* is [: :], and is used thus, $3 : 6 :: 4 : 8$; this may be read, 3 is to 6 *as* 4 is to 8; another reading, the ratio of 3 to 6 *is equal to* the ratio of 4 to 8.

34. The signs [( ), ——], are *signs of Aggregation*—the first is the *Parenthesis*, the second the *Vinculum*. They are used for the same purpose; thus, $24 - (8 + 7)$, or $24 - \overline{8 + 7}$, means that the sum of $8 + 7$ is to be subtracted from 24. The numbers within the parenthesis, or under the vinculum, are considered as one quantity.

35. The dots [. . . .], used to guide the eye from words at the left to the right, are called *Leaders*, or the *sign of Continuation*, and are read, *and so on*.

36. The *sign of Deduction* is [∴], and is read *therefore*, *hence*, or *consequently*.

37. The signs, =, :, ::, ( ), ——, ..., ∴, are symbols of relation.

38. Arithmetic depends upon this primary proposition: that any number may be increased or diminished. "Increased" comprehends Addition, Multiplication, and Involution; "decreased," Subtraction, Division, and Evolution.

39. The fundamental operations of Arithmetic in the order of their arrangement, are: Numeration and Notation, Addition, Subtraction, Multiplication, and Division.

Topical Outline.

INTRODUCTION.

1. **Definition.**
2. **Quantity.**
3. **Science.**
4. **Principles.**
5. **Art.**
6. **Mathematics.**
7. **Proposition.**
8. **Demonstration.**
9. **Solution.**
10. **Operation.**
11. **Rule.**
12. **Formula.**
13. **Unit.**
14. **Number.**
15. **Sign.**
16. **Signs most used.**
17. **Primary Proposition.**
18. **Fundamental Operations**

Topical Outline of Arithmetic.

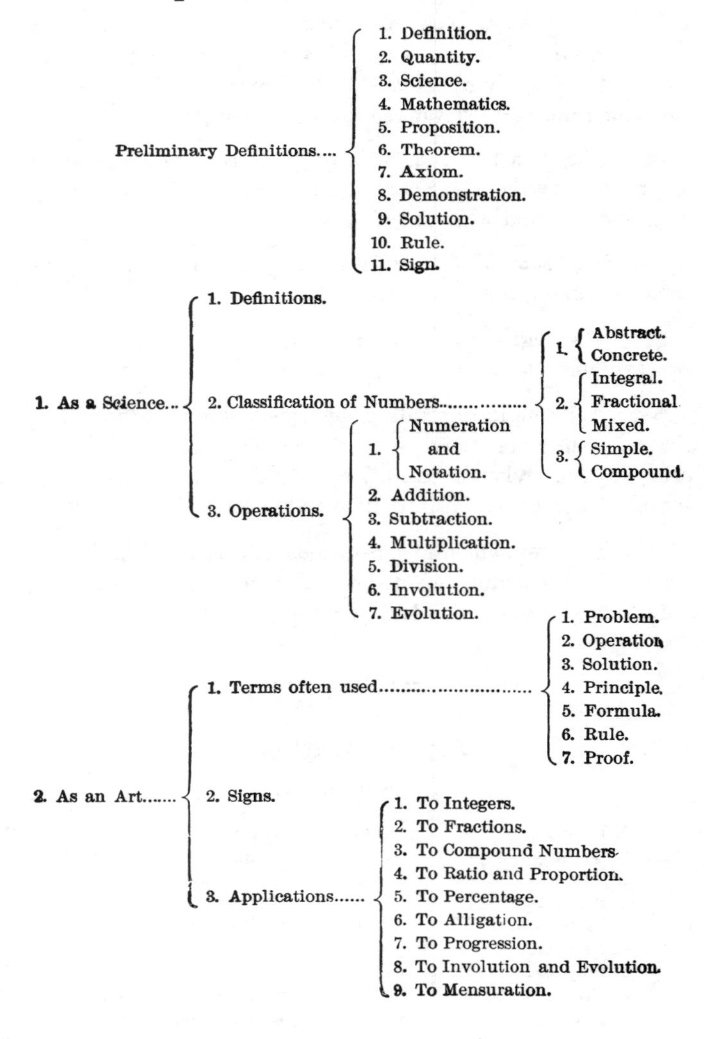

- Preliminary Definitions....
  1. Definition.
  2. Quantity.
  3. Science.
  4. Mathematics.
  5. Proposition.
  6. Theorem.
  7. Axiom.
  8. Demonstration.
  9. Solution.
  10. Rule.
  11. Sign.

1. As a Science...
   1. Definitions.
   2. Classification of Numbers..................
      1. Abstract. / Concrete.
      2. Integral. / Fractional. / Mixed.
      3. Simple. / Compound.
   3. Operations.
      1. Numeration and Notation.
      2. Addition.
      3. Subtraction.
      4. Multiplication.
      5. Division.
      6. Involution.
      7. Evolution.
2. As an Art.......
   1. Terms often used.............................
      1. Problem.
      2. Operation.
      3. Solution.
      4. Principle.
      5. Formula.
      6. Rule.
      7. Proof.
   2. Signs.
   3. Applications......
      1. To Integers.
      2. To Fractions.
      3. To Compound Numbers.
      4. To Ratio and Proportion.
      5. To Percentage.
      6. To Alligation.
      7. To Progression.
      8. To Involution and Evolution.
      9. To Mensuration.

II. NUMERATION AND NOTATION

40. Numeration is the method of reading numbers.

Notation is the method of writing numbers. Numbers are expressed in three ways; viz., by *words*, *letters*, and *figures*.

41. The first nine numbers are each represented by a single figure, thus:

1	2	3	4	5	6	7	8	9
one.	two.	three.	four.	five.	six.	seven.	eight.	nine.

All other numbers are represented by combinations of these and another figure, 0, called *zero*, *naught*, or *cipher*.

REMARK.—The cipher, 0, is used to indicate *no value*. The other figures are called *significant* figures, because they indicate *some value*.

42. The number next higher than 9 is named *ten*, and is written with two figures, thus, 10: in which the cipher, 0, merely serves to show that the unit, 1, on its left, is different from the unit, 1, standing alone, which represents a *single thing*, while this, 10, represents a *single group of ten things*.

The nine numbers succeeding *ten* are written and named as follows:

11	12	13	14	15	16
eleven.	twelve.	thirteen.	fourteen.	fifteen.	sixteen.

17	18	19
seventeen.	eighteen.	nineteen.

In each of these, the 1 on the left represents a *group of ten things*, while the figure on the right expresses the units or *single things* additional, required to make up the number.

REMARK.—The words *eleven* and *twelve* are supposed to be derived from the Saxon, meaning *one left after ten*, and *two left after ten*. The words *thirteen*, *fourteen*, etc., are contractions of *three and ten*, *four and ten*, etc.

The next number above nineteen (*nine* and *ten*), is *ten* and *ten*, or two *groups of ten*, written 20, and called *twenty*.

The next numbers are *twenty-one*, 21; *twenty-two*, 22; etc., up to *three tens*, or *thirty*, 30; *forty*, 40; *fifty*, 50; *sixty*, 60; *seventy*, 70; *eighty*, 80; *ninety*, 90.

The highest number that can be written with two figures is 99, called *ninety-nine;* that is, nine tens and nine units.

The next higher number is 9 tens and ten, or *ten tens*, which is called *one hundred*, and written with *three* figures, 100; in which the two ciphers merely show that the unit on their left is neither a single thing, 1, nor a group of ten things, 10, but a *group of ten tens*, being a unit of a higher order than either of those already known.

In like manner, 200, 300, etc., express *two hundreds, three hundreds*, and so on, up to *ten hundreds*, called a *thousand*, and written with *four* figures, 1000, being a unit of a still higher order.

43. The **Order** of a figure is the place it occupies in a number.

From what has been said, it is clear that a figure in the 1st place, with no others to the right of it, expresses *units* or single things; but standing on the left of another figure, that is, in the 2d place, expresses *groups of tens;* and standing at the left of two figures, or in the 3d place, expresses *tens of tens*, or *hundreds;* and in the 4th place, expresses *tens of hundreds* or *thousands*. Hence, counting from the right hand,

The order of *Units*	is in the 1st place,	1
The order of *Tens*	is in the 2d place,	10
The order of *Hundreds*	is in the 3d place,	100
The order of *Thousands*	is in the 4th place,	1000

By this arrangement, *the same figure has different values according to the place, or order*, in which it stands. Thus, 3 in the first place is 3 *units;* in the second place 3 *tens*, or thirty; in the third place 3 *hundreds;* and so on.

44. The word *Units* may be used in naming all the orders, as follows:

Simple units	are called	*Units of the* 1*st order.*
Tens	" "	*Units of the* 2*d order.*
Hundreds	" "	*Units of the* 3*d order.*
Thousands	" "	*Units of the* 4*th order.*
etc.		etc.

45. The following table shows the place and name of each order up to the fifteenth.

TABLE OF ORDERS.

15th.	14th.	13th.	12th.	11th.	10th.	9th.	8th.	7th.	6th.	5th.	4th.	3d.	2d.	1st.
Hundreds of Trillions	Tens of Trillions	Trillions	Hundreds of Billions	Tens of Billions	Billions	Hundreds of Millions	Tens of Millions	Millions	Hundreds of Thousands	Tens of Thousands	Thousands	Hundreds	Tens	Units

46. For convenience in reading and writing numbers, orders are divided into groups of three each, and each group is called a **period.** The following table shows the grouping of the first fifteen orders into five periods:

TABLE OF PERIODS.

of Trillions.			of Billions.			of Millions.			of Thousands.			of Units.		
Hundreds	Tens	Units	Hundreds	Tens	Units	Hundreds	Tens	Units	Hundreds	Tens	Units	Hundreds	Tens	Units
6	5	4	3	2	1	9	8	7	6	5	4	3	2	1
5th Period.			4th Period.			3d Period.			2d Period.			1st Period.		

47. It will be observed that each period is composed of units, tens, and hundreds of *the same denomination.*

48. List of the Periods, according to the common or French method of Numeration.

First	Period,	Units.	Sixth	Period,	Quadrillions.
Second	"	Thousands.	Seventh	"	Quintillions.
Third	"	Millions.	Eighth	"	Sextillions.
Fourth	"	Billions.	Ninth	"	Septillions.
Fifth	"	Trillions.	Tenth	"	Octillions.

The next twelve periods are, Nonillions, Decillions, Undecillions, Duodecillions, Tredecillions, Quatuordecillions, Quindecillions, Sexdecillions, Septendecillions, Octodecillions, Novendecillions, Vigintillions.

PRINCIPLES.—1. *Ten units of any order always make one of the next higher order.*

2. *Removing a significant figure one place to the left increases its value tenfold; one place to the right, decreases its value tenfold.*

3. *Vacant orders in a number are filled with ciphers.*

PROBLEM.—Express in words the number which is represented by 608921045.

SOLUTION.—The number, as divided into periods, is 608·921·045; and is read six hundred and eight million nine hundred and twenty-one thousand and forty-five.

Rule for Numeration.—1. *Begin at the right, and point the number into periods of three figures each.*

2. *Commence at the left, and read in succession each period with its name.*

REMARK.—Numbers may also be read by merely *naming each figure with the name of the place in which it stands.* This method, however, is rarely used except in teaching beginners. Thus, the numbers expressed by the figures 205, may be read *two hundred and five,* or *two* hundreds *no* tens and *five* units.

EXAMPLES IN NUMERATION.

7	4053	204026	4300201
40	7009	500050	29347283
85	12345	730003	45004024
278	70500	1375482	343827544
1345	165247	6030564	830070320

832045682327825000000321
8007006005004003002001000000
60030020090080070050060030070
504030209102800703240703250207

PROBLEM.—Express in figures the number four million twenty thousand three hundred and seven. 4020307.

SOLUTION.—Write 4 in *millions* period; place a dot after it to separate it from the next period: then write 20 in *thousands* period; place another dot: then write 307 in *units* period. This gives 4·20·307. As there are but *two* places in the thousands period, a cipher must be put before 20 to complete its orders, and the number correctly written, is 4020307.

NOTE.—Every period, except the highest, must have *three* figures; and if any period is not mentioned in the given number, supply its place with three ciphers.

Rule for Notation.—*Begin at the left, and write each period in its proper place—filling the vacant orders with ciphers.*

PROOF.—Apply to the number, as written, the rule for Numeration, and see if it agrees with the number given.

EXAMPLES IN NOTATION.

1. Seventy-five.
2. One hundred and thirty-four.
3. Two hundred and four.
4. Three hundred and seventy.
5. One thousand two hundred and thirty-four.
6. Nine thousand and seven.
7. Forty thousand five hundred and sixty-three.
8. Ninety thousand and nine.
9. Two hundred and seven thousand four hundred and one.

10. Six hundred and forty thousand and forty.
11. Seven hundred thousand and seven.
12. One million four hundred and twenty-one thousand six hundred and eighty-five.
13. Seven million and seventy.
14. Ten million one hundred thousand and ten.
15. Sixty million seven hundred and five thousand.
16. Eight hundred and seven million forty thousand and thirty-one.
17. Two billion and twenty million.
18. Nineteen quadrillion twenty trillion and five hundred billion.
19. Ten quadrillion four hundred and three trillion ninety billion and six hundred million.
20. Eighty octillion sixty sextillion three hundred and twenty-five quintillion and thirty-three billion.
21. Nine hundred decillion seventy nonillion six octillion forty septillion fifty quadrillion two hundred and four trillion ten million forty thousand and sixty.

English Method of Numeration.

49. In the **English Method** of Numeration six figures make a period. The first period is *units*, the second *millions*, the third *billions*, the fourth *trillions*, etc.

The following table illustrates this method:

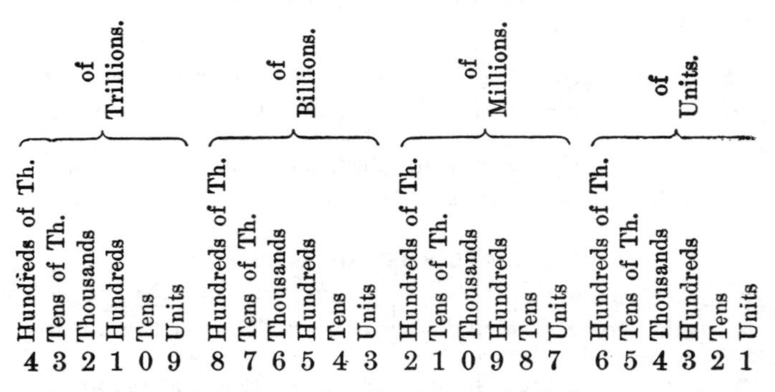

By this system the twelve figures at the right are read, two hundred and ten thousand nine hundred and eighty-seven million six hundred and fifty-four thousand three hundred and twenty-one. By the French method they would be read, two hundred and ten billion nine hundred and eighty-seven million six hundred and fifty-four thousand three hundred and twenty-one.

Roman Notation.

50. In the Roman Notation, numbers are represented by seven *letters*. The letter I represents *one*; V, *five*; X, *ten*; L, *fifty*; C, *one hundred*; D, *five hundred*; and M, *one thousand*. The other numbers are represented according to the following principles:

1st. Every time a letter is repeated, its value is repeated. Thus, II denotes *two*; XX denotes *twenty*.

2d. Where a letter of less value is placed *before* one of greater value, the less is taken from the greater; thus, IV denotes *four*.

3d. Where a letter of less value is placed *after* one of greater value, the less is added to the greater; thus, XI denotes *eleven*.

4th. Where a letter of less value stands *between* two letters of greater value, it is taken from the following letter, not added to the preceding; thus, XIV denotes *fourteen*, not *sixteen*.

5th. A bar [—] placed over a letter increases its value a *thousand times*. Thus, $\overline{\text{V}}$ denotes *five thousand*; $\overline{\text{M}}$ denotes *one million*.

Roman Table.

I	One.	XI	Eleven.
II	Two.	XIV	Fourteen.
III	Three.	XV	Fifteen.
IV	Four.	XVI	Sixteen.
V	Five.	XIX	Nineteen.
VI	Six.	XX	Twenty.
IX	Nine.	XXI	Twenty-one.
X	Ten.	XXX	Thirty.

ROMAN TABLE. (CONTINUED.)

XL	Forty.	DC	Six hundred.
L	Fifty.	DCC	Seven hundred.
LX	Sixty.	DCCC	Eight hundred.
XC	Ninety.	DCCCC	Nine hundred.
C	One hundred.	M	One thousand.
CCCC	Four hundred.	MM	Two thousand.
D	Five hundred.	MDCCCLXXXI	1881.

Topical Outline.

NUMERATION AND NOTATION.

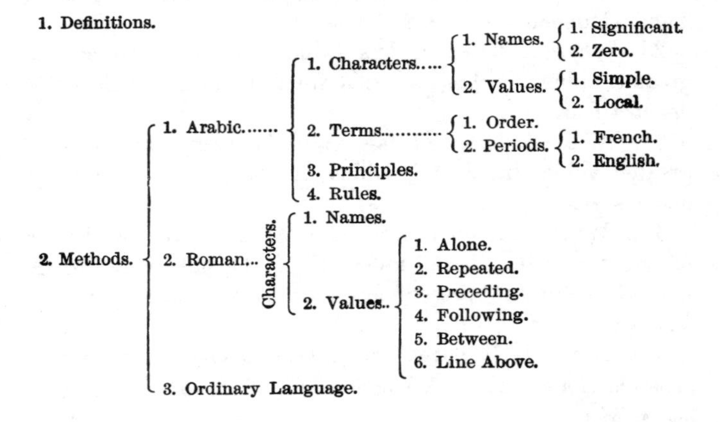

III. ADDITION.

51. Addition is the process of uniting two or more like numbers into one equivalent number.

Sum or **Amount** is the result of Addition.

52. Since a number is a collection of units of the *same* kind, *two or more numbers can be united into one sum, only when their units are of the same kind. Two apples* and 3 *apples* are 5 *apples;* but 2 *apples* and 3 *peaches* can not be united into *one* number, either of *apples* or of *peaches.*

Nevertheless, numbers of different names may be added together, if they can be brought under a *common denomination.*

PRINCIPLES.—1. *Only like numbers can be added.*

2. *The sum is equal to all the units of all the parts.*

3. *The sum is the same in kind as the numbers added.*

4. *Units of the same order, and only such, can be added directly.*

5. *The sum is the same in whatever succession the numbers are added.*

REMARK.—*Like numbers, similar numbers,* and *numbers of the same kind* are those having the same unit.

PROBLEM.—What is the sum of 639, 82, and 543?

SOLUTION.—Writing the numbers as in the margin, say, 3, 5, 14 *units,* which are 1 ten and 4 units. Write the 4 units beneath, and carry the 1 ten to the next column. Then 1, 5, 13, 16 *tens,* which are 6 tens to be written beneath, and 1 hundred to be carried to the next column. Lastly, 1, 6, 12 *hundreds,* which is set beneath, there being no other columns to carry to or add.

OPERATION
```
 639
  82
 543
----
1264
```

DEMONSTRATION.—1. Figures of the same order are written in the same column for *convenience*, since none but units of the same name can be added. (Art. 52.)

2. Commence at the right to add, so that when the sum of any column is greater than nine, the *tens* may be carried to the next column, and, thereby, units of the same name added together.

3. Carry *one* for every ten, since ten units of each order make one unit of the order next higher.

Rule for Adding Simple Numbers.—1. *Write the numbers to be added, so that figures of the same order may stand in the same column, and draw a line directly beneath.*

2. *Begin at the right and add each column separately, writing the units under the column added, and carrying the tens, if any, to the next column. At the last column write the last whole amount.*

METHODS OF PROOF.—1. Add the figures *downward* instead of *upward*; or

2. Separate the numbers into two or more divisions; find the sum of the numbers in each division, and then add the several sums together; or

3. Commence at the left; add each column separately; place each sum under that previously obtained, but extending one figure further to the right, and then add them together.

In each of these methods the result should be the same as when the numbers are added upward.

NOTE.—For proof by casting out the 9's, see Art. 105.

EXAMPLES FOR PRACTICE.

Find the sum,

1. Of 76767; 7654; 50121; 775.
2. Of 97674; 686; 7676; 9017.
3. Of 971; 7430; 97476; 76734.
4. Of 999; 3400; 73; 47; 452; 11000; 193; 97; 9903; 42; and 5100.

5. Of four hundred and three; 5025; sixty thousand and seven; eighty-seven thousand; two thousand and ninety; and 100.

6. Of 20050; three hundred and seventy thousand two hundred; four million and five; two million ninety thousand seven hundred and eighty; one hundred thousand and seventy; 98002; seven million five thousand and one; and 70070.

7. Of 609505; 90070; 90300420; 9890655; 789; 37599; 19962401; 5278; 2109350; 41236; 722; 8764; 29753; and 370247.

8. Of two hundred thousand two hundred; three hundred million six thousand and thirty; seventy million seventy thousand and seventy; nine hundred and four million nine thousand and forty; eighty thousand; ninety million nine thousand; six hundred thousand and sixty; five thousand seven hundred.

In each of the 7 following examples, find the sum of the consecutive numbers from A to B, including these numbers:

	A.	B.
9.	119	131.
10.	987	1001.
11.	3267	3281.
12.	4197	4211.
13.	5397	5416.
14.	7815	7831.
15.	31989	32028.

16. Paid for coffee, $375; for tea, $280; for sugar, $564; for molasses, $119; and for spices, $75: what did the whole amount to?

17. I bought three pieces of cloth: the first cost $87; the second, $25 more than the first; and the third, $47 more than the second: what did all cost?

18. A man bought three bales of cotton. The first cost

$325; the second cost $16 more than the first; and the third, as much as both the others: what sum was paid for the three bales?

19. A has $75; B has $19 more than A; C has as much as A and B, and $23 more; and D has as much as A, B, and C together: what sum do they all possess?

SUGGESTIONS.—Two things are of the greatest importance in arithmetical operations,—absolute accuracy and rapidity. The figures should always be plain and legible. Frequent exercises in adding long columns of figures are recommended; also, practice in grouping numbers at sight into *tens* and *twenties* is a useful exercise.

Accountants usually resort to artifices in Addition to save time and extra labor, such as writing the number to be carried in small figures beneath the column to which it belongs, also writing the whole amount of each column separately, etc.

Topical Outline.

ADDITION.

1. Definitions.
2. Sign.
3. Principles.
4. Operation..
   1. Writing the Numbers.
   2. Drawing Line Beneath.
   3. Adding, Reducing, etc.
5. Rule.
6. Methods of Proof.

IV. SUBTRACTION.

53. **Subtraction** is the process of finding the difference between two numbers of the same kind.

The larger number is the *Minuend;* the less, the *Subtrahend;* the number left, the *Difference* or *Remainder.*

Minuend means *to be diminished*; subtrahend, *to be subtracted.*

54. Subtraction is the reverse of Addition, and since none but numbers of the same kind can be added together (Art. **52**), it follows, therefore, that a number can be subtracted only from another of the same kind: 2 *cents* can not be taken from 5 *apples,* nor 3 *cows* from 8 *horses.*

PRINCIPLES.—1. *The minuend and subtrahend must be of the same kind.*

2. *The difference is the same in kind as the minuend and subtrahend.*

3. *The difference equals the minuend minus the subtrahend*

4. *The minuend equals the difference plus the subtrahend.*

5. *The subtrahend equals the minuend minus the difference.*

PROBLEM.—From 827 dollars take 534 dollars.

OPERATION.
$827
534
$293 *Rem.*

SOLUTION.—After writing figures of the same order in the same column, say, 4 units from 7 units leave 3 units. Then, as 3 tens can not be taken from 2 tens, add 10 tens to the 2 tens, which make 12 tens, and 3 tens from 12 tens leave 9 tens. To compensate for the 10 tens added to the 2 tens, add one hundred (10 tens) to the 5 hundreds, and say, 6 hundreds from 8 hundreds leaves 2 hundreds; and the whole remainder is 2 hundreds 9 tens and 3 units, or 293.

DEMONSTRATION.—1. Since a number can be subtracted only from another of the same kind (Art. **51**), figures of the same name are placed in the same column to be convenient to each other, the less number being placed below as a matter of custom.

2. Commence at the right to subtract, so that if any figure is greater than the one above it, the upper may be increased by 10, and the next in the subtrahend increased by 1, or, as some prefer, the next in the minuend decreased by 1.

Rule for Subtracting Simple Numbers.—1. *Write the less number under the greater, placing units under units, tens under tens, etc., and draw a line directly beneath.*

2. *Begin at the right, subtract each figure from the one above it, placing the remainder beneath.*

3. *If any figure exceeds the one above it, add ten to the upper, subtract the lower from the sum, increase by* 1 *the units of the next order in the subtrahend, and proceed as before.*

PROOF.—Add the remainder to the less number. If the work be correct, the sum will be equal to the greater.

NOTE.—For proof by casting out the 9's, see Art. **105**.

EXAMPLES FOR PRACTICE.

1. From 30020037 take 50009.

REMARK.—When there are no figures in the lower number to correspond with those in the upper, consider the vacant places occupied by zeros.

OPERATION.
```
30020037
   50009
--------
29970028 Rem.
```

2. From 79685 take 30253.
3. From 1145906 take 39876.
4. From 2900000 take 777888.
5. From 71086540 take 64179730.
6. From 101067800 take 100259063.
7. How many years from the discovery of America in 1492, to the Declaration in 1776?
8. A farm that cost $7253, was sold at a loss of $395: for how much was it sold?

9. The difference of two numbers is 19034, and the greater is 75421: what is the less?

10. How many times can the number 285 be subtracted from 1425?

11. Which is the nearer number to 920736; 1816045 or 25427?

12. From a tract of land containing 10000 acres, the owner sold to A 4750 acres; and to B 875 acres less than A: how many acres had he left?

13. A, B, C, D, are 4 places in order in a straight line. From A to D is 1463 miles; from A to C, 728 miles; and from B to D, 1317 miles. How far is it from A to B, from B to C, and from C to D?

BUSINESS TERMS AND EXPLANATIONS.

55. Book-keeping is the science and art of recording business transactions.

56. Business records are called **Accounts,** and are kept in books called Account Books. The books mostly used are the *Day-book* and *Ledger.* In the Day-book are recorded the daily transactions in business, and the Ledger is used to classify and arrange the results of all transactions under distinct heads.

57. Each account has two sides: **Dr.**—Debits, and **Cr.**—Credits. Sums a person *owes* are his Debits; sums *owing to him* are his Credits. The difference between the sum of the Debits and the sum of the Credits, is called the **Balance.** Debits are preceded by "To," and Credits by "By."

58. Finding the difference between the sum of the Debits and the sum of the Credits, and writing it under the less side as Balance, is called **Balancing.**

Practical Exercises.

Dr.		JAMES CRAIG.			**Cr.**
1878.			1878.		
July 4	To Merchandise . .	560 50	July 5	By Cash	550 50
„ 6	„ Interest	24 90	„ 11	„ Bills Payable . .	890 70
„ 10	„ Sundries . . .	870 60	„ 26	„ Sundries . . .	310 80
„ 25	„ Merchandise . .	320 10	„ 31	„ Cash	100 00
„ 31	„ Ditto . . .	125 40		" Balance	49 50
1878.		1901 50			1901 50
Aug. 1	To Balance	49 50			

Balance the following account:

Dr.		THOMAS BALDWIN.			**Cr.**
1879.			1879.		
Jan. 3	To Merchandise . .	810 30	Jan. 7	By Sundries . . .	1000 00
„ 16	„ Sundries . . .	580 20	„ 20	„ Cash	300 00
„ 25	„ Cash	381 25	„ 31	„ Merchandise . .	225 20
„ 31	„ Merchandise . .	60 75		„ Balance	
1879.					
Feb. 1	To Balance				

Topical Outline.

Subtraction.

1. **Definition.**
2. **Terms**..............
   1. Minuend.
   2. Subtrahend.
   3. Difference or Remainder.
3. **Sign.**
4. **Principles.**
5. **Operation**........
   1. Writing the Numbers.
   2. Drawing Line Beneath.
   3. Subtracting and Writing Difference.
6. **Rule.**
7. **Proof.**
8. **Applications.**

V. MULTIPLICATION.

59. 1. **Multiplication** is taking one number as many times as there are units in another; or,

2. Multiplication is a short method of adding numbers that are equal.

60. The number to be taken, is called the *Multiplicand;* the other number, the *Multiplier;* and the result obtained, the *Product.* The Multiplicand and Multiplier are together called *Factors* (makers), because they *make* the Product.

PROBLEM.—How many trees in 3 rows, each containing 42 trees?

SOLUTION.—Since 3 rows contain 3 *times* as many trees as *one* row, take 42 *three* times. This may be done by writing 42 three times, and then *adding.* This gives 126 trees for the whole number of trees.

Instead, however, of writing 42 *three* times, write it *once;* then placing under it the figure 3, the *number of times* it is to be taken, say, 3 times 2 are 6, and 3 times 4 are 12. This process is *Multiplication.*

OPERATION.

First row,	42 trees.
Second row,	42 trees.
Third row,	42 trees.
	126 trees.

42 trees.
3
126 trees.

PRINCIPLES.—1. *The multiplicand may be either concrete or abstract.*

2. *The multiplier must always be an abstract number.*

3. *The product is the same in kind as the multiplicand.*

4. *The product is the same, whichever factor is taken as the multiplier.*

5. *The partial products are the same in kind as the multiplicand.*

6. *The sum of the partial products is equal to the total product.*

MULTIPLICATION TABLE.

1	2	3	4	5	6	7	8	9	10	11	12	13	14	15	16	17	18	19	20
2	4	6	8	10	12	14	16	18	20	22	24	26	28	30	32	34	36	38	40
3	6	9	12	15	18	21	24	27	30	33	36	39	42	45	48	51	54	57	60
4	8	12	16	20	24	28	32	36	40	44	48	52	56	60	64	68	72	76	80
5	10	15	20	25	30	35	40	45	50	55	60	65	70	75	80	85	90	95	100
6	12	18	24	30	36	42	48	54	60	66	72	78	84	90	96	102	108	114	120
7	14	21	28	35	42	49	56	63	70	77	84	91	98	105	112	119	126	133	140
8	16	24	32	40	48	56	64	72	80	88	96	104	112	120	128	136	144	152	160
9	18	27	36	45	54	63	72	81	90	99	108	117	126	135	144	153	162	171	180
10	20	30	40	50	60	70	80	90	100	110	120	130	140	150	160	170	180	190	200
11	22	33	44	55	66	77	88	99	110	121	132	143	154	165	176	187	198	209	220
12	24	36	48	60	72	84	96	108	120	132	144	156	168	180	192	204	216	228	240
13	26	39	52	65	78	91	104	117	130	143	156	169	182	195	208	221	234	247	260
14	28	42	56	70	84	98	112	126	140	154	168	182	196	210	224	238	252	266	280
15	30	45	60	75	90	105	120	135	150	165	180	195	210	225	240	255	270	285	300
16	32	48	64	80	96	112	128	144	160	176	192	208	224	240	256	272	288	304	320
17	34	51	68	85	102	119	136	153	170	187	204	221	238	255	272	289	306	323	340
18	36	54	72	90	108	126	144	162	180	198	216	234	252	270	288	306	324	342	360
19	38	57	76	95	114	133	152	171	190	209	228	247	266	285	304	323	342	361	380
20	40	60	80	100	120	140	160	180	200	220	240	260	280	300	320	340	360	380	400

61. Multiplication is divided into two cases:

1. *When the multiplier does not exceed* 12.
2. *When the multiplier exceeds* 12.

CASE I.

62. When the multiplier does not exceed 12.

PROBLEM.—At the rate of 53 miles an hour, how far will a railroad car run in four hours?

SOLUTION.—Here say, 4 times 3 (units) are 12 (units); write the 2 in units' place, and carry the 1 (ten); then, 4 times 5 are 20, and 1 carried makes 21 (tens), and the work is complete.

OPERATION.

```
 5 3 miles.
   4
 -----
2 1 2 miles.
```

DEMONSTRATION.—The multiplier being written under the multiplicand for convenience, begin with units, so that if the product should contain tens, they may be carried to the tens; and so on for each successive order.

Since every figure of the multiplicand is multiplied, therefore, the *whole* multiplicand is multiplied.

Rule.—1. *Write the multiplicand, and place the multiplier under it, so that units of the same order shall stand in the same column, and draw a line beneath.*

2. *Begin with units; multiply each figure of the multiplicand by the multiplier, carrying as in Addition.*

PROOF.—Separate the multiplier into any two parts; multiply by these separately. The sum of the products must be equal to the first product.

EXAMPLES FOR PRACTICE.

1. 195×3.
2. 3823×4.
3. 8765×5.
4. 98374×6.
5. 64382×7.
6. 58765×8.
7. 837941×9.
8. 645703×10
9. 407649×11.

10. If 4 men can perform a certain piece of work in 15 days, how long will it require 1 man?

SOLUTION.—One man must work four times as long as four men.
4×15 days $= 60$ days.

11. How many pages in a half-dozen books, each containing 336 pages?

12. How far can an ocean steamer travel in a week, at the rate of 245 miles a day?

13. What is the yearly expense of a cotton-mill, if $32053 are paid out every month?

14. A receives from his business an average of $45 a day. He pays three clerks $3; three, $9; and three, $12 a week; other expenses amount to $4 a day; what are his profits for one week?

CASE II.

63. When the multiplier exceeds 12.

PROBLEM.—Multiply 246 by 235.

SOLUTION.—First multiply by 5 (units), and place the first figure of the product, 1230, under the 5 (units). Then multiply by 3 (tens), and place the first figure of the product, 738, under the 3 (tens). Lastly, multiply by 2 (hundreds), and place the first figure of the product, 492, under the 2 (hundreds). Then add these several products for the entire product.

OPERATION

```
  246
  235
 1230 product by   5
 738  product by  30
492   product by 200
57810 product by 235
```

DEMONSTRATION.—The 0 of the first product, 1230, is *units* (Art. 62). The 8 of the second product, 738, is *tens*, because 3 (tens) times 6 = 6 times 3 (tens) = 18 (tens); giving 8 (tens) to be written in the tens' column. The 2 of the third product, 492, is hundreds, because 2 (hundreds) times 6 = 6 times 2 (hundreds) = 12 (hundreds), giving 2 (hundreds) to be written in the hundreds' column. The right-hand figure of each product being in its proper column, the other figures will fall in their proper columns; and each line being the product of the multiplicand by a *part* of the multiplier, their sum will be the product by *all the parts* or the *whole* of the multiplier.

Rule.—1. *Write the multiplier under the multiplicand, placing figures of the same order in the same column, and draw a line beneath.*

2. *Multiply each figure of the multiplicand by each figure of the multiplier successively; first by the units' figure, then by the tens' figure, etc.; placing the right-hand figure of each product under that figure of the multiplier which produces it, then draw a line beneath.*

3. *Add the several partial products together; their sum will be the required product.*

METHODS OF PROOF.—1. Multiply the multiplier by the multiplicand; this product must be the same as the first product.

2. The same as when the multiplier does not exceed 12.

NOTE.—For proof by casting out the 9's, see Art. **105**.

REMARK.—Although it is customary to use the figures of the multiplier in regular order beginning with units, it will give the same product to use them in any order, observing that *the right-hand figure of each partial product must be placed under the figure of the multiplier which produces it.*

OPERATION.

```
  246
  235
 ----
  738   product by  30
 492    product by 200
 1230   product by   5
-----              ---
57810   product by 235
```

EXAMPLES FOR PRACTICE.

1. 7198 × 216.
2. 8862 × 189.
3. 7575 × 7575.
4. 15607 × 3094.
5. 93186 × 4455.
6. 135790 × 24680.
7. 3523725 × 2583.
8. 4687319 × 1987.
9. 9264397 × 9584.
10. 9507340 × 7071.
11. 1644405 × 7749.
12. 1389294 × 8900.
13. 2778588 × 9867.
14. 204265 × 562402.

PRACTICAL PROBLEMS.

1. In a mile are 63360 inches: how many inches are there in the circumference of the earth at the equator if the distance be 25000 miles?

2. The flow of the Mississippi at Memphis is about 434000 cubic feet a second: required the weight of water passing that point in one day of 86400 seconds, if a cubic foot of water weigh 62 pounds?

3. John Sexton sold 25625 bushels of wheat, at $1.20 a bushel, and received in payment 320 acres of land, valued at $50 an acre; 60 head of horses, valued at $65 a head; 10 town lots, worth $150 each; and the remainder in money: how much money did he receive?

4. If light comes from the sun to the earth in 495 seconds, what is the distance from the earth to the sun light moving 192500 miles a second?

5. If 3702754400 cubic feet of solid matter is deposited in the Gulf of Mexico by the Mississippi every year, what is the deposit for 6000 years?

6. The area of Missouri is 65350 square miles: how many acres are there in the State, allowing 640 acres to each square mile?

7. In the United States, at the close of 1878, there were 81841 miles of railroad: if the average cost of building be $50000 a mile, what has been the total cost of building the railroads in this country?

8. The number of pounds of tobacco produced in this country in 1870 was 260000000. If this were manufactured into plugs one inch wide and six inches long, and four plugs weigh a pound, what would be the length in inches of the entire crop?

BUSINESS TERMS AND EXPLANATIONS.

64. A **Bill** is an account of goods sold or delivered, services rendered, or work done. Usually the price or value is annexed to each article, and the date of purchase given.

It is customary to write the total amount off to the right, and not directly under the column of amounts added.

65. A **Receipt** is a written acknowledgment of payment. The common form consists in signing the name after the words "Received Payment" written at the foot of the bill.

1. Joseph Allen bought of Seth Ward, at Springfield, Ill., Jan. 2, 1879, 30 barrels of flour, at $3.60 a barrel; 48 barrels of mess pork, at $16.25 a barrel; 16 boxes of candles, at $3.50 a box; 23 barrels of molasses, at $28.75 a barrel; and 64 sacks of coffee, at $47.50 a sack. Place the purchases in bill form.

SOLUTION.

SPRINGFIELD, ILL., *Jan.* 2, 1879.

JOSEPH ALLEN,

Bought of SETH WARD.

1879.						
Jan.	2	To 30 bl. flour, @ $ 3.60 a bl.	108	00		
,,	2	,, 48 ,, mess pork, ,, 16.25 ,,	780	00		
,,	2	,, 16 boxes candles, ,, 3.50 ,, box	56	00		
,,	2	,, 23 bl. molasses, ,, 28.75 ,, bl.	661	25		
,,	2	,, 64 sacks coffee, ,, 47.50 ,, sack	3040	00		
					$4645	25

2. At St. Louis, March 1, 1879, Chester Snyder bought of Thomas Glenn, 4 lb. of tea, at 40 ct.; 21 lb. of butter, at 21 ct.; 58 lb. of bacon, at 13 ct.; 16 lb. of lard, at 9 ct.; 30 lb. of cheese, at 12 ct.; 4 lb. of raisins, at 20 ct.; and 9 doz. of eggs, at 15 ct. Place these purchases in the form of a receipted bill?

66. A **Statement of Account** is a written form rendered to a customer, showing his debits and credits as they appear on the books. The following is an example:

CINCINNATI, *Feb.* 2, 1880.

JOHN SMITH,

In Account with VAN ANTWERP, BRAGG & CO.

1880.						
Jan.	2	To 525 McGuffey's Revised First Readers, @ 16c.	84			
,,	10	,, 50 Ray's New Higher Arithmetics, ,, 75c.	37	50		
		Cr.			121	50
,,	29	By Cash	20			
,,	31	,, Merchandise	12	75	32	75
					$88	75

3. James Wilson & Co. bought of the Alleghany Coal Co., March 2, 1880, five hundred tons of coal, at $2.75 a ton, and sold the same Company during the month, as follows: March 3d, 14 barrels of flour, at $6.55 a barrel; March 10th, 6123 pounds of sugar, at 8c. a pound; they also paid them on account, on March 15th, cash, $687.50. Make out a statement of account in behalf of the Alleghany Coal Co. under date of April 1, 1880.

CONTRACTIONS IN MULTIPLICATION.

CASE I.

67. When the multiplier is a composite number.

A **Composite Number** is the product of two or more whole numbers, each greater than 1, called its *factors*. Thus, 10 is a composite number, whose factors are 2 and 5; and 30 is one whose factors are 2, 3, and 5.

PROBLEM.—At 7 cents a piece, what will 6 melons cost?

ANALYSIS.—Three times 2 times are 6 times. Hence, it is the same to take 2 times 7, and then take this product 3 times, as to take 6 times 7. The same may be shown of any other composite number.

OPERATION.

```
 7 cents, cost of 1 melon.
 2
--
14 cents, cost of 2 melons.
 3
--
42 cents, cost of 6 melons.
```

Rule.—*Separate the multiplier into two or more factors. Multiply first by one of the factors, then this product by another factor, and so on till each factor has been used as a multiplier. The last product will be the result required.*

EXAMPLES FOR PRACTICE.

1. At the rate of 37 miles a day, how far will a man walk in 28 days?

2. Sound moves about 1130 feet per second: how far will it move in 54 seconds?

3. If an engine travel at an average speed of 25 miles an hour, how far can it travel in a week, or 168 hours?

CASE II.

68. When the multiplier is 1 with ciphers annexed, as 10, 100, 1000, etc.

DEMONSTRATION.—By the principles of Notation (Art. 43), placing *one* cipher on the right of a number, changes the units into tens, the tens into hundreds, and so on, and, therefore, *multiplies the number by* 10.

Annexing *two* ciphers to a number changes the units into hundreds, the tens into thousands, and so on, and, therefore, *multiplies the number by* 100. Annexing *three* ciphers multiplies the number by 1000, etc.

Rule.—*Annex to the multiplicand as many ciphers as there are in the multiplier; the result will be the required product.*

EXAMPLES FOR PRACTICE.

1. Multiply 743 by 10.
2. Multiply 375 by 100.
3. Multiply 207 by 1000.

CASE III.

69. When ciphers are on the right in one or both factors.

PROBLEM.—Find the product of 5400 by 130.

SOLUTION.—Find the product of 54 by 13, and then annex *three* ciphers; that is, as many as there are on the right in both the factors.

OPERATION

```
  5400
   130
  ----
  162
  54
------
702000
```

ANALYSIS.—Since 13 times 54 = 702, it follows that 13 times 54 hundreds (5400) = 702 hundreds (70200); and 130 times 5400 = 10 times 13 times 5400 = 10 times 70200 = 702000.

Rule.—*Multiply as if there were no ciphers on the right in the numbers; then annex to the product as many ciphers as there are on the right in both the factors.*

EXAMPLES FOR PRACTICE.

1. 15460 × 3200.
2. 30700 × 5904000.

CASE IV.

70. When the multiplier is a little less or a little greater than 10, 100, 1000, etc.

PROBLEM.—Multiply 3046 by 997.

ANALYSIS.—Since 997 is equal to 1000 diminished by 3, to multiply by it is the same as to multiply by 1000 (that is, to annex 3 ciphers) and by 3, and take the difference of the products; and the same can be shown in any similar case.

OPERATION.

```
   3046
    997
-------
3046000
   9138
-------
3036862
```

NOTE.—Where the number is a little *greater* than 10, 100, 1000, etc., the two products must be *added.*

Rule.—*Annex to the multiplicand as many ciphers as there are in the* 100, 1000, *etc., approximating the multiplier; multiply the multiplicand by the difference between the multiplier and* 100, 1000, *etc., and add or subtract the smaller result as the multiplier is greater or less than* 100, 1000, *etc.*

EXAMPLES FOR PRACTICE.

1. 7023 × 99.
2. 16642 × 996.
3. 372051 × 1002.

CASE V.

71. When one part taken as units, in the multiplier, is a factor of another part so taken.

PROBLEM.—Multiply 387295 by 216324.

SOLUTION.—Commence with the 3 of the multiplier, and obtain the first partial product, 1161885; then multiply this product by 8, which gives the product of the multiplicand by 24 at once (since 8 times 3 times any number make 24 times it). Set the right-hand figure under the right-hand figure 4 of the multiplier in use. Multiply the second partial product by 9, which gives the product of the multiplicand by 216 (since 9 times 24 times a number make 216 times that number). Set the right-hand figure of this partial product under the 6 of the multiplicand; and, finally, add to obtain the total product.

OPERATION.

```
      387295
      216324
    --------
    1161885
      9295080
  83655720
  -----------
  83781203580
```

Rule.—1. *Multiply the multiplicand by some figure or figures of the multiplier, which are a factor of one or more parts of the multiplier.*

2. *Multiply this partial product by a factor of some other figure or figures of the multiplier, and write the right-hand figure thus obtained under the right-hand figure of the multiplier thus used.*

3. *Continue thus until the entire multiplier is used, and then add the partial products.*

EXAMPLES FOR PRACTICE.

1. 38057 × 48618.
2. 267388 × 14982.
3. 481063 × 63721.
4. 66917 × 849612.
5. 102735 × 273162.
6. 536712 × 729981.

Topical Outline.

MULTIPLICATION.

1. Definitions.
2. Terms...........
   1. Multiplicand.
   2. Multiplier.
   3. Partial Product.
   4. Product.
3. Sign.
4. Principles.
5. Operation.....
   1. Writing Numbers.
   2. Drawing Line Beneath.
   3. Finding Partial Products.
   4. Drawing a Line Beneath Partial Products.
   5. Adding the Partial Products.
6. Rule.
7. Proof.
8. Applications.
9. Contractions.

VI. DIVISION.

72. 1. **Division** is the process of finding how many times one number is contained in another; or,

2. Division is a short method of making several subtractions of the same number.

3. Division is also an operation in which are given the product of two factors, and one of the factors, to find the other factor.

73. The product is the *Dividend;* the given factor is the *Divisor;* and the required factor is the *Quotient.* The *Remainder* is the number which is sometimes left after dividing.

NOTE.—Dividend signifies *to be divided.* Quotient is derived from the Latin word *quoties*, which signifies *how often.*

PROBLEM.—How many times is 24 cents contained in 73 cents?

SOLUTION.—Twenty-four cents from 73 cents leaves 49 cents; 24 cents from 49 cents leaves 25 cents; 24 cents from 25 cents leaves 1 cent.

Here, 24 cents is taken 3 times from (*out of*) 73 cents, and 1 cent remains; hence, 24 cents is *contained in* 73 cents 3 times, with a remainder of 1 cent.

OPERATION.

$$\begin{array}{rl} 73 & \text{cents.} \\ \underline{24} & \\ 49 & \text{cents remaining.} \\ \underline{24} & \\ 25 & \text{cents remaining.} \\ \underline{24} & \\ 1 & \text{cent remaining.} \end{array}$$

74. The divisor and quotient in Division, correspond to the factors in Multiplication, and the dividend corresponds to the product. Thus:

$$\text{Factors}\left\{\begin{array}{l} 5 \times 3 = 15 \\ 3 \times 5 = 15 \end{array}\right\}\text{Product.}$$

$$\text{Dividend}\left\{\begin{array}{l} 15 \div 5 = 3 \\ 15 \div 3 = 5 \end{array}\right\}\text{Divisors and Quotients.}$$

75. There are three methods of expressing division; thus,

$$12 \div 3, \qquad \frac{12}{3}, \qquad \text{or } 3)12.$$

Each indicates that 12 is to be divided by 3.

PRINCIPLES.—1. *When the dividend and divisor are like numbers, the quotient is abstract.*

2. *When the divisor is an abstract number, the quotient is like the dividend.*

3. *The remainder is like the dividend.*

4. *The dividend is equal to the product of the quotient by the divisor, plus the remainder.*

76. Multiplication is a short method of making several *additions* of the same number; Division is a short method of making several *subtractions* of the same number; hence, Division is the reverse of Multiplication.

77. All problems in Division are divided into two classes:

1. *To find the number of equal parts of a number.*
2. *To divide a number into equal parts.*

78. Two methods are employed in solving problems in Division: **Long Division,** when the work is written in full in solving the problem; and **Short Division,** when the result only is written, the work being performed in the mind.

The following illustrates the methods:

PROBLEM.—Divide 820 by 5.

LONG DIVISION.

```
5)820(164 Quotient.
  5
  32 tens.
  30
   20 units.
   20
```

SHORT DIVISION.

```
5)820
  164 Quotient.
```

Both operations are performed on the same principle. In the first, the subtraction is written; in the second, it is performed mentally.

LONG DIVISION.

Problem.—Divide $4225 equally among 13 men.

Solution.—As 13 is not contained in 4 (thousands), therefore, the quotient has no thousands. Next, take 42 (hundreds) as a partial dividend; 13 is contained in it 3 (hundreds) times; after multiplying and subtracting, there are 3 hundreds left. Then bring down 2 tens, and 32 tens is the next partial dividend. In this, 13 is contained 2 (tens) times, with a remainder of 6 tens. Lastly, bringing down the 5 units, 13 is contained in 65 (units) exactly 5 (units) times. The entire quotient is 3 hundreds 2 tens and 5 units.

OPERATION.

```
     thous. hund. tens. units.   hund. tens. units.
13 ) 4 2 2 5 ) 3 2 5
     3 9         hundreds.
     ---
       3 2
       2 6       tens.
       ---
         6 5
         6 5     units.
         ---
```

This may be further shown by separating the dividend into parts, each exactly divisible by 13, as follows:

```
DIVISOR.        DIVIDEND.             QUOTIENT.
  13 ) 3900 + 260 + 65 ( 300 + 20 + 5
       3900
       ----
            + 260
            + 260
            -----
                  + 65
                  + 65
                  ----
```

Rule for Long Division.—1. *Draw curved lines on the right and left of the dividend, placing the divisor on the left.*

2. *Find how often the divisor is contained in the left-hand figure, or figures, of the dividend, and write the number in the quotient at the right of the dividend.*

3. *Multiply the divisor by this quotient figure, and write the product under that part of the dividend from which it was obtained.*

4. *Subtract this product from the figures above it; to the remainder bring down the next figure of the dividend, and divide as before, until all the figures of the dividend are brought down.*

5. *If at any time after a figure is brought down, the number thus formed is too small to contain the divisor, a cipher must be placed in the quotient, and another figure brought down, after which divide as before.*

6. *If there is a final remainder after the last division, place the divisor under it and annex it to the quotient.*

PROOF.—Multiply the Divisor by the Quotient, and to this product add the Remainder, if any; the sum is equal to the Dividend when the work is correct.

NOTES.—1. The product must never be greater than the partial dividend from which it is to be subtracted; if so, the quotient figure is too large, and must be diminished.

2. The remainder after each subtraction must be *less* than the divisor; if not, the last quotient figure is too *small*, and must be increased.

3. The order of each quotient figure is the same as the lowest order in the partial dividend from which it was derived.

EXAMPLES FOR PRACTICE.

1. 1004835 ÷ 33.
2. 5484888 ÷ 67.
3. 4326422 ÷ 961.
4. 1457924651 ÷ 1204.
5. 65358547823 ÷ 2789.
6. 33333333333 ÷ 5299.
7. 245379633477 ÷ 1263.
8. 555555555555 ÷ 123456.
9. 555555555555 ÷ 654321.

In the following, multiply A by itself, also B by itself: divide the difference of the products by the sum of A and B.

	A.	B.
10.	2856	3765.
11.	33698	42856.
12.	47932	152604.

In the following, multiply A by itself, also B by itself: divide the difference of the products by the difference of A and B.

	A.	B.
13.	4986	5369.
14.	3973	4308.
15.	23798	59635.
16.	47329	65931.

17. If 25 acres produce 1825 bushels of wheat, how much is that per acre?

18. How many times 1024 in 1048576?

19. How many sacks, each containing 55 pounds, can be filled with 2035 pounds of flour?

20. How many pages in a book of 7359 lines, each page containing 37 lines?

21. In what time will a vat of 10878 gallons be filled, at the rate of 37 gallons an hour?

22. In what time will a vat of 3354 gallons be emptied, at the rate of 43 gallons an hour?

23. The product of two numbers is 212492745; one is 1035; what is the other?

24. What number multiplied by 109, with 98 added to the product, will give 106700?

SHORT DIVISION.

PROBLEM.—How often is 2 cents contained in 652 cents?

SOLUTION.—Two in 6 (hundreds) is contained 3 (hundreds) times; 2 in 5 (tens) is contained 2 (tens) times, with a remainder of 1 (ten); lastly, 1 (ten) prefixed to 2 makes 12, and 2 in 12 (units) is contained 6 times, making the entire quotient 326.

OPERATION.

```
2)652
  ---
  326
```

REMARKS.—Commence at the left to divide, so that if there is a remainder it may be carried to the next lower order.

By the operation of the rule, the dividend is separated into parts corresponding to the different orders. Having found the number of times the divisor is contained in each of these parts, the *sum* of these must give the number of times the divisor is contained in the *whole* dividend. Analyze the preceding dividend thus:

652=600+40+12
2 in 600 is contained 300 times.
2 in 40 is contained 20 times.
2 in 12 is contained 6 times.
Hence, 2 in 652 is contained 326 times.

Rule for Short Division.—1. *Write the divisor on the left of the dividend with a curved line between them, and draw a line directly beneath the dividend. Begin at the left, divide successively each figure or figures of the dividend by the divisor, and set the quotient beneath.*

2. *Whenever a remainder occurs, prefix it to the figure in the next lower order, and divide as before.*

3. *If the figure, except the first, in any order does not contain the divisor, place a cipher beneath it, prefix it to the figure in the next lower order, and divide as before.*

4. *If there is a remainder after dividing the last figure, place the divisor under it and annex it to the quotient.*

PROOF.—The same as in Long Division.

EXAMPLES FOR PRACTICE.

1. Divide 512653 by 5.
2. Divide 534959 by 7.
3. Divide 986028 by 8.
4. Divide 986974 by 11.
5. At $6 a head, how many sheep can be bought for $222?
6. At $5 a barrel, how many barrels of flour can be bought for $895?

CONTRACTIONS IN DIVISION.

CASE I.

79. When the divisor is a composite number.

This case presents no difficulty except when remainders occur.

PROBLEM.—Divide 217 by 15.

SOLUTION.—$15 = 3 \times 5$, hence $217 \div 3 = 72$ and 1 remainder; $72 \div 5 = 14$ and 2 remainder. Dividing 217 by 3, the quotient is 72 *threes*, and 1 *unit* remainder. Dividing by 5, the quotient is 14 (*fifteens*), and a remainder of 2 *threes;* hence the quotient is 14, and the *true* remainder is $2 \times 3 + 1 = 7$.

Rule.—1. *Divide the dividend by one factor of the divisor, and divide this quotient by another factor, and so on, till each factor has been used; the last quotient will be the required result.*

2. *Multiply each remainder by all of the divisors preceding the one which produced it. The sum of the products, plus the first remainder, will be the true remainder.*

REMARK.—This rule is not much used.

CASE II.

80. When the divisor is 1 with ciphers annexed.

This case presents no difficulty. Proceed thus:

PROBLEM.—Divide 23543 by 100.

OPERATION.

SOLUTION.—1|00) 235|43

$235\frac{43}{100}$

Rule.—*Cut off as many figures in the dividend as there are ciphers in the divisor; the figures cut off will be the remainder, and the other figure or figures the quotient.*

CASE III.

81. When ciphers are on the right of the divisor.

PROBLEM.—Divide 3846 by 400.

SOLUTION.—To divide by 400 is the same as to divide by 100 and then by 4 (Art. 79). Dividing by 100 gives 38, and 46 remainder (Art. 80); then, dividing by 4 gives 9, and 2 remainder: the true remainder is $2 \times 100 + 46 = 246$ (Art. 79).

OPERATION.
4|00)38|46
9 Quotient,
$200 + 46 = 246$, Rem.

Rule.—1. *Cut off the ciphers at the right of the divisor, and as many figures from the right of the dividend.*

2. *Divide the remaining part of the dividend by the remaining part of the divisor.*

3. *Annex to the remainder the figures cut off, and thus obtain the true remainder.*

ARITHMETICAL SIGNS.

82. If a number be multiplied, it is simply repeated as many times as there are units in the multiplier; if a number be divided, it is simply decreased by the divisor as many times as there are units in the quotient. It is thus evident, that Addition and Subtraction are the fundamental conceptions in all the operations of Arithmetic; and, hence, all numbers may be classified as follows:

1. Numbers to be added; or, **positive numbers.**
2. Numbers to be subtracted; or, **negative numbers.**

83. Positive numbers are distinguished by the sign $+$, negative numbers by the sign $-$; thus, $+8$ is a positive 8, and -8 a negative 8.

REMARK.—When a number is preceded by no sign, as, for example, the number 4 in the first of the following exercises, it is to be considered positive.

84. The signs $\times$ and $\div$ do not show whether *their results* are to be added or to be subtracted; they simply show what operations are to be performed on the positive or negative numbers which they follow.

Thus, in the statement, $+12-5\times 2$, the sign $\times$ shows that 5 is to be taken twice, but it does not show what is to be done with the resulting 10; *that* is shown by the —. We are to take two 5's from 12. So, in $18+9\div 3$, the sign $\div$ shows that 9 is to be divided by 3; what is to be done with the quotient, is shown by the + before the 9.

85. In every such numerical statement, the + or the — must be understood to affect the *whole result* of the operations indicated *between* it and the *next* + or —, or between it and the close of the expression.

Thus, in $5+7\times 2\times 9-2\times 6$, the + indicates the addition of 126, not of 7 only; and the — indicates the subtraction of 12. The same meaning is conveyed by $5+(7\times 2\times 9)-(2\times 6)$.

86. When the signs $\times$ and $\div$ occur in succession, they are to have their particular effects in the exact order of their occurrence.

Thus, we would indicate by $96\div 12\times 4$, that the operator is first to divide by 12, and then multiply the quotient by 4. The result intended is 32, not 2; if the latter were intended, we should write $96\div(12\times 4)$. Usage has been divided on this point, however.

REMARK.—It will be observed that in no case can the sign $\times$ or $\div$ affect any number before the preceding + or —, or beyond the following + or —.

EXERCISES.

1. $4\times 3+7\times 2-9\times 3+6\times 4-3\times 3=?$

SOLUTION. $+4\times 3=12$, $7\times 2=14$, $-9\times 3=-27$, $6\times 4=24$, $-3\times 3=-9$. Grouping and adding according to the signs, we have, $12+14+24=50$; and $-27-9=-36$. Therefore, $50-36=14$, *Ans.*

2. $2\times 2-1\times 2-2\times 2-5\times 3-5\times 3-4\times 2-4\times 2-8\times 3-5\times 2-9\times 3-7\times 2-12\times 4-7\times 2=?$

SOLUTION.$+4-2-4-15-15-8-8-24-10-27-14-48--14$. Grouping and adding, we have, $4-189=-185$, *Ans.*

3. $21\div3\times7-1\times1\div1\times4\div2+18\div3\times6\div(2\times2)+(4-2+6-7)\times4\times6\div8=?$

REMARK.—Whenever several numbers are included within the marks of parenthesis, brackets, or vinculum, they are regarded as *one* number. Note the advantage of this in example 3.

4. $16\times4\div8-7+48\div16-3-7\times4\times0\times9\times16+24\times6\div48-4\times9\div12=?$

5. $(16\div16\times96\div8-7-5+3)\times[(27\div9)\div3-1]+(91\div13\times7-45-3)\times9=?$

GENERAL PRINCIPLES.

87. The following are the **General Principles** of Multiplication and Division.

PRINCIPLE I.—*Multiplying either factor of a product, multiplies the product by the same number.*

Thus, $5\times4=20$, and $5\times4\times2=40$, whence $20\times2=40$.

II.—*Dividing either factor of a product, divides the product by the same number.*

Thus, $5\times4=20$, and $5\times4\div2=10$, whence $20\div2=10$.

III.—*Multiplying one factor of a product, and dividing the other factor by the same number, does not alter the product.*

Thus, $6\times4=24$, and $6\times2\times4\div2=24$, whence $6\times2\times2=24$.

IV.—*Multiplying the dividend, or dividing the divisor, by any number, multiplies the quotient by that number.*

If 24 be the dividend and 6 the divisor, then 4 is the quotient; hence $24\times2\div6=8$, and $24\div(6\div2)=8$.

V.—*Dividing the dividend, or multiplying the divisor, by any number, divides the quotient by that number.*

Thus, if 24 be the dividend and 6 the divisor, then $24 \div 2 = 12$ and $12 \div 6 = 2$; whence $24 \div (6 \times 2) = 2$. Therefore, $4 \div 2 = 2$.

VI.—*Multiplying or dividing both dividend and divisor by the same number, does not change the quotient.*

Thus, $24 \times 2 = 48$, and $6 \times 2 = 12$; consequently, $48 \div 12 = 4$, and $24 \div 6 = 4$.

CONTRACTIONS IN MULTIPLICATION AND DIVISION.

CASE I.

88. To multiply by any simple part of 100, 1000, etc.

NOTE.—Let the pupil study carefully the following table of equivalent parts:

PARTS OF 100.	PARTS OF 1000.
$12\frac{1}{2} = \frac{1}{8}$ of 100.	$125 = \frac{1}{8}$ of 1000.
$16\frac{2}{3} = \frac{1}{6}$ of 100.	$166\frac{2}{3} = \frac{1}{6}$ of 1000.
$25 = \frac{1}{4}$ of 100.	$250 = \frac{1}{4}$ of 1000.
$33\frac{1}{3} = \frac{1}{3}$ of 100.	$333\frac{1}{3} = \frac{1}{3}$ of 1000.
$37\frac{1}{2} = \frac{3}{8}$ of 100.	$375 = \frac{3}{8}$ of 1000.
$62\frac{1}{2} = \frac{5}{8}$ of 100.	$625 = \frac{5}{8}$ of 1000.
$66\frac{2}{3} = \frac{2}{3}$ of 100.	$666\frac{2}{3} = \frac{2}{3}$ of 1000.
$75 = \frac{3}{4}$ of 100.	$750 = \frac{3}{4}$ of 1000.
$87\frac{1}{2} = \frac{7}{8}$ of 100.	$875 = \frac{7}{8}$ of 1000.

PROBLEM.—Multiply 246 by $87\frac{1}{2}$.

SOLUTION.—Since $87\frac{1}{2}$ is $\frac{7}{8}$ of 100, annex two ciphers to the multiplicand, which multiplies it by 100, and then take $\frac{7}{8}$ of the result.

OPERATION.

```
      24600
          7
  8)172200
Ans. 21525
```

Rule.—*Multiply by* 100, 1000, *etc., and take such a part of the result as the multiplier is of* 100, 1000, *etc.*

EXAMPLES FOR PRACTICE.

1. $422 \times 33\frac{1}{3}$.
2. $6564 \times 62\frac{1}{2}$.
3. $10724 \times 16\frac{2}{3}$.

CASE II.

89. To multiply by any number whose digits are all alike.

PROBLEM.—Multiply 592643 by 66666.

SOLUTION.—Multiply 592643 by 99999 (Art. 70), the product is 59263707357; take $\frac{2}{3}$ of this product, since 6 is $\frac{2}{3}$ of 9; the result is 39509138238.

Rule.—*Multiply as if the digits were* 9's, *and take such a part of the product as the digit is of* 9.

EXAMPLES FOR PRACTICE.

1. 451402×3333.
2. 281257×555555.
3. 630224×4444000.

CASE III.

90. To divide by a number ending in any simple part of 100, 1000, etc.

PROBLEM.—Divide 6903141128 by 21875.

SOLUTION.—Multiply both by 8 and 4 successively. The divisor becomes 700000, and the dividend 220900516096, while the quotient remains the same. (Art. 87, VI.) Performing the division as in Art. 81, the quotient is 315572, and remainder 116096. The remainder being a part of the dividend, has been made too large by the multiplication by 8 and 4, and is, therefore, reduced to its true dimensions by dividing by 8 and 4. This gives 3628 for the true remainder.

Rule.—*Multiply both dividend and divisor by such a number as will convert the final figures of the divisor into ciphers, and then divide the former product by the latter.*

NOTES.—1. If there be a remainder, it should be divided by the multiplier, to get the true remainder.

2. The multiplier is 3, 4, 6, etc., according as the final portion of the divisor is *thirds, fourths, sixths,* etc., of 100, 1000.

EXAMPLES FOR PRACTICE.

1. $300521761 \div 225$.
2. $1510337264 \div 43750$.
3. $22500712361 \div 1406250$.
4. $620712480 \div 20833\frac{1}{3}$.
5. $742851692 \div 2916\frac{2}{3}$.

GENERAL PROBLEMS.

NOTE.—Let the pupil make a special problem under each general problem, and solve it.

1. When the separate cost of several things is given, how is the entire cost found?

2. When the sum of two numbers, and one of them, are given, how is the other found?

3. When the less of two numbers and the difference between them are given, how is the greater found?

4. When the greater of two numbers and the difference between them are given, how is the less found?

5. When the cost of one article is given, how do you find the cost of any number at the same price?

6. If the total cost of a given number of articles of equal value is stated, how do you find the value of one article?

7. When the divisor and quotient are given, how do you find the dividend?

8. How do you divide a number into parts, each containing a certain number of units?

9. How do you divide a number into a given number of equal parts?

10. If the product of two numbers, and one of them, are given, how do you find the other?

11. If the dividend and quotient are given, how do you find the divisor?

12. If you have the product of three numbers, and two of them are given, how do you find the third?

13. If the divisor, quotient, and remainder are given, how do you find the dividend?

14. If the dividend, quotient, and remainder are given, how do you find the divisor?

Miscellaneous Exercises.

1. A grocer gave 153 barrels of flour, worth $6 a barrel, for 54 barrels of sugar: what did the sugar cost per barrel?

2. When the divisor is 35, quotient 217, and remainder 25, what is the dividend?

3. What number besides 41 will divide 4879 without a remainder?

4. Of what number is 103 both the divisor and the quotient?

5. What is the nearest number to 53815, that can be divided by 375 without a remainder?

6. A farmer bought 25 acres of land for $2675: what did 19 acres of it cost?

7. I bought 15 horses, at $75 a head: at how much per head must I sell them to gain $210?

8. A locomotive has 391 miles to run in 11 hours: after running 139 miles in 4 hours, at what rate per hour must the remaining distance be run?

9. A merchant bought 235 yards of cloth, at $5 per yard: after reserving 12 yards, what will he gain by selling the remainder at $7 per yard?

10. A grocer bought 135 barrels of pork for $2295; he sold 83 barrels at the same rate at which he purchased, and the remainder at an advance of $2 per barrel: how much did he gain?

11. A drover bought 5 horses, at $75 each, and 12 at $68 each; he sold them all at $73 each: what did he gain?

At what price per head must he have sold them to have gained $118?

12. A merchant bought 3 pieces of cloth of equal length, at $4 a yard; he gained $24 on the whole, by selling 2 pieces for $240: how many yards were there in each piece?

13. If 18 men can do a piece of work in 15 days, in how many days will one man do it?

14. If 13 men can build a wall in 15 days, in how many days can it be done if 8 men leave?

15. If 14 men can perform a job of work in 24 days, in how many days can they perform it with the assistance of 7 more men?

16. A company of 45 men have provisions for 30 days: how many men must depart, that the provisions may last the remainder 50 days?

17. A horse worth $85, and 3 cows at $18 each, were exchanged for 14 sheep and $41 in money: at how much each were the sheep valued?

18. A drover bought an equal number of sheep and hogs for $1482: he gave $7 for a sheep, and $6 for a hog: what number of each did he buy?

19. A trader bought a lot of horses and oxen for $1260; the horses cost $50, and the oxen $17, a head; there were twice as many oxen as horses: how many were there of each?

20. In a lot of silver change, worth 1050 cents, one seventh of the value is in 25-cent pieces; the rest is made up of 10-cent, 5-cent, and 3-cent pieces, of each an equal number: how many of each coin are there?

21. A speculator had 140 acres of land, which he might have sold at $210 an acre, and gained $6300; but after holding, he sold at a loss of $5600: how much an acre did the land cost him, and how much an acre did he sell it for?

Topical Outline.

Division.

- **1. Definitions.**
- **2. Terms** .... { 1. Dividend. 2. Divisor. 3. Quotient. 4. Remainder.
- **3. Sign.**
- **4. Principles.**
- **5. Operation** .... {
  - 1. Writing the Numbers.
  - 2. Drawing Curved Lines.
  - 3. Finding Quotient Figure.
  - 4. Multiplying Divisor and Writing Product.
  - 5. Drawing Line.
  - 6. Subtracting.
  - 7. Annexing Lower Order.
  - 8. Repeating the Process from 3.
  - 9. Writing Remainder.
- **6 Rules.** { Long Division. Short Division.
- **7 Proof.**
- **8. Applications.**

- **9. Contractions** .... { Division. Multiplication and Division.
- **10. Arithmetical Signs.** {
  - 1. Of Numbers... { Positive. Negative.
  - 2. Of Operation... { Multiplying. Dividing.
- **11. General Principles.**
- **12. Applications.**

VII. PROPERTIES OF NUMBERS.

DEFINITIONS.

91. 1. The Properties of Numbers are those qualities which belong to them.

2. Numbers are classified (1), as Integral, Fractional, and Mixed (Art. **22**); (2), as Abstract and Concrete (Art. **21**); (3), Prime and Composite; (4), Even and Odd; (5), Perfect and Imperfect.

3. An **integer** is a whole number; as, 1, 2, 3, etc.

4. Integers are divided into two classes—*prime* numbers and *composite* numbers.

5. A **prime number** is one that can be exactly divided by no other whole number but itself and unity, (1); as, 1, 2, 3, 5, 7, 11, etc.

6. A **composite number** is one that can be exactly divided by some other whole number besides itself and unity; as, 4, 6, 8, 9, 10, etc.

REMARK.—Every composite number is the product of two or more other numbers, called its *factors* (Art. **60**).

7. Two numbers are *prime to each other*, when unity is the only number that will exactly divide both; as, 4 and 5.

REMARK.—Two prime numbers are always prime to each other: sometimes, also, two composite numbers; as, 4 and 9.

8 An **even number** is one which can be divided by 2 without a remainder; as, 2, 4, 6, 8, etc.

9. An **odd number** is one which can not be divided by 2 without a remainder; as, 1, 3, 5, 7, etc.

REMARK.—All even numbers except 2 are composite the odd numbers are partly prime and partly composite.

10. A **perfect number** is one which is equal to the sum of all its integral divisors, except itself; as, $6 = 1 + 2 + 3$; $28 = 1 + 2 + 4 + 7 + 14$.

11. An **imperfect number** is one not equal to the sum of all its integral divisors, except itself. Imperfect numbers are *Abundant* or *Defective:* Abundant when the number is less than the sum of the divisors; as, 18, less than $1 + 2 + 3 + 6 + 9$; and Defective when the number is greater than the sum; as, 16, greater than $1 + 2 + 4 + 8$.

12. A **divisor** of a number, is a number that will exactly divide it.

13. One number is *divisible* by another when it contains that other without a remainder; 8 is divisible by 2.

14. A **multiple** of a number is the product obtained by taking it a certain number of times; 15 is a multiple of 5, being equal to 5 taken 3 times; hence,

1st. *A multiple of a number can always be divided by it without a remainder.*

2d. *Every multiple is a composite number.*

15. Since every composite number is the product of factors, each factor must divide it exactly; hence, every factor of a number is a divisor of it.

16. A **prime factor** of a number is a prime number that will exactly divide it: 5 is a *prime* factor of 20; while 4 is a *factor* of 20, not a prime factor; hence,

1st. *The prime factors of a number are all the prime numbers that will exactly divide it.*

EXAMPLE.—1, 2, 3, and 5 are the prime factors of 30.

2d. *Every composite number is equal to the product of all its prime factors.*

EXAMPLE.—All the prime factors of 15 are 1, 3, and 5; and $1 \times 3 \times 5 = 15$.

17. Any factor of a number is called an **aliquot part** of it.

EXAMPLE.—1, 2, 3, 4, and 6, are aliquot parts of 12.

FACTORING.

92. Factoring is resolving composite numbers into factors; it depends on the following principles and propositions.

PRINCIPLE 1.—*A factor of a number is a factor of any multiple of that number.*

DEMONSTRATION.—Since $6 = 2 \times 3$, therefore, any multiple of $6 = 2 \times 3 \times$ some number; hence, every factor of 6 is also a factor of the multiple. The same may be proved of the multiple of any composite number.

PRINCIPLE 2.—*A factor of any two numbers is also a factor of their sum.*

DEMONSTRATION.—Since each of the numbers contains the factor a certain number of times, their sum must contain it as often as both the numbers; 2, which is a factor of 6 and 10, must be a factor of their sum, for 6 is 3 *twos*, and 10 is 5 *twos*, and their sum is 3 *twos* + 5 *twos* = 8 *twos*.

93. From these principles are derived the six following propositions:

PROP. I.—*Every number ending with* 0, 2, 4, 6, *or* 8, *is divisible by* 2.

DEMONSTRATION.—Every number ending with a 0, is either 10 or some number of *tens;* and since 10 is divisible by 2, therefore, by Principle 1st, Art. **92**, any number of tens is divisible by 2.

Again, any number ending with 2, 4, 6, or 8, may be considered as a certain number of *tens* plus the figure in the units' place; and since each of the two parts of the number is divisible by 2, therefore, by Principle 2d, Art. **92**, the number itself is divisible by 2; thus, $36 = 30 + 6 = 3$ tens $+ 6$; each part is divisible by 2; hence, 36 is divisible by 2.

Conversely, *No number is divisible by* 2, *unless it ends with* 0, 2, 4, 6, *or* 8.

PROP. II.—*A number is divisible by* 4, *when the number denoted by its two right-hand digits is divisible by* 4.

DEMONSTRATION.—Since 100 is divisible by 4, any number of hundreds will be divisible by 4 (Art. **92**, Principle 1st); and any number consisting of more than two places may be regarded as a certain number of hundreds plus the number expressed by the digits in tens' and units' places (thus, 384 is equal to 3 hundreds + 84); then, if the latter part (84) is divisible by 4, both parts, or the number itself, will be divisible by 4 (Art. **92**, Prin. 2d).

Conversely, *No number is divisible by* 4, *unless the number denoted by its two right-hand digits is divisible by* 4.

PROP. III.—*A number ending in* 0 *or* 5 *is divisible by* 5.

DEMONSTRATION.—Ten is divisible by 5, and every number of two or more figures is a certain number of tens, plus the right-hand digit; if this is 5, both parts of the number are divisible by 5, and, hence, the number itself is divisible by 5 (Art. **92**, Prin. 2d).

Conversely, *No number is divisible by* 5, *unless it ends in* 0 *or* 5.

PROP. IV.—*Every number ending in* 0, 00, *etc., is divisible by* 10, 100, *etc.*

DEMONSTRATION.—If the number ends in 0, it is either 10 or a multiple of 10; if it ends in 00, it is either 100, or a multiple of 100, and so on; hence, by Prin. 1st, Art. **92**, the proposition is true.

PROP. V.—*A composite number is divisible by the product of any two or more of its prime factors.*

DEMONSTRATION.—Since $2\times3\times5=30$, it follows that 2×3 taken 5 times, makes 30; hence, 30 contains 2×3 (6) exactly 5 times. In like manner, 30 contains 3×5 (15) exactly 2 times, and 2×5 (10), exactly 3 times.

Hence, *If any even number is divisible by* 3, *it is also divisible by* 6.

DEMONSTRATION.—An even number is divisible by 2; and if also by 3, it must be divisible by their product 2×3, or 6.

PROP. VI.—*Every prime number, except* 2 *and* 5, *ends with* 1, 3, 7, *or* 9.

DEMONSTRATION.—This is in consequence of Props. I. and III.

Prop. VII.—*Any integer is divisible by 9 or by 3, if the sum of its digits be thus divisible.*

94. To find the prime factors of a composite number.

Problem.—Find the prime factors of 42.

Solution.—42 is divisible by 2, and 21 is divisible by 3 or 7, which is found by trial; hence, the prime factors of 42 are 2, 3, 7, ∴ $2 \times 3 \times 7 = 42$.

OPERATION.

$$\begin{array}{r} 2\,)\,\underline{42} \\ 3\,)\,\underline{21} \\ 7 \end{array}$$

Rule.—*Divide the given number by any prime number that will exactly divide it; divide the quotient in like manner, and so continue until the quotient is a prime number; the several divisors and the last quotient are the prime factors.*

Remarks.—1. Divide first by the smallest prime factor.

2. The *least* divisor of any number is a *prime* number; for, if it were a composite number, its factors, which are less than itself, would also be divisors (Art. **92**), and then it would not be the *least* divisor. Therefore, the prime factors of any number may be found by dividing it first by the *least* number that will exactly divide it, then dividing this quotient in like manner, and so on.

3. Since 1 is the factor of every number, either prime or composite, it is not usually specified as a factor.

Find the prime factors of:

1. 45.
2. 54.
3. 72.
4. 75.
5. 96.
6. 98.
7. Factor 210.
8. Factor 1155.
9. Factor 10010.
10. Factor 36414.
11. Factor 58425.

95. The prime factors common to several numbers may be found by resolving each into its prime factors, then taking the prime factors alike in all.

Find the prime factors common to:

1. 42 and 98.
2. 45 and 105.
3. 90 and 210.
4. 210 and 315.

96. To find all the divisors of any composite number.

Any composite number is divisible, not only by each of its prime factors, but also by the product of any two or more of them (Art. **93**, Prop. V.); thus,

$42 = 2 \times 3 \times 7$; and all its divisors are 2, 3, 7, and 2×3, 2×7, and 3×7; or, 2, 3, 7, 6, 14, 21. Hence,

Rule.—*Resolve the number into its prime factors, and then form from these factors all the different products of which they will admit; the prime factors and their products will be all the divisors of the given number.*

Find all the divisors:

1. Of 70.
2. Of 196.
3. Of 231.
4. Of 496; and name the properties of 496.

GREATEST COMMON DIVISOR.

97. A **common divisor** (C. D.) of two or more numbers, is a number that exactly divides each of them.

98. The **greatest common divisor** (G. C. D.) of two or more numbers is the *greatest number* that exactly divides each of them.

PRINCIPLES.—1. *Every prime factor of a number is a divisor of that number.*

2. *Every product of two or more prime factors of a number, is a divisor of that number.*

3. *Every number is equal to the continued product of all its prime factors.*

4. *A divisor of a number is a divisor of any number of times that number.*

5. *A common divisor of two or more numbers is a divisor of their sum, and also of their difference.*

6. *The product of all the prime factors, common to two or more numbers, is their greatest common divisor.*

7. *The greatest common divisor of two numbers, is a divisor of their difference.*

TO FIND THE GREATEST COMMON DIVISOR.

CASE I.

99. By simple factoring.

PROBLEM.—Find the G. C. D. of 30 and 105.

OPERATION.

$$\left.\begin{array}{r} 30 = 2 \times 3 \times 5. \\ 105 = 3 \times 5 \times 7. \end{array}\right\} 3 \times 5 = 15, \text{ G. C. D.}$$

DEMONSTRATION.—The product 3×5 is a divisor of both the numbers, since each contains it, and it is their greatest common divisor, since it contains all the factors common to both.

PROBLEM.—Find the G. C. D. of 36, 63, 144, and 324.

OPERATION.

$$\text{SOLUTION.—}\begin{array}{r|l} 3 & 36,\ 63,\ 144,\ 324 \\ 3 & 12,\ 21,\ \ 48,\ 108 \\ & \ \ 4,\ \ \ 7,\ \ 16,\ \ 36 \\ & \therefore 3 \times 3 = 9, \text{ G. C. D.} \end{array}$$

Rule.—*Resolve the given numbers into their prime factors and take the product of the factors common to all the numbers*

Find the greatest common divisor:

1. Of 30 and 42.
2. Of 42 and 70.
3. Of 63 and 105.
4. Of 66 and 165.
5. Of 90 and 150.
6. Of 60 and 84.
7. Of 90 and 225.
8. Of 112 and 140.
9. Of 30, 45, and 75.
10. Of 84, 126, and 210.
11. Of 16, 40, 88, and 96.
12. Of 21, 42, 63, and 126.

CASE II.

100. By successive divisions.

PROBLEM.—Find the G. C. D. of 348 and 1024.

OPERATION.

```
348)1024(2
     696
     328)348(1
         328
          20)328(16
             20
             128
             120
               8)20(2
                 16
     G. C. D. 4)8
                2
```

DEMONSTRATION.—If 348 will divide 1024, it is the G. C. D.; but it will not divide it.

If 328 (Art. **98**, Prin. 5,) will divide 348, it is the G. C. D. of 328, 348, and 1024; but it will not divide 348 and 1024 exactly.

If 20 will divide 328 (by same process of reasoning), it is the G. C. D.; but there is a remainder of 8; hence, if 8 will divide 20, it is the G. C. D.; but there is also a remainder of 4.

Now, 4 divides 8 without a remainder. Therefore, 4 is the greatest number that will divide 4, 8, 20, 328, 348, and 1024, and is the G. C. D.

Rule.—*Divide the greater number by the less, and the divisor by the remainder, and so on; always dividing the last divisor by the last remainder, till nothing remains; the last divisor will be the greatest common divisor sought.*

NOTE.—A condensed form of operation may be used after the pupils are familiar with the preceding process.

REMARK.—To find the greatest common divisor of more than two numbers, find the G. C. D. of any two; then of that G. C. D., and any one of the remaining numbers, and so on for all of the numbers; the last C. D. will be the G. C. D. of all the numbers.

CONDENSED OPERATION.

348	1024	2
	696	
328	328	1
20	320	16
16	8	2
4	8	2
	0	

REMARK.—If in any case it be *obvious* that one of the numbers has a prime factor not found in the other, *that* factor may be suppressed by division before applying the rule. Thus, let the two numbers be 715 and 11011. It is plain that the prime 5 divides the first but not the second; and since that prime can be no factor of any common divisor of the two, their G. C. D. is the same as the G. C. D. of 143 and 11011.

Find the greatest common divisor of:

1. 85 and 120.
2. 91 and 133.
3. 357 and 525.
4. 425 and 493.
5. 324 and 1161.
6. 589 and 899.
7. 597 and 897.
8. 825 and 1287.
9. 423 and 2313.
10. 18607 and 24587.
11. 105, 231, and 1001.
12. 165, 231, and 385.
13. 816, 1360, 2040, and 4080.
14. 1274, 2002, 2366, 7007, and 13013.

LEAST COMMON MULTIPLE.

101. **A common multiple** (C. M.) of two or more numbers, is a number that can be divided by *each* of them without a remainder.

102. The **least common multiple** (L. C. M.) of two or more numbers, is the *least* number that is divisible by each of them without a remainder.

PRINCIPLES.—1. *A multiple of a number is divisible by that number.*

2. *A multiple of a number must contain all of the prime factors of that number.*

3. *A common multiple of two or more numbers is divisible by each of those numbers.*

4. *A common multiple of two or more numbers contains all of the prime factors of each of those numbers.*

5. *The least common multiple of two or more numbers must contain all of the prime factors of each of those numbers, and no other factors.*

6. *If two or more numbers are prime to each other, their continued product is their least common multiple.*

TO FIND THE LEAST COMMON MULTIPLE.

CASE I.

103. **By factoring the numbers separately.**

PROBLEM.—Find the L. C. M. of 10, 12, and 15.

OPERATION.

$10 = 2 \times 5$

$12 = 2 \times 2 \times 3$

$15 = 3 \times 5$

L. C. M. $= 2 \times 2 \times 3 \times 5 = 60$.

SOLUTION.—Resolve each number into its prime factors. A multiple of 10 contains the prime factors 2 and 5; of 12, the prime factors 2, 2, and 3; of 15, the prime factors 3 and 5. But the L. C. M. of 10, 12, and 15 must contain all of the different prime factors of these numbers, and no other factors; hence, the L. C. M. $=2\times2\times3\times5=60$ (Art. **102**, Prin. 2 and 5).

Rule.—*Resolve each number into its prime factors, and then take the continued product of all the different prime factors, using each factor the greatest number of times it occurs in any one of the given numbers.*

REMARKS.—1. Each factor must be taken in the least common multiple the greatest number of times it occurs in either of the numbers. In the preceding solution, 2 must be taken *twice*, because it occurs twice in 12, the number containing it most.

2. To avoid mistakes, after resolving the numbers into their prime factors, strike out the needless factors.

Find the L. C. M. of:

1. 8, 10, 15.
2. 6, 9, 12.
3. 12, 18, 24.
4. 8, 14, 21, 28.
5. 10, 15, 20, 30.
6. 15, 30, 70, 105.

CASE II.

104. By dividing the numbers successively by their common primes.

PROBLEM.—Find the L. C. M. of 10, 20, 25, and 30.

OPERATION.

2	~~10~~	20	25	30
5		10	25	15
		2	5	3

$2\times5\times2\times5\times3=300$ L. C. M.

SOLUTION.—Write the numbers as in the margin. Strike out 10, because it is contained in 20 and 30. Next, divide 20 and 30 by the prime factor 2; write the quotients 10 and 15, and the undivided number 25 in a line beneath. Divide these numbers by the common prime factor 5. The three quotients—2, 5, 3, are prime to one another; whence, the L. C. M. is the product of the divisors 2, 5, and the quotients 2, 5, 3. By division, all needless factors are suppressed.

Rule.—1. *Write the numbers in a horizontal line; strike out any number that will exactly divide any of the others; divide by any prime number that will divide two or more of them without a remainder; write the quotients and undivided numbers in a line beneath.*

2. *Proceed with this line as before, and continue the operation till no number greater than 1 will exactly divide two or more of the numbers.*

3. *Multiply together the divisors and the numbers in the last line; their product will be the least common multiple required.*

REMARK.—Prime factors not obvious may be found by Art. 100.

Find the least common multiple of:

1. 6, 9, 20.
2. 15, 20, 30.
3. 7, 11, 13, 5.
4. 35, 45, 63, 70.
5. 8, 15, 20, 25, 30.
6. 30, 45, 48, 80, 120, 135.
7. 174, 485, 4611, 14065, 15423.
8. 498, 85988, 235803, 490546.
9. 2183, 2479, 3953.
10. 1271, 2573, 3403.

SOME PROPERTIES OF THE NUMBER NINE.

105. Addition, Subtraction, Multiplication, and Division may be proved by "casting out the 9's." To cast the 9's out of any number, is to divide the sum of the digits by 9, and find the excess.

PROBLEM.—Find the excess of 9's in 768945.

EXPLANATION.—Begin at the left, thus: $7+6$ are 13; drop the 9; $4+8$ are 12; drop the 9; $3+4+5$ are 12; drop the 9; the excess is 3. The 9 in the number was not counted.

PRINCIPLE.—*Any number divided by 9, will leave the same remainder as the sum of its digits divided by 9.*

ILLUSTRATION.

$$768945 = \begin{cases} 700000 = 7 \times 100000 = 7 \times (99999 + 1) = 7 \times 99999 + 7 \\ 60000 = 6 \times 10000 = 6 \times (9999 + 1) = 6 \times 9999 + 6 \\ 8000 = 8 \times 1000 = 8 \times (999 + 1) = 8 \times 999 + 8 \\ 900 = 9 \times 100 = 9 \times (99 + 1) = 9 \times 99 + 9 \\ 40 = 4 \times 10 = 4 \times (9 + 1) = 4 \times 9 + 4 \\ 5 = 5 \times 1 = \quad = 5 \end{cases}$$

Whence, $7 \times 99999 + 6 \times 9999 + 8 \times 999 + 9 \times 99 + 4 \times 9 + 7 + 6 + 8 + 9 + 4 + 5 = 768945$.

SOLUTION.—An examination of the above shows that 768945 has been separated into multiples of 9, and the sum of the digits composing the number; the same may be shown of any other number. There can be no remainder in the multiples, except in the sum of the digits.

PROOF OF ADDITION.—*The sum of the excess of 9's in the several numbers must equal the excess of 9's in their sum.*

ILLUSTRATION.—The excesses in the numbers are 8, 2, 4, and 3, and the excess in the sum of these excesses is 8. The excess in the sum of the numbers is 8, the two excesses being the same, as they ought to be when the work is correct.

OPERATION.

7352	8
5834	2
6241	4
7302	3
26729	8

PROOF OF SUBTRACTION.—*The excess of 9's in the minuend must equal the sum of the excess of 9's in the subtrahend and remainder.*

ILLUSTRATION.—As the minuend is the sum of the subtrahend and remainder, the reason of this proof is seen from that of Addition.

OPERATION.

Minuend,	7640	8
Subtrahend,	1234	1
Remainder,	6406	7

PROOF OF MULTIPLICATION.—*Find the excess of 9's in the factors and in the product. The excess of 9's in the product of the excesses of the factors, should equal the excess in the product of the factors themselves.*

ILLUSTRATION.—Multiply 835 by 76; the product is 63460. The excess in the multiplicand is 7, in the multiplier 4, and in the product 1; the two former multiplied, give 28; and the excess in 28 is also 1, as it should be.

OPERATION.

$835 \times 76 = 63460$
835, excess $= 7$
76, " $= 4$
$7 \times 4 = 28$, " $= 1$
63460, " $= 1$

PROOF OF DIVISION.—*Find the excess of 9's in each of the terms. To the excess of 9's in the product of the excesses in the divisor and quotient, add the excess in the remainder; the excess in the sum should equal the excess in the dividend.*

ILLUSTRATION.—Divide 8915 by 25; the quotient is 356, and the remainder, 15. The excess of 9's in the divisor is 7; in the quotient, 5; their product is 35, the excess of which is 8. The excess in the remainder is 6. $6 + 8 = 14$, of which the excess is 5. The excess of 9's in the dividend is also 5.

OPERATION.

356, excess 5
25, " 7
35

35, " 8
15, " 6
14

14, " 5
8915, " 5

CANCELLATION.

106. Cancellation is the process of crossing out equal factors from dividend and divisor.

The **sign of Cancellation** is an oblique line drawn across a figure; thus, 2̸, 6̸, 9̸.

PRINCIPLES.—1. *Canceling a factor in any number, divides the number by that factor.*

2. *Canceling a factor in both dividend and divisor, does not change the quotient.* (Art. **129**, III.)

PROBLEM.—Multiply 75, 153, and 28 together, and divide by the product of 63 and 36.

Solution.—Indicate the operations as in the margin. Cancel 4 out of 28 and 36, leaving 7 above and 9 below. Cancel this 7 out of the dividend and out of the 63 in the divisor, leaving 9 below. Cancel a 9 out of the divisor and out of 153 in the dividend, leaving 17 above. Cancel 3 out of 9 and 75, leaving 25 above and 3 below. No further canceling is possible; the factors remaining in the dividend are 25 and 17, whose product, 425, divided by the 3 in the divisor, gives $141\frac{2}{3}$.

OPERATION.

$$\frac{\overset{25}{\cancel{75}} \times \overset{17}{\cancel{153}} \times \overset{\cancel{7}}{\cancel{28}}}{\underset{\underset{3}{\cancel{9}}}{\cancel{63}} \times \underset{\cancel{9}}{\cancel{36}}} =$$

$$\frac{25 \times 17}{3} = 141\frac{2}{3}$$

Rule for Cancellation.—1. *Indicate the multiplications which produce the dividend, and those, if any, which produce the divisor.*

2. *Cancel equal factors from dividend and divisor; multiply together the factors remaining in the dividend, and divide the product by the product of the factors left in the divisor.*

Note.—If no factor remains in the divisor, the product of the factors remaining in the dividend will be the quotient; if only one factor is left in the dividend, it will be the answer.

Examples for Practice.

1. How many cows, worth $24 each, can I get for 9 horses, worth $80 each?

2. I exchanged 8 barrels of molasses, each containing 33 gallons, at 40 cents a gallon, for 10 chests of tea, each containing 24 pounds: how much a pound did the tea cost me?

3. How many bales of cotton, of 400 pounds each, at 12 cents a pound, are equivalent to 6 hogsheads of sugar, 900 pounds each, at 8 cents a pound?

4. Divide $15 \times 24 \times 112 \times 40 \times 10$ by $25 \times 36 \times 56 \times 90$.

Topical Outline.

Properties of Numbers.

1. **Definitions**......
   - Properties.
   - Numbers Classified.......
     - Integer ......
       - Prime.
       - Composite.
       - Even.
       - Odd.
       - Perfect.
       - Imperfect.
     - Fraction.
     - Mixed.
   - Divisor; Divisible; Multiple; Factors; Prime Factors; Aliquot Parts.
2. **Factoring** ....
   1. Definition.
   2. Principles.
   3. Propositions.
   4. Operations.
   5. Rules.
   6. Applications.
3. G. C. D...........
   1. Definitions....
      1. Common Divisor.
      2. Greatest Common Divisor.
   2. Principles.
   3. Operations.....
      - Case I.....
        - Rule.
        - Applications.
      - Case II...
        - Rule.
        - Applications.
4. L. C. M..........
   1. Definitions......
      1. Common Multiple.
      2. Least Common Multiple.
   2. Principles.
   3. Operations......
      - Case I.....
        - Rule.
        - Applications.
      - Case II...
        - Rule.
        - Applications.
5. Some Properties of the No. 9..
   1. Principle.
   2. Application to...
      1. Addition.
      2. Subtraction.
      3. Multiplication.
      4. Division.
6. Cancellation...
   1. Definition.
   2. Sign, ⁄.
   3. Principles.
   4. Operation.
   5. Rule.
   6. Applications.

VIII. COMMON FRACTIONS.

DEFINITIONS.

107. A **Fraction** is an expression for one or more of the equal parts of a divided whole.

108. Fractions are divided into two classes; viz., *common* fractions and *decimal* fractions.

109. A **Common Fraction** is expressed by two numbers, one above and one below a horizontal line; thus, $\frac{2}{3}$, which is read *two thirds*.

110. The **Denominator** is the number below the line. It shows the number of parts into which the whole is divided, and thus the size of the parts.

111. The **Numerator** is the number above the line. It shows how many of the parts are taken.

NOTE.—The denominator *denominates*, or names, the parts; the numerator *numbers* the parts.

112. The **Terms** of a fraction are the numerator and the denominator.

ILLUSTRATION.—The expression $\frac{4}{5}$, four fifths, shows that the whole is divided into five equal parts, and that four of those parts are taken. 5 is the denominator, 4 is the numerator, and the terms of the fraction are 4 and 5.

113. Every fraction implies: 1. That a number is divided; 2. That the parts are equal; 3. That one or more of the parts are taken.

114. There are two ways of considering a fraction whose numerator is greater than 1. Four fifths may be 4 fifths of one thing, or 1 fifth of four things; therefore,

The numerator of a fraction may be regarded as showing the number of units to be divided; the denominator, the number of parts into which the numerator is to be divided; the fraction itself being the value of one of those parts.

Hence, a fraction may be considered as an *indicated division* (Art. 75) in which,

1. *The dividend is the numerator.*
2. *The divisor is the denominator.*
3. *The quotient is the fraction itself.*

115. The **value** of a fraction is its relation to a unit.

116. Common fractions are divided into classes with respect to their *value* and *form*.

(1). As to value, into *Proper* and *Improper*.

(2). As to form, into *Simple*, *Complex*, and *Compound*.

117. A **Proper Fraction** is one whose numerator is less than its denominator; as, $\frac{1}{2}$.

118. An **Improper Fraction** is one whose numerator is equal to, or greater than, its denominator; as, $\frac{4}{3}$.

119. A **Mixed Number** is a number composed of an integer and a fraction; as, $3\frac{2}{5}$.

120. A **Simple Fraction** is a single fraction whose terms are integral; as, $\frac{3}{8}$, $\frac{8}{3}$, $\frac{8}{8}$.

121. A **Complex Fraction** is one which has one or both of its terms fractional; as, $\frac{4}{3\frac{1}{2}}$, $\frac{3\frac{1}{2}}{4}$, or $\frac{3\frac{1}{2}}{5\frac{1}{3}}$.

122. A **Compound Fraction** is a fraction of a fraction; as, $\frac{1}{4}$ of $\frac{2}{3}$.

123. An Integer may be expressed as a fraction by writing 1 under it as a denominator; thus, $\frac{7}{1}$, which is read *seven ones*.

124. The **Reciprocal** of a number is 1 divided by that number; thus, the reciprocal of 5 is $\frac{1}{5}$.

125. Similar Fractions are those that have the same denominator; as, $\frac{3}{8}$ and $\frac{5}{8}$.

126. Dissimilar Fractions are those that have unlike denominators; as, $\frac{3}{4}$ and $\frac{3}{5}$.

REMARK.—The word "fraction" is from the Latin, *frango,* I break, and literally means a broken number. In mathematics, however, the word "fraction," as a general term, means simply the indicated quotient of a required division.

NUMERATION AND NOTATION OF FRACTIONS.

127. Numeration of Fractions is the art of reading fractional numbers.

128. Notation of Fractions is the art of writing fractional numbers.

Rule for Reading Common Fractions.—*Read the number of parts taken as expressed by the numerator, and then the size of the parts as expressed by the denominator.*

EXAMPLE.—$\frac{7}{9}$ is read *seven ninths.*

REMARK.—Seven ninths ($\frac{7}{9}$), signifies 7 ninths of one, or $\frac{1}{9}$ of 7, or 7 divided by 9.

Rule for Writing Common Fractions.—*Write the number of parts; place a horizontal line below it, under which write the number which indicates the size of the parts.*

Fractions to be written in figures:

Seven eighths. Four elevenths. Five thirteenths. One seventeenth. Three twenty-ninths. Eight twenty-firsts. Nine forty-seconds. Nineteen ninety-thirds. Thirteen one-hundredths. Twenty-four one-hundred-and-fifteenths.

129. Since a fraction is an indicated division (Art. **114**); therefore,

PRINCIPLES.—I. *A Fraction is multiplied,*
1st. *By multiplying the numerator.*
2d. *By dividing the denominator.*

II. *A Fraction is divided,*
1st. *By dividing the numerator.*
2d. *By multiplying the denominator.*

III. *The value of a Fraction is not changed,*
1st. *By multiplying both terms by the same number.*
2d. *By dividing both terms by the same number.*

REMARK.—The proof of I is found in Art. **87**, Principle IV; the proof of II is in Principle V; and of III, in Principle VI.

REDUCTION OF FRACTIONS.

130. Reduction of Fractions consists in changing their form without altering their value.

CASE I.

131. To reduce a fraction to its lowest terms.

REMARKS.—1. Reducing a fraction to lower terms, is changing it to an equivalent fraction whose terms are smaller numbers.

2. A fraction is in its *lowest* terms when the numerator and denominator are prime to each other; as, $\frac{2}{3}$, but not $\frac{4}{6}$.

PROBLEM.—Reduce $\frac{20}{30}$ to its lowest terms.

SOLUTION.—Dividing both terms by the common factor 2, the result is $\frac{10}{15}$; dividing this by 5 (**129**, III), the result is $\frac{2}{3}$, which can not be reduced lower.

FIRST OPERATION.

$2)\frac{20}{30} = \frac{10}{15}$

$5)\frac{10}{15} = \frac{2}{3}$, *Ans.*

Or, dividing at once by 10, the greatest common divisor of both terms, the result is $\frac{2}{3}$, as before.

SECOND OPERATION.

$10)\frac{20}{30} = \frac{2}{3}$, *Ans.*

Rule.—*Reject all factors common to both terms of the fraction. Or, divide both terms of the fraction by their greatest common divisor.*

Reduce to their lowest terms:

1. $\frac{30}{45}$.
2. $\frac{32}{56}$.
3. $\frac{42}{189}$.
4. $\frac{105}{195}$.
5. $\frac{154}{210}$.
6. $\frac{156}{221}$.
7. $\frac{253}{414}$.
8. $\frac{667}{783}$.
9. $\frac{1767}{4557}$.
10. $\frac{9702}{18522}$.

Express the following in their simplest forms:

11. $923 \div 1491$.
12. $890 \div 1691$.
13. $2261 \div 4123$.
14. $6160 \div 40480$.

CASE II.

132. To reduce a fraction to higher terms.

REMARK.—Reducing a fraction to higher terms, is changing it to an equivalent fraction whose terms are larger numbers.

PROBLEM.—Reduce $\frac{5}{8}$ to fortieths.

SOLUTION.—Divide 40 by 8, the quotient is 5; multiply both terms of the given fraction, $\frac{5}{8}$, by 5 (**129**, III), and the result is $\frac{25}{40}$, the equivalent fraction required.

OPERATION.

$$40 \div 8 = 5$$

$$\frac{5}{8} = \frac{5 \times 5}{8 \times 5} = \frac{25}{40}, \textit{ Ans.}$$

Rule.—*Divide the required denominator by the denominator of the given fraction; multiply both terms of the given fraction by this quotient; the result is the equivalent fraction required.*

1. Reduce $\frac{3}{11}$ and $\frac{4}{33}$ to ninety-ninths.
2. Reduce $\frac{4}{9}$, $\frac{3}{7}$, and $\frac{3}{21}$ to sixty-thirds.
3. Reduce $\frac{8}{17}$, $\frac{9}{19}$, and $\frac{11}{21}$ to equivalent fractions having 6783 for a denominator.

CASE III.

133. To reduce a whole or mixed number to an improper fraction.

PROBLEM.—Reduce $3\frac{3}{4}$ to an improper fraction; to fourths.

SOLUTION.—In 1 (unit), there are 4 fourths; in 3 (units), there are 3 times 4 fourths, = 12 fourths: and 12 fourths + 3 fourths = 15 fourths.

Rule.—*Multiply together the whole number and the denominator of the fraction: to the product add the numerator, and write the sum over the denominator.*

1. In $\$7\frac{3}{8}$, how many eighths of a dollar?
2. In $19\frac{3}{4}$ gallons, how many fourths?
3. In $13\frac{37}{60}$ hours, how many sixtieths?

Reduce to improper fractions:

4. $11\frac{2}{3}$.
5. $15\frac{8}{11}$.
6. $127\frac{11}{17}$.
7. $109\frac{9}{19}$.
8. $5\frac{207}{211}$.
9. $13\frac{51}{73}$.

REMARK.—To reduce a whole number to a fraction having a given denominator, is a special case under the preceding.

PROBLEM.—Reduce 8 to a fraction whose denominator is 7.

SOLUTION.—Since $\frac{7}{7}$ equals one, 8 equals 8 times $\frac{7}{7}$, or $\frac{56}{7}$.

CASE IV.

134. To reduce an improper fraction to a whole or mixed number.

PROBLEM.—Reduce $\frac{13}{5}$ of a dollar to dollars.

SOLUTION.—Since 5 fifths make 1 dollar, there will be as many dollars in 13 fifths as 5 fifths are contained times in 13 fifths; that is, $2\frac{3}{5}$ dollars.

Rule.—*Divide the numerator by the denominator; the quotient will be the whole or mixed number.*

REMARK.—If there be a fraction in the answer, reduce it to its lowest terms.

1. In $\frac{37}{8}$ of a dollar, how many dollars?
2. In $\frac{137}{4}$ of a bushel, how many bushels?
3. In $\frac{785}{60}$ of an hour, how many hours?

Reduce to whole or mixed numbers:

4. $\frac{23}{23}$.
5. $\frac{1295}{37}$.
6. $\frac{800}{9}$.
7. $\frac{1162}{11}$.
8. $\frac{4260}{13}$.
9. $\frac{15780}{31}$.

CASE V.

135. To reduce compound to simple fractions.

PROBLEM.—Reduce $\frac{3}{4}$ of $\frac{5}{7}$ to a simple fraction.

SOLUTION.—$\frac{1}{4}$ of $\frac{1}{7}=\frac{1}{28}$; $\frac{1}{4}$ of $\frac{5}{7}=\frac{5}{28}$; and $\frac{3}{4}$ of $\frac{5}{7}=\frac{3\times5}{28}=\frac{15}{28}$.

OPERATION.

$\frac{3}{4}$ of $\frac{5}{7}=\frac{3\times5}{4\times7}=\frac{15}{28}$, *Ans.*

Rule.—*Multiply the numerators together for the numerator and the denominators together for the denominator of the fraction, canceling common factors if they occur in both terms.*

REMARK.—Whole or mixed numbers must be reduced to improper fractions before applying the rule.

PROBLEM.—Reduce $\frac{2}{3}$ of $\frac{9}{10}$ of $\frac{7}{12}$ to a simple fraction.

SOLUTION.—Indicate the work, and employ cancellation, as shown in the accompanying operation.

OPERATION.

$$\frac{\overset{}{\not{2}}\times\overset{3}{\not{9}}\times7}{\not{3}\times\underset{5}{\not{10}}\times\underset{4}{\not{12}}}=\frac{7}{20},\ \textit{Ans.}$$

Reduce to simple fractions:

1. $\frac{1}{3}$ of $\frac{3}{4}$ of $\frac{4}{7}$.
2. $\frac{2}{5}$ of $\frac{4}{7}$ of $2\frac{5}{8}$.
3. $\frac{4}{5}$ of $\frac{15}{16}$ of $2\frac{2}{3}$.
4. $\frac{1}{2}$ of $\frac{4}{5}$ of $3\frac{3}{4}$.
5. $\frac{3}{4}$ of $\frac{8}{9}$ of $\frac{4}{7}$ of $8\frac{3}{4}$.
6. $\frac{1}{3}$ of $\frac{3}{5}$ of $\frac{6}{7}$ of $\frac{3}{4}$ of $4\frac{2}{3}$.
7. $\frac{8}{11}$ of $\frac{3}{7}$ of $\frac{4}{19}$ of $\frac{77}{24}$ of $7\frac{1}{8}$.
8. $\frac{12}{13}$ of $\frac{9}{16}$ of $\frac{7}{18}$ of $\frac{10}{21}$ of $1\frac{4}{35}$.

COMMON DENOMINATOR.

136. **A common denominator** of two or more fractions, is a denominator by which they express *like* parts of a unit.

137. The **least common denominator** (L. C. D.) of two or more fractions, is the *least* denominator by which they can express like parts of a unit.

PRINCIPLES—1. *Only a common multiple of different denominators can become a common denominator.*

2. *Only a least common multiple can become a least common denominator.*

CASE VI.

138. To reduce fractions to equivalent fractions having a common denominator.

PROBLEM.—Reduce $\frac{1}{2}$, $\frac{2}{3}$, and $\frac{3}{4}$ to a common denominator.

OPERATION.

$$\frac{1\times3\times4}{2\times3\times4}=\frac{12}{24}$$
$$\frac{2\times2\times4}{3\times2\times4}=\frac{16}{24}$$
$$\frac{3\times2\times3}{4\times2\times3}=\frac{18}{24}$$

SOLUTION.—Since $2\times3\times4=24$, 24 is a common multiple of all the denominators. The terms of $\frac{1}{2}$ must be multiplied by 3×4; the terms of $\frac{2}{3}$, by 2×4; and the terms of $\frac{3}{4}$, by 2×3. The values of the fractions are unaltered. (**129**, III.)

Rule.—*Multiply both terms of each fraction by the denominators of the other fractions.*

REMARK.—Since the denominator of each new fraction is the product of the same numbers—viz., all the denominators of the given fractions—it is unnecessary to find this product more than once. The operation is generally performed as in the following example:

PROBLEM.—Reduce $\frac{1}{2}$, $\frac{3}{5}$, and $\frac{6}{7}$ to a common denominator.

OPERATION.

$2\times5\times7=70$, common denominator.
$1\times5\times7=35$, first numerator.
$3\times2\times7=42$, second numerator.
$6\times2\times5=60$, third numerator.

$$\left.\begin{matrix}\frac{1}{2}=\frac{35}{70}\\ \frac{3}{5}=\frac{42}{70}\\ \frac{6}{7}=\frac{60}{70}\end{matrix}\right\}\textit{Ans.}$$

NOTE.—Mixed numbers and compound fractions must first be reduced to simple fractions; the lowest terms are preferable.

Reduce to a common denominator:

1. $\frac{1}{2}$, $\frac{2}{3}$, $\frac{3}{5}$.
2. $\frac{1}{4}$, $\frac{1}{5}$, $\frac{1}{6}$.
3. $\frac{2}{3}$, $\frac{8}{7}$, $\frac{5}{8}$.
4. $\frac{1}{2}$, $\frac{3}{5}$, $\frac{5}{6}$, $\frac{7}{8}$.
5. $\frac{2}{3}$, $\frac{1}{2}$ of $3\frac{1}{2}$, $\frac{2}{3}$ of $\frac{3}{5}$.
6. $\frac{2}{3}$ of $\frac{6}{7}$, $\frac{3}{4}$ of $\frac{8}{9}$, $\frac{1}{2}$ of $\frac{4}{5}$ of $\frac{3}{7}$ of $2\frac{5}{8}$.

REMARK.—When the terms of the fractions are small, and one denominator is a multiple of the others, reduce the fractions to a common denominator, by multiplying both terms of each by such a number as will render its denominator the same as the largest denominator. *This number will be found by dividing the largest denominator by the denominator of the fraction to be reduced.*

PROBLEM.—Reduce $\frac{1}{3}$ and $\frac{5}{6}$ to a common denominator.

SOLUTION.—The largest denominator, 6, is a multiple of 3; therefore, if we multiply both terms of $\frac{1}{3}$ by 6 divided by 3, which is 2, it is reduced to $\frac{2}{6}$.

OPERATION.

$$\frac{1\times2}{3\times2}=\frac{2}{6}$$
$$\frac{5}{6}=\frac{5}{6}$$

Reduce to a common denominator:

1. $\frac{1}{2}$, $\frac{3}{4}$, and $\frac{5}{8}$.
2. $\frac{2}{3}$, $\frac{5}{6}$ and $\frac{7}{12}$.
3. $\frac{3}{4}$, $\frac{4}{5}$, $\frac{9}{10}$ and $\frac{11}{20}$.

CASE VII.

139. To reduce fractions of different denominators, to equivalent fractions having the least common denominator.

PROBLEM.—Reduce $\frac{5}{8}$, $\frac{7}{9}$, and $\frac{9}{24}$ to equivalent fractions, having the least common denominator.

SOLUTION.—Find the L. C. M. of 8, 9, and 24, which is 72; divide 72 by the given denominators 8, 9, and 24, respectively; multiply both terms of each fraction by the quotient obtained by dividing 72 by its denominator; the L. C. M. is the denominator of the equivalent fractions. In practice it is not necessary to multiply each denominator *in form*.

OPERATION.

$$\frac{5}{8}=\frac{5\times9}{8\times9}=\frac{45}{72}$$
$$\frac{7}{9}=\frac{7\times8}{9\times8}=\frac{56}{72}$$
$$\frac{9}{24}=\frac{9\times3}{24\times3}=\frac{27}{72}$$

Rule.—1. *Find the L. C. M. of the denominators of the given fractions, for the L. C. D.*

2. *Divide this L. C. D. by the denominator of each fraction, and multiply the numerator by the quotient.*

3. *Write the product after each multiplication as a numerator above the L. C. D.*

REMARK.—All expressions should be in the simplest form.

Reduce to the least common denominator:

1. $\frac{1}{3}$, $\frac{3}{4}$, $\frac{5}{6}$.
2. $\frac{1}{2}$, $\frac{3}{5}$, $\frac{9}{10}$, $\frac{3}{4}$.
3. $\frac{3}{7}$, $\frac{5}{8}$, $\frac{11}{14}$,
4. $\frac{3}{4}$, $\frac{6}{8}$, $\frac{9}{12}$, $\frac{15}{20}$.
5. $\frac{6}{9}$, $\frac{9}{12}$, $\frac{12}{20}$, $\frac{7}{10}$.
6. $1\frac{3}{4}$, $3\frac{2}{3}$, and $\frac{3}{10}$ of $3\frac{4}{7}$.

ADDITION OF FRACTIONS.

140. Addition of Fractions is the process of uniting two or more fractional numbers in one sum.

REMARK.—As integers to be added must express *like* units (Art. **52**), so fractions to be added must express *like parts* of like units.

PROBLEM.—What is the sum of $\frac{3}{4}$, $\frac{5}{8}$, and $\frac{7}{12}$?

SOLUTION.—Reducing the given fractions to equivalent fractions having a common denominator, we have $\frac{18}{24}$, $\frac{15}{24}$, and $\frac{14}{24}$. Since these are now of the same kind, they can be added by adding their numerators. Their sum is $\frac{47}{24}=1\frac{23}{24}$.

OPERATION.

$\frac{3}{4}=\frac{18}{24}$ $\quad \frac{5}{8}=\frac{15}{24}$

$\frac{7}{12}=\frac{14}{24}$

$\frac{18}{24}+\frac{15}{24}+\frac{14}{24}=\frac{47}{24}$

$\frac{47}{24}=1\frac{23}{24}$, *Ans.*

Rule.—*Reduce the fractions to a common denominator, add their numerators, and write the sum over the common denominator.*

REMARKS.—1. Each fractional expression should be in its simplest form before applying the rule.

2. Mixed numbers and fractions may be added separately and their sums united.

3. After adding, reduce the sum to its lowest terms.

EXAMPLES FOR PRACTICE.

1. $\frac{1}{6}$, $\frac{2}{9}$, and $\frac{5}{12}$.
2. $\frac{2}{3}$, $\frac{3}{5}$, $\frac{7}{9}$, and $\frac{4}{15}$.
3. $1\frac{2}{3}$ and $2\frac{3}{5}$.
4. $2\frac{1}{4}$, $3\frac{2}{7}$, and $4\frac{5}{6}$.
5. $\frac{6}{10}$, $\frac{4}{14}$, $\frac{8}{12}$, and $2\frac{1}{3}$.
6. $1\frac{1}{2}$, $2\frac{1}{3}$, $3\frac{1}{4}$, and $4\frac{1}{5}$.
7. $\frac{2}{3}$ of $\frac{4}{5}$, and $\frac{3}{7}$ of $\frac{5}{8}$ of $2\frac{1}{3}$.
8. $\frac{3}{8}+\frac{1}{22}+\frac{7}{24}+\frac{29}{88}$.
9. $\frac{7}{8}+\frac{11}{12}+\frac{17}{18}+\frac{23}{24}+\frac{20}{27}$.
10. $\frac{4}{7}$ of $96\frac{1}{4}+\frac{8}{9}$ of $\frac{11}{12}$ of $5\frac{1}{6}$.
11. $\frac{2}{3}+\frac{8}{9}+\frac{26}{27}+\frac{80}{81}+\frac{242}{243}+\frac{728}{729}$.

SUBTRACTION OF FRACTIONS.

141. Subtraction of Fractions is the process of finding the difference between two fractional numbers.

REMARK.—In subtraction of integers, the numbers must be of *like* units (Art. **54**); in subtraction of fractions, the minuend and subtrahend must express *like parts* of like units.

PROBLEM.—Find the difference between $\frac{5}{6}$ and $\frac{8}{15}$.

SOLUTION.—Reducing the given fractions to equivalent fractions having a common denominator, we have $\frac{5}{6}=\frac{25}{30}$, and $\frac{8}{15}=\frac{16}{30}$; the difference is $\frac{9}{30}=\frac{3}{10}$.

OPERATION.

$\frac{5}{6}=\frac{25}{30}$

$\frac{8}{15}=\frac{16}{30}$

$\frac{25}{30}-\frac{16}{30}=\frac{9}{30}=\frac{3}{10}$, *Ans.*

Rule.—*Reduce the fractions to a common denominator, and write the difference of their numerators over the common denominator.*

REMARK.—Before applying the rule, the fractions should be in their simplest form. The difference should be reduced to its lowest terms.

EXAMPLES FOR PRACTICE.

1. $\frac{4}{5}-\frac{5}{12}$.
2. $\frac{8}{11}-\frac{3}{17}$ of $\frac{1}{2}$.
3. $\frac{11}{54}-\frac{3}{28}$ of $\frac{5}{6}$.
4. $\frac{5}{11}-\frac{1}{13}$ of 4.
5. $\frac{11}{14}-\frac{4}{63}$.
6. $\frac{9}{55}-\frac{1}{15}$.
7. $\frac{7}{45}-\frac{11}{75}$.
8. $\frac{10}{39}-\frac{8}{65}$.

REMARK.—When the mixed numbers are small, reduce them to improper fractions before subtracting; if they are not small, subtract whole numbers and fractions separately, and then unite the results. Thus,

PROBLEM.—Subtract $23\frac{15}{16}$ from $31\frac{3}{4}$.

SOLUTION.—Reducing the fractions to a common denominator, $\frac{3}{4}=\frac{12}{16}$. But $\frac{15}{16}$ can not be taken from $\frac{12}{16}$. Take a unit, $\frac{16}{16}$, from the integer of the minuend and add it to $\frac{12}{16}$. Then $\frac{16}{16}+\frac{12}{16}=\frac{28}{16}$, and $\frac{28}{16}-\frac{15}{16}=\frac{13}{16}$. $30-23=7$. Therefore the answer is $7\frac{13}{16}$.

OPERATION.

$$\begin{array}{r} 31\frac{3}{4} \\ \underline{23\frac{15}{16}} \\ 7\frac{13}{16} \end{array}, \textit{ Ans.} \qquad \begin{array}{l} \frac{12}{16}+\frac{16}{16}=\frac{28}{16} \\ \frac{28}{16}-\frac{15}{16}=\frac{13}{16} \end{array}$$

9. $12\frac{3}{4}-10\frac{13}{16}$.
10. $12\frac{23}{28}-9\frac{27}{35}$.
11. $5\frac{23}{32}-2\frac{2}{7}$.
12. $7\frac{5}{12}-3\frac{1}{2}$.
13. $15-\frac{3}{7}$.
14. $18-5\frac{3}{8}$.
15. $\frac{5}{3}$ of $2\frac{7}{9}-3\frac{17}{18}$.
16. $3\frac{1}{3}-\frac{4}{5}$ of $1\frac{7}{8}$.
17. $\frac{16}{3}$ of $4\frac{1}{2}-\frac{13}{4}$ of $3\frac{1}{5}=$ what?
18. $11\frac{2}{3}+8\frac{7}{9}-9\frac{19}{22}=$ what?
19. A man owned $\frac{25}{38}$ of a ship, and sold $\frac{3}{5}$ of his share: how much had he left?
20. After selling $\frac{4}{7}$ of $\frac{5}{8}+\frac{1}{5}$ of $\frac{3}{7}$ of a farm, what part of it remains?
21. $3\frac{1}{4}+4\frac{2}{5}-5\frac{1}{2}+16\frac{5}{8}-7\frac{11}{24}+10-14\frac{5}{6}$, is equal to what?
22. $5\frac{1}{5}-2\frac{5}{6}+\frac{13}{2}-3\frac{3}{10}+3\frac{1}{12}+8\frac{1}{9}-16\frac{1}{4}$, is equal to what?
23. $1-\frac{3}{8}$ of $\frac{5}{6}-\frac{2}{3}$ of $\frac{3}{7}=$ what?

MULTIPLICATION OF FRACTIONS.

142. Multiplication of Fractions is finding the product when either or when each of the factors is a fractional number. There are three cases:

1. *To multiply a fraction by an integer.*
2. *To multiply an integer by a fraction.*
3. *To multiply one fraction by another*

NOTE.—Since any whole number may be expressed in the form of a fraction, the first and second cases are special cases of the third.

PROBLEM.—Multiply $\frac{3}{7}$ by $\frac{5}{8}$.

SOLUTION.—Once $\frac{3}{7}$ is $\frac{3}{7}$. $\frac{1}{8}$ times $\frac{3}{7}$ is $\frac{1}{8}$ of $\frac{3}{7}=\frac{3}{56}$. $\frac{5}{8}$ times $\frac{3}{7}$, then, is 5 times $\frac{3}{56}=\frac{15}{56}$.

OPERATION.

$\frac{3}{7}\times\frac{5}{8}=\frac{15}{56}$, *Ans.*

PROBLEM.—Multiply $\frac{3}{4}$ by 6.

SOLUTION.—Six times 3 fourths is 18 fourths. Reducing to its simplest form, we have $4\frac{1}{2}$.

OPERATION.

$\frac{3}{4}\times\frac{6}{1}=\frac{18}{4}=4\frac{1}{2}$, *Ans.*

or, $\frac{3}{4}\times 6=\frac{18}{4}=4\frac{1}{2}$

PROBLEM.—Multiply 8 by $\frac{3}{5}$.

SOLUTION.—One fifth times 8 is $\frac{8}{5}$, 3 fifths times 8 is 3 times $\frac{8}{5}=\frac{24}{5}$. Reducing to a mixed number, we have $4\frac{4}{5}$.

OPERATION.

$\frac{8}{1}\times\frac{3}{5}=\frac{24}{5}=4\frac{4}{5}$, *Ans.*

or, $8\times\frac{3}{5}=\frac{24}{5}=4\frac{4}{5}$

NOTE.—The three operations are alike, and from them we may derive the rule.

Rule.—*Multiply the numerators together for the numerator of the product, and the denominators for the denominator of the product*

REMARK.—Whole numbers may be expressed in the form of fractions.

EXAMPLES FOR PRACTICE.

1. $\frac{10}{13}\times 12$.
2. $\frac{11}{24}\times 18$.
3. $\frac{29}{48}\times 24$.
4. $\frac{9}{16}\times 28$.
5. $\frac{13}{15}\times 30$.
6. $3\frac{2}{3}\times 5$.

REMARK.—In multiplying a mixed number by a whole number, multiply the whole number and the fraction separately, and add the products; or, reduce the mixed number to an improper fraction, and multiply it; as, in the last example

OPERATIONS.

$$\begin{array}{r} 3\frac{2}{3} \\ 5 \\ \hline 15 \\ 3\frac{1}{3} \\ \hline 18\frac{1}{3} \end{array}, \textit{Ans.}$$

$$3\frac{2}{3}=\frac{11}{3}$$
$$\frac{11}{3}\times 5=\frac{55}{3}$$
$$\frac{55}{3}=18\frac{1}{3}\ \textit{Ans.}$$

7. $45 \times \frac{7}{9}$.
8. $50 \times \frac{11}{14}$.
9. $25 \times \frac{3}{4}$.
10. $32 \times 2\frac{3}{8}$.
11. $28 \times 3\frac{3}{8}$.
12. $\frac{12}{35} \times \frac{7}{16}$.
13. $\frac{15}{16} \times \frac{24}{25}$.
14. $\frac{42}{55} \times \frac{22}{35}$.

15. What will $3\frac{1}{3}$ yards of cloth cost, at $\$4\frac{1}{2}$ per yard?

REMARK.—In finding the product of two mixed numbers, it is generally best to reduce them to improper fractions; thus,

$$4\tfrac{1}{2}=\tfrac{9}{2};\ 3\tfrac{1}{3}=\tfrac{10}{3};\ \$\tfrac{9}{2}\times\tfrac{10}{3}=\tfrac{90}{6}=\$15,\ \textit{Ans.}$$

SOLUTION.—The operation may be performed without reducing to improper fractions; thus, 3 yards will cost $\$13\frac{1}{2}$, and $\frac{1}{3}$ of a yard will cost $\frac{1}{3}$ of $\$4\frac{1}{2}=\$1\frac{1}{2}$; hence, the whole will cost $15.

OPERATION.

$$\begin{array}{r} \$4\tfrac{1}{2} \\ \underline{3\tfrac{1}{3}} \\ 13\tfrac{1}{2} \\ \underline{1\tfrac{1}{2}} \\ \textit{Ans}, \$15 \end{array}$$

16. $6\frac{2}{3} \times 4\frac{1}{2}$.
17. $4\frac{4}{5} \times 2\frac{3}{8}$.
18. $12\frac{3}{8} \times 3\frac{3}{11}$.
19. $7\frac{11}{12} \times 3\frac{7}{19}$.

20. Multiply $\frac{1}{5}$ of 8 by $\frac{1}{4}$ of 10.
21. Multiply $\frac{2}{3}$ of $5\frac{2}{5}$ by $\frac{3}{5}$ of $3\frac{1}{3}$.
22. Multiply $\frac{3}{4}$ of $\frac{2}{3}$ of $5\frac{3}{5}$ by $\frac{3}{7}$ of $3\frac{3}{8}$.
23. Multiply 5, $4\frac{1}{4}$, $2\frac{1}{3}$, and $\frac{3}{7}$ of $4\frac{4}{9}$.
24. Multiply $\frac{4}{5}$, $\frac{3}{7}$, $\frac{5}{11}$, $\frac{1}{3}$ of $2\frac{1}{2}$, and $\frac{4}{7}$ of $3\frac{1}{7}$.
25. Multiply $\frac{3}{5}$, $\frac{7}{9}$, $\frac{9}{11}$, $3\frac{1}{3}$, and $3\frac{1}{7}$.
26. Multiply $3\frac{1}{2}$, $4\frac{2}{3}$, $5\frac{3}{5}$, $\frac{2}{9}$ of $\frac{5}{14}$, and $6\frac{3}{4}$.
27. At $\frac{7}{8}$ of a dollar per yard, what will 25 yards of cloth cost?
28. A quantity of provisions will last 25 men $12\frac{3}{4}$ days: how long will the same last one man?
29. At $3\frac{1}{2}$ cents a yard, what will $2\frac{3}{4}$ yards of tape cost?
30. What must be paid for $\frac{3}{5}$ of $\frac{2}{3}$ of a lot of groceries that cost $\$18\frac{3}{4}$?
31. K owns $\frac{5}{8}$ of a ship, and sells $\frac{3}{4}$ of his share to L: what part has he left?

DIVISION OF FRACTIONS.

143. Division of Fractions is finding the quotient when the dividend or divisor is fractional, or when both are fractional. There are three cases:

1. *To divide a fraction by an integer.*
2. *To divide an integer by a fraction.*
3. *To divide one fraction by another.*

NOTE.—Since any whole number may be expressed in the form of a fraction, the first and second cases reduce to the third case.

PROBLEM.—Divide $\frac{5}{6}$ by $\frac{4}{7}$.

OPERATION.

$\frac{5}{6} \div \frac{4}{7} = \frac{5}{6} \times \frac{7}{4} = \frac{35}{24}$

SOLUTION.—$\frac{1}{7}$ is contained in 1, seven times; $\frac{1}{7}$ is contained in $\frac{1}{6}$, $\frac{1}{6}$ of $7 = \frac{7}{6}$ times; $\frac{1}{7}$ is contained in $\frac{5}{6}$, $5 \times \frac{7}{6} = \frac{35}{6}$ times; $\frac{4}{7}$ is contained in $\frac{5}{6}$, $\frac{1}{4}$ of $\frac{35}{6} = \frac{35}{24}$ times. It will be seen that the terms of the dividend have been *multiplied,* and that the terms of the divisor have exchanged places. Writing the terms thus is called "inverting the terms of the divisor," or simply, "inverting the divisor."

PROBLEM.—Divide 3 by $\frac{2}{5}$.

OPERATION.

$\frac{3}{1} \div \frac{2}{5} = \frac{3}{1} \times \frac{5}{2} = \frac{15}{2} = 7\frac{1}{2}$, *Ans.*

Or, $3 \div \frac{2}{5} = \frac{15}{2} = 7\frac{1}{2}$.

SOLUTION.—$\frac{1}{5}$ is contained in 1, five times; $\frac{1}{5}$ is contained in 3, three times 5 times $= 15$ times; $\frac{2}{5}$ is contained in 3, $\frac{1}{2}$ of 15 times $= \frac{15}{2} = 7\frac{1}{2}$ times.

PROBLEM.—Divide $\frac{4}{7}$ by 2.

OPERATION.

$\frac{4}{7} \div \frac{2}{1} = \frac{4}{7} \times \frac{1}{2} = \frac{4}{14} = \frac{2}{7}$, *Ans*

Or, $\frac{4}{7} \div 2 = \frac{2}{7}$.

SOLUTION.—Two is contained in 1, $\frac{1}{2}$ times; 2 is contained in $\frac{1}{7}$, $\frac{1}{2}$ of $\frac{1}{7} = \frac{1}{14}$ times; 2 is contained in $\frac{4}{7}$, 4 times $\frac{1}{14} = \frac{4}{14} = \frac{2}{7}$ times.

NOTE.—From these solutions we may derive the following rule.

Rule.—*Multiply the dividend by the divisor with its terms inverted.*

Remark.—The terms of the divisor are inverted because the solution requires it. The same may be shown by a different solution, as below.

Problem.—Divide $\frac{2}{9}$ by $\frac{3}{5}$.

Solution.—Reduce both dividend and divisor to a common denominator. The quotient of $\frac{10}{45} \div \frac{27}{45}$ is the same as $10 \div 27 = \frac{10}{27}$. The same result is obtained by multiplying the dividend by the divisor, with its terms inverted. thus, $\frac{2}{9} \times \frac{5}{3} = \frac{10}{27}$.

OPERATION.

$\frac{2}{9} = \frac{10}{45}$ $\frac{3}{5} = \frac{27}{45}$

$\frac{10}{45} \div \frac{27}{45} = \frac{10}{27}$, *Ans.*

Remark.—Mixed numbers must be reduced to improper fractions. Use cancellation when applicable.

Examples for Practice.

1. $\frac{9}{16} \div 3$.
2. $\frac{14}{23} \div 7$.
3. $\frac{3}{5} \div 8$.
4. $6 \div \frac{2}{3}$.
5. $21 \div \frac{9}{10}$.
6. $\frac{3}{4} \div \frac{1}{2}$.
7. $\frac{2}{3} \div \frac{1}{40}$.
8. $\frac{21}{25} \div \frac{14}{15}$.
9. $\frac{12}{35} \div \frac{30}{77}$.
10. $1\frac{3}{4} \div 5$.
11. $8\frac{1}{6} \div \frac{2}{3}$.
12. $19\frac{1}{2} \div 1\frac{7}{8}$.
13. $73\frac{1}{2} \div 9\frac{4}{5}$.
14. $54\frac{43}{48} \div 25\frac{5}{6}$.
15. Divide $1\frac{1}{2}$ by $\frac{1}{2}$ of $\frac{3}{5}$ of $7\frac{1}{2}$.
16. Divide $\frac{3}{10}$ of $\frac{3}{16}$ of $\frac{5}{12}$ by $\frac{7}{24}$ of $\frac{12}{35}$.
17. Divide $\frac{3}{4}$ of $\frac{1}{2}$ by $\frac{1}{4}$ of $\frac{2}{3}$.
18. Divide $\frac{7}{9}$ of $3\frac{3}{5}$ by $\frac{13}{14}$ of 7.
19. Divide $\frac{1}{3}$ of $\frac{4}{5} \times \frac{39}{320}$ by $\frac{2}{75}$ of $3\frac{1}{4}$.
20. Divide $\frac{2}{7}$ of $5\frac{1}{2}$ by $\frac{2}{5}$ of $\frac{9}{14}$ of $3\frac{1}{3}$.
21. Divide $\frac{1}{3}$ of $\frac{2}{7}$ of $\frac{4}{11}$ by $\frac{2}{5}$ of $\frac{1}{3}$ of $\frac{4}{7}$.
22. Divide $1\frac{7}{8}$ times $4\frac{2}{3}$ by $1\frac{7}{11}$ times $3\frac{8}{9}$.
23. Divide $3\frac{1}{7}$ by $\frac{4}{9}$ of $8\frac{1}{4}$ times $\frac{5}{11}$ of $3\frac{3}{10}$.
24. Divide $\frac{9}{11}$ of $\frac{2}{3}$ of $27\frac{1}{2}$ by $\frac{4}{9}$ of $\frac{3}{17}$ of $5\frac{1}{2}$.
25. What is $2\frac{5}{8} \times \frac{4}{5}$ of $19\frac{1}{3} \div (4\frac{5}{6} \times \frac{3}{10}$ of 8)?

144. To reduce complex to simple fractions.

PROBLEM.—Reduce $\frac{1\frac{5}{6}}{2\frac{1}{4}}$ to a simple fraction.

OPERATION.

$1\frac{5}{6} = \frac{11}{6}$ $\quad 2\frac{1}{4} = \frac{9}{4}$

$\frac{11}{6} \div \frac{9}{4} = \frac{11}{6} \times \frac{4}{9} = \frac{44}{54} = \frac{22}{27}$, *Ans.*

EXPLANATION.—The mixed numbers are reduced to improper fractions, and the numerator is divided by the denominator. (Art. 114.)

Rule.—*Divide the numerator by the denominator, as in division of fractions.*

Reduce to simple fractions:

1. $\frac{\frac{3}{8}}{\frac{11}{4}}$.
2. $\frac{1\frac{3}{4}}{2\frac{4}{5}}$.
3. $\frac{4\frac{2}{3}}{2\frac{11}{12}}$.
4. $\frac{2\frac{5}{14}}{2\frac{2}{21}}$.
5. $\frac{12\frac{3}{8}}{18}$.
6. $\frac{62}{16\frac{10}{11}}$.

REMARK.—Complex fractions may be multiplied or divided, by reducing them to simple fractions. The operation may often be shortened by cancellation.

7. $\frac{7}{10} \times \frac{4\frac{1}{4}}{8}$.
8. $\frac{31}{25} \times \frac{8\frac{1}{2}}{10\frac{1}{5}}$.
9. $\frac{5\frac{1}{3}}{9\frac{1}{5}} \times \frac{11\frac{11}{12}}{24\frac{1}{5}}$.
10. $\frac{6\frac{1}{4}}{2\frac{2}{5}} \div \frac{7}{12\frac{1}{2}}$.
11. $\frac{7\frac{3}{7}}{40\frac{5}{9}} \div \frac{17\frac{1}{3}}{73}$.
12. $\frac{2\frac{4}{11}}{2\frac{3}{5}} \div \frac{2\frac{7}{11}}{8\frac{7}{10}}$.

THE GREATEST COMMON DIVISOR OF FRACTIONS.

145. **The greatest common divisor** of two or more fractions is the greatest fraction that will exactly divide each of them.

One fraction is *divisible* by another when the numerator of the divisor is a factor of the numerator of the dividend, and the denominator of the divisor is a multiple of the denominator of the dividend.

Thus, $\frac{9}{16}$ is divisible by $\frac{3}{32}$; for $\frac{9}{16}=\frac{18}{32}$; $\frac{18}{32}\div\frac{3}{32}=6$.

The *greatest* common divisor of two or more fractions, must be that fraction whose numerator is the G. C. D. of all the numerators, and whose denominator is the L. C. M. of the denominators.

Thus, the G. C. D. of $\frac{8}{35}$ and $\frac{18}{49}$ is $\frac{2}{245}$.

PROBLEM.—Find the G. C. D. of $\frac{5}{8}$, $\frac{25}{32}$, and $\frac{7}{12}$.

SOLUTION.—Since 5 and 7 are both prime numbers, 1 is the G. C. D. of all the numerators; 96 is the L. C. M. of 8, 32, and 12; therefore, the G. C. D. of the fraction is $\frac{1}{96}$.

OPERATION.

1 = G. C. D. of 5, 25, 7
96 = L. C. M. of 8, 32, 12
$\frac{1}{96}$, *Ans.*

Rule.—*Find the G. C. D. of the numerators of the fractions, and divide it by the L. C. M. of their denominators.*

REMARK.—The fractions should be in their simplest forms before the rule is applied.

Find the greatest common divisor:

1. Of $83\frac{1}{3}$ and $268\frac{3}{4}$.
2. Of $14\frac{7}{12}$ and $95\frac{3}{8}$.
3. Of $59\frac{1}{9}$ and $735\frac{14}{15}$.
4. Of $23\frac{7}{16}$ and $213\frac{13}{24}$.
5. Of $418\frac{3}{5}$ and $1772\frac{1}{3}$.
6. Of $261\frac{13}{14}$ and $652\frac{11}{21}$.
7. Of $44\frac{4}{9}$, $546\frac{2}{3}$, and 3160.
8. A farmer sells $137\frac{1}{2}$ bushels of yellow corn, $478\frac{1}{8}$ bushels of white corn, and $2093\frac{3}{4}$ bushels of mixed corn: required the size of the largest sacks that can be used in shipping, so as to keep the corn from being mixed; also the number of sacks for each kind.

9. A owns a tract of land, the sides of which are $134\frac{3}{4}$, $128\frac{1}{3}$, and $115\frac{1}{2}$ feet long: how many rails of the greatest length possible will be needed to fence it in straight lines, the fence to be 6 rails high, and the rails to lap 6 inches at each end?

THE LEAST COMMON MULTIPLE OF FRACTIONS.

146. The **least common multiple** of two or more fractions is the least number that each of them will divide exactly.

NOTE.—The G. C. D. of several fractions must be a fraction, but the L. C. M. of several fractions may be an integer or a fraction.

A fraction is a *multiple* of a given fraction when its numerator is a multiple, and its denominator is a divisor, of the corresponding terms of the given fraction.

ILLUSTRATION.—$\frac{8}{11}$ is a multiple of $\frac{2}{33}$. 8 is a multiple of 2, and 11 is a divisor of 33; hence, $\frac{8}{11} \div \frac{2}{33} = \frac{8}{11} \times \frac{33}{2} = 12$. The same result is otherwise obtained; thus, $\frac{8}{11} = \frac{24}{33}$, and $\frac{24}{33} \div \frac{2}{33} = 12$.

A fraction is a *common* multiple of two or more given fractions when its numerator is a common multiple of the numerators of the given fractions, and its denominator is a common divisor of the denominators of the given fractions.

A fraction is the *least* common multiple of two or more fractions when its numerator is the least common multiple of the given numerators, and its denominator is the greatest common divisor of the given denominators.

PROBLEM.—Find the L. C. M. of $\frac{1}{3}$, $\frac{3}{4}$, and $\frac{5}{6}$.

SOLUTION.—The L. C. M. of the numerators is 15. The G. C. D. of the denominators is 1; therefore, the L. C. M. of the fractions is $\frac{15}{1}$, or 15.

OPERATION.

L. C. M. of 1, 3, 5 $= 3 \times 5 = \mathbf{15}$

G. C. D. of 3, 4, 6 $= 1$

$\therefore \frac{15}{1}$, *Ans.*

Rule.—*Divide the L. C. M. of the numerators by the G. C. D. of the denominators.*

REMARK.—The fractions must be in their simplest forms before the rule is applied.

Find the least common multiple:

1. Of $\frac{2}{3}$, $\frac{3}{4}$, $\frac{4}{5}$, $\frac{5}{6}$, and $\frac{6}{7}$.
2. Of $4\frac{1}{2}$, $6\frac{3}{4}$, $5\frac{5}{6}$, and $10\frac{1}{2}$.
3. Of $3\frac{1}{3}$, $4\frac{3}{8}$, $\frac{5}{12}$, $5\frac{5}{9}$, and $12\frac{1}{2}$.
4. A can walk around an island in $14\frac{2}{7}$ hours; B, in $9\frac{1}{11}$ hours; C, in $16\frac{2}{3}$ hours; and D, in 25 hours. If they start from the same point, and at the same time, how many hours after starting till they are all together again?

PROMISCUOUS EXERCISES.

NOTE TO TEACHERS.—All problems marked thus [*], are to be solved mentally by the class. In the solution of such problems, the following is earnestly recommended:

1. The teacher will read the problem slowly and distinctly, and not repeat it.
2. The pupil designated by the teacher, will then give the answer to the question.
3. Some pupil, or pupils, will now reproduce the question in the exact language in which it was first given to the class.
4. The pupil, or pupils, called upon by the teacher, will give a short, logical analysis of the problem.

1. What is the sum of $3\frac{1}{2}$, $4\frac{1}{3}$, $5\frac{1}{4}$, $\frac{3}{4}$ of $\frac{7}{8}$, and $\frac{1}{2}$ of $\frac{1}{3}$ of $\frac{5}{8}$?

2. The sum of $1\frac{1}{26}$ and $\dfrac{1}{1\frac{4}{9}}$ is equal to how many times their difference?

3. What is $(2\frac{3}{4} + \dfrac{5}{2}$ of $\dfrac{7}{3\frac{4}{5}} - \dfrac{1\frac{2}{3}}{2\frac{1}{2}}) \div 1\frac{77}{228}$?

4. Reduce $\dfrac{4\frac{4}{15} \text{ of } 2\frac{5}{8}}{5\frac{1}{5} - 4\frac{1}{2}}$ and $\dfrac{5}{7} \times (100 - \dfrac{200}{3} + \dfrac{7\frac{1}{3}}{2\frac{1}{4}})$ to their simplest forms.

5. What is $\frac{1}{4}$ of $5\frac{1}{4} - \frac{1}{3}$ of $3\frac{7}{8}$?

6. What is $\frac{32}{51} \times \frac{85}{112} \times \frac{189}{207} \times \frac{23}{36}$ equal to?

7.* $\dfrac{1}{2} \times \dfrac{1 - \frac{1}{2}}{2} \times \dfrac{2 - \frac{1}{2}}{3}$?

8. $\dfrac{1}{3} \times \dfrac{1 - \frac{1}{3}}{2} \times \dfrac{2 - \frac{1}{3}}{3} \times \dfrac{3 - \frac{1}{3}}{5} \times \dfrac{4 - \frac{1}{3}}{4}$?

9. $\dfrac{(2 + \frac{1}{5}) \div (3 + \frac{1}{7})}{(2 - \frac{1}{3}) \times (4 - 3\frac{3}{7})} =$ what?

10. $\dfrac{4\frac{1}{3} \times 4\frac{1}{3} \times 4\frac{1}{3} - 1}{4\frac{1}{3} \times 4\frac{1}{3} - 1} =$ what?

11. Add $\frac{2}{3}$ of $\frac{3}{4}$ of $\frac{7}{8}$, $\frac{1}{5} \times \frac{2}{3}$ of $1\frac{7}{8}$, and $\frac{1}{4}$.

12. $\frac{3}{5}$ of $\frac{10}{9}$ of what number, diminished by leaves $\frac{25}{64}$?

13.* James's money equals $\frac{3}{5}$ of Charles's money; and $\frac{3}{4}$ of James's money + \$33 equals Charles's money: how much has each?

14. A leaves L for N at the same time that B leaves N for L. The two places are exactly 109 miles apart: A travels $7\frac{1}{2}$ miles per hour, and B, $8\frac{1}{4}$ miles per hour; in how many hours will they meet, and how far will each have traveled?

15. What number multiplied by $\frac{5}{9}$ of $\frac{3}{7}$ of $3\frac{11}{15}$ will produce $2\frac{1}{2}$?

16. What, divided by $1\frac{3}{5}$, gives $14\frac{3}{4}$?

17. What, added to $14\frac{5}{8}$, gives $29\frac{23}{35}$?

18.* I spend $\frac{5}{9}$ of my income in board, $\frac{1}{6}$ of it in clothes, and save \$60 a year: what is my income?

19.* Divide 51 into two such parts that $\frac{2}{3}$ of the first is equal to $\frac{3}{4}$ of the second.

20.* $\frac{1}{2}$ is what part of $\frac{2}{3}$?

21. Divide $\frac{7}{9}$ of $3\frac{3}{5}$ by $\frac{13}{14}$ of 7; and $\frac{9}{11}$ of $\frac{2}{3}$ of $27\frac{1}{2}$ by $\frac{4}{9}$ of $\frac{3}{17}$ of $5\frac{1}{2}$.

22. Multiply $\frac{7}{11}$ of $2\frac{1}{2}$ by $\frac{8}{13}$ of $19\frac{1}{2}$; and divide $\frac{7}{8}$ of $\frac{5}{9}$ of $14\frac{1}{7}$ by $\frac{3}{11}$ of $\frac{2}{7}$ of $13\frac{4}{9}$.

23. $\left(\frac{3131}{4331} + \frac{1470}{2911} + \frac{1981}{2501}\right) \div \dfrac{2\frac{1}{2}}{7} =$ what?

24.* A bequeathed $\frac{11}{20}$ of his estate to his elder son; the rest to his younger, who received $525 less than his brother. What was the estate?

25. Find the sum, difference, and product of $3\frac{7}{8}$ and $2\frac{1}{9}$; also, the quotient of their sum by the difference.

26.* A cargo is worth 7 times the ship: what part of the cargo is $\frac{5}{16}$ of the ship and cargo?

27. By what must the sum of $\frac{757}{2747}$, $\frac{951}{2479}$, and $\frac{517}{1517}$ be multiplied to produce 1000?

28. Multiply the sum of all the divisors of 8128, excluding itself, by the number of its prime factors excluding 1, and divide by $149\frac{1}{3}$.

29. $6\frac{1}{2}$ is what part of $10\frac{7}{11}$? Reduce to its simplest form $\frac{8}{9} - \frac{1}{7} + \frac{5}{8} - 1\frac{1}{4}$.

30. Multiply $\dfrac{4}{5\frac{1}{3}}$, $14\frac{1}{7}$, $\dfrac{2\frac{3}{4}}{4}$, $\dfrac{5}{7\frac{1}{3}}$, $\dfrac{1\frac{1}{5}}{2\frac{3}{4}}$, and 6.

31. $\frac{7}{8}$ of $\frac{5}{9}$ of what number equals $9\frac{13}{18}$?

32.* A 63-gallon cask is $\frac{5}{8}$ full: $9\frac{1}{2}$ gallons being drawn off, how full will it be?

33. If a person going $3\frac{3}{4}$ miles per hour, perform a journey in $14\frac{3}{4}$ hours, how long would it take him if he traveled $5\frac{1}{4}$ miles per hour?

34. A man buys $32\frac{3}{4}$ pounds of coffee, at $17\frac{5}{8}$ cents a pound: if he had got it $4\frac{2}{3}$ cents a pound cheaper, how many more pounds would he have received?

35. Henry spent $\frac{2}{11}$ of his money and then received $65; he then lost $\frac{3}{4}$ of all his money, and had in hand $10 less than at first. How much had he at first?

Topical Outline.

Common Fractions.

1. Definition.
2. Classes
   1. As to Kinds
      1. Common.
      2. Decimal.
   2. As to Value
      1. Proper.
      2. Improper.
      3. Mixed.
   3. As to Form
      1. Simple.
      2. Complex.
      3. Compound.
3. Terms
   1. Numerator.
   2. Denominator.
   3. Similar.
   4. Dissimilar.
4. Principles.
5. Reduction
   1. Cases
      1. Lowest Terms.
      2. Higher Terms.
      3. Mixed Numbers to Improper Fractions.
      4. Improper Fractions to Mixed Numbers.
      5. Compound to Simple Fractions.
      6. Common Denominator.
      7. Least Common Denominator.
   2. Principles.
   3. Rules.
6. Practical Applications
   1. Addition
      1. Definition.
      2. Principles.
      3. Rule.
   2. Subtraction
      1. Definition.
      2. Principles.
      3. Rule.
   3. Multiplication
      1. Definition.
      2. Principles.
      3. Rule.
   4. Division
      1. Definition.
      2. Principles.
      3. Rule.
   5. Divisors, Multiples, etc.

IX. DECIMAL FRACTIONS.

147. **A Decimal Fraction** is a fraction whose denominator is 10, or some product of 10, expressed by 1 with ciphers annexed.

REMARK 1.—A decimal fraction is also defined as a *fraction whose denominator is some power of* 10. By the "*power*" of a quantity, is usually understood, either that quantity itself, or the product arising from taking *only* that quantity a certain number of times as a factor. Thus, $9 = 3 \times 3$, or the *second power* of 3.

REMARK 2.—Since decimal fractions form only one of the *classes* (Art. **108**) under the term *fractions*, the general principles relating to common fractions relate also to decimals.

148. The orders of integers decrease from left to right in a tenfold ratio (Art. **48**). The orders may be continued from the place of units toward the right by the same law of decrease.

149. The places at the right of units are called **decimal places,** and decimal fractions when so written, without a denominator expressed, are called **decimals.**

150. The **decimal point,** or **separatrix,** is a dot [.] placed at the left of decimals to distinguish them from integers.

Thus, $\frac{1}{10}$	is written		.1
$\frac{1}{100}$	"	"	.01
$\frac{1}{1000}$	"	"	.001
$\frac{1}{10000}$	"	"	.0001

From this, it is evident that,

The denominator of any decimal is 1 *with as many ciphers annexed as there are places in the decimal.*

151. A **pure decimal** consists of decimal places only; as, .325

152. A **mixed decimal** consists of a whole number and a decimal written together; as, 3.25

REMARK.—A mixed decimal may be read as an improper fraction, since $3\frac{25}{100}=\frac{325}{100}$.

153. A **complex decimal** has a common fraction in its right-hand place; as, $.033\frac{1}{3}$

154. From the general law of notation (Arts. **48** and **148**) may be derived the following principles:

PRINCIPLE I.—*If, in any decimal, the point be moved to the right, the decimal is multiplied by* 10 *as often as the point is removed one place.*

ILLUSTRATION.—If, in the decimal .032, we move the point one place to the right, we have .32. The first has three decimal places, and represents *thousandths;* while the second has two places, and represents *hundredths.* (Art. **129**, Prin. I.)

PRINCIPLE II.—*If, in any decimal, the point be moved to the left, the decimal is divided by* 10 *as often as the point is removed one place.*

ILLUSTRATION.—If, in the decimal .35 we move the point one place to the left, we have .035. The first represents *hundredths;* the second, *thousandths,* while the numerator is not changed. (Art. **129**, Prin. II.)

PRINCIPLE III.—*Decimal ciphers may be annexed to, or omitted from, the right of any number without altering its value.*

ILLUSTRATION.—.5 is equal to .500; for $\frac{5\times100}{10\times100}=\frac{500}{1000}$. The reverse may be shown in the same way. (Art. **129**, Prin. III.)

NUMERATION AND NOTATION OF DECIMALS.

155. Since $.6=\frac{6}{10}$; $.06=\frac{6}{100}$; and $.006=\frac{6}{1000}$, any figure expresses *tenths, hundredths,* or *thousandths,* according as it is in the 1st, 2d, or 3d decimal place; hence, these places are named respectively the *tenths'*, the *hundredths'*, the *thousandths'* place; other places are named in the same way as seen in the following table:

TABLE OF DECIMAL ORDERS.

Tenths. Hundredths. Thousandths. Ten-thousandths. Hundred-thousandths. Millionths. Ten-millionths. Hundred-millionths. Billionths.

1st	place	.2	read	2 Tenths.
2d	"	.08	"	8 Hundredths.
3d	"	.005 . . .	"	5 Thousandths.
4th	"	.0007 . . .	"	7 Ten-thousandths.
5th	"	.00003 . .	"	3 Hundred-thousandths.
6th	"	.000001 . .	"	1 Millionth.
7th	"	.0000009 .	"	9 Ten-millionths.
8th	"	.00000004 .	"	4 Hundred-millionths.
9th	"	.000000006	"	6 Billionths.

NOTE.—The names of the decimal orders are derived from the names of the orders of whole numbers. The table may therefore be extended to trillionths, quadrillionths, etc.

PROBLEM.—Read the decimal .0325

SOLUTION.—The numerator is 325; the denomination is ten-thousandths since there are four decimal places. It is read three hundred and twenty-five ten-thousandths.

Rule.—*Read the number expressed in the decimal places as the numerator, give it the denomination expressed by the right-hand figure.*

EXAMPLES TO BE READ.

1. .9
2. $.6\frac{2}{3}$
3. $.0\frac{3}{7}$
4. .035

5. .7200
6. .5060
7. 1.008
8. 9.00$\frac{1}{3}$
9. 105.0$\frac{7}{8}$
10. .0003
11. 00.100
12. 180.010
13. 2030.0
14. 40.68031
15. 200.002
16. .0900001
17. 61.010001
18. 31.0200703
19. .000302501
20. .03672113

Exercises in Notation.

156. The numerator is written as a simple number; the denomination is then expressed by the use of the decimal point, and, if necessary, by the use of ciphers in vacant places.

Problem.—Write eighty-three thousand and one billionths.

Explanation.—First write the numerator, 83001. If the point were placed immediately at the left of the 8, the denominator would be hundred-thousandths; it is necessary to fill four places with ciphers so that the final figure may be in billionths' place.

OPERATION.
.000083001

Rule.—*Write the numerator as a whole number; then place the decimal point so that the right-hand figure shall be of the same name as the decimal.*

Examples to be Written.

1. Five tenths.
2. Twenty-two hundredths.
3. One hundred and four thousandths.
4. Two units and one hundredth.
5. One thousand six hundred and five ten-thousandths.
6. Eighty-seven hundred-thousandths.
7. Twenty-nine and one half ten-millionths.

8. Nineteen million and one billionths.
9. Seventy thousand and forty-two units and sixteen hundredths.
10. Two thousand units and fifty-six and one third millionths.
11. Four hundred and twenty-one tenths.
12. Six thousand hundredths.
13. Forty-eight thousand three hundred and five thousandths.
14. Eight units and one half a hundredth.
15. Thirty-three million ten-millionths.
16. Four hundred thousandths.
17. Four hundred-thousandths.
18. One unit and one half a billionth.
19. Sixty-six thousand and three millionths.
20. Sixty-six million and three thousandths.
21. Thirty-four and one third tenths.

REDUCTION OF DECIMALS.

157. Reduction of decimals is changing their form without altering their value.

CASE I.

158. To reduce a decimal to a common fraction.

PROBLEM.—Reduce .24 to a common fraction.

SOLUTION.—.24 is equal to $\frac{24}{100}$, which, reduced, is $\frac{6}{25}$.

OPERATION.

$.24 = \frac{24}{100} = \frac{6}{25}$, *Ans.*

PROBLEM.—Reduce $.12\frac{1}{2}$ to a common fraction.

SOLUTION.—Write $12\frac{1}{2}$ as a numerator, and under it place 100 as a denominator. Reduce the complex fraction according to Art. 144.

OPERATION.

$.12\frac{1}{2} = \frac{12\frac{1}{2}}{100} = \frac{\frac{25}{2}}{100} = \frac{25}{200} = \frac{1}{8}$, *Ans.*

Rule.—*Write the decimal as a common fraction; then reduce the fraction to its lowest terms.*

REMARK.—If the decimal contains many decimal places, an approximate value is sometimes used. For example, 3.14159 = nearly $3\frac{1}{7}$.

Reduce to common fractions

1. .25625
2. .15234375
3. 2.125
4. 19.01750
5. $16.00\frac{1}{5}$
6. $350.028\frac{4}{7}$
7. $.666666\frac{2}{3}$
8. .003125
9. $11.0\frac{5}{9}$
10. .390625
11. $.1944\frac{4}{9}$
12. $.24\frac{4}{9}$
13. $.33\frac{1}{3}$
14. $.66\frac{2}{3}$
15. .25
16. .75

CASE II.

159. To reduce common fractions to decimals.

PROBLEM.—Reduce $\frac{7}{8}$ to a decimal.

OPERATION.

$$\begin{array}{r} 8\,)\,7.000 \\ \hline .875,\ \textit{Ans.} \end{array}$$

SOLUTION.—The fraction $\frac{7}{8} = \frac{1}{8}$ of 7. 7 = 7.0, $\frac{1}{8}$ of 7.0 = .8, with 6 tenths remaining; .6 = .60, $\frac{1}{8}$ of .60 = .07, with 4 hundredths remaining; .04 = .040, $\frac{1}{8}$ of .040 = .005. The answer is .875.

NOTE.—Another form of solution may be obtained by multiplying both terms of the fraction by 1000; dividing both terms of the resulting fraction by the first denominator and writing the answer as a decimal. Thus, $\frac{7}{8} = \frac{7000}{8000}$, $\frac{7000}{8000} = \frac{875}{1000} = .875$. Both solutions depend on Art. **129**, Prin. III.

Rule.—*Annex ciphers to the numerator and divide it by the denominator. Then point off as many decimal places in the quotient as there are ciphers annexed.*

REMARK.—Any fraction in its lowest terms having in its denominator any prime factor other than 2 and 5, can not be reduced exactly to a decimal. Thus, $\frac{1}{12} = .08333+$. The sign + is used at the end of a decimal to indicate that the result is less than the true quotient. The sign — is also sometimes used to indicate that the last figure is too great. Thus, $\frac{1}{7}=.1428+$, or, by abbreviating, $\frac{1}{7}=.143-$.

Reduce to decimals:

1. $\frac{3}{4}$.
2. $\frac{1}{8}$.
3. $\frac{1}{20}$.
4. $\frac{15}{32}$.
5. $\frac{9}{1600}$.
6. $\frac{4}{5}$.
7. $\frac{99}{200}$.
8. $\frac{5}{64}$.
9. $\frac{13}{256}$.
10. $\frac{1}{1024}$.

NOTE.—The rule converts a mixed number into a mixed decimal, and a complex into a pure decimal; thus, $9\frac{3}{8}=9.375$, since $\frac{3}{8}=.375$; and $.26\frac{3}{25}=.2612$, since $\frac{3}{25}=.12$.

11. $16\frac{1}{2}$
12. $42\frac{3}{16}$
13. $.015\frac{1}{4}$
14. $101.01\frac{3}{4}$
15. $75119\frac{3}{80}$
16. $2.00\frac{1}{320}$

ADDITION OF DECIMALS.

160. Addition of Decimals is finding the sum of two or more decimals.

NOTE.—Complex decimals, if there are any, must be made pure as far, at least, as the decimal places extend in the other numbers.

PROBLEM.—Add 23.8 and $17\frac{1}{2}$ and .0256 and $.41\frac{2}{3}$.

SOLUTION.—Write the numbers as in the operation, and add as in simple addition. Write the decimal point in the sum to the left of tenths.

OPERATION.

$$\begin{array}{rr} & 23.8 \\ 17\frac{1}{2}= & 17.5 \\ & .0256 \\ .41\frac{2}{3}= & .4166\frac{2}{3} \\ \hline & 41.7422\frac{2}{3}, \textit{Ans.} \end{array}$$

Rule.—1. *Write the numbers so that figures of the same order shall stand in the same column.*

2. *Then add as in simple numbers, and put the decimal point to the left of tenths.*

REMARK.—The proof of each fundamental operation in decimals is the same as in simple numbers.

1. Find the sum of $1 + .9475$
2. Of $1.33\frac{1}{3}$ added to $2.66\frac{2}{3}$
3. Of 14.034, 25, $.000062\frac{1}{2}$, .0034
4. Of 83 thousandths, 2101 hundredths, 25 tenths, and $94\frac{1}{2}$ units.
5. Of $.16\frac{2}{3}$, $.37\frac{1}{2}$, 5, $3.4\frac{3}{8}$, $.000\frac{7}{8}$
6. Of 4 units, 4 tenths, 4 hundredths.
7. Of $.11\frac{1}{9} + .6666\frac{2}{3} + .222222\frac{2}{9}$
8. Of $.14\frac{2}{7}$, $.018\frac{3}{5}$, 920, $.0139\frac{3}{7}$
9. Of $16.008\frac{7}{9}$, $.0074\frac{2}{3}$, $.2\frac{5}{6}$, $.000190421\frac{1}{5}$
10. Of 675 thousandths, 2 millionths, $64\frac{1}{8}$, 3.489107, and .00089407
11. Of four times $4.067\frac{7}{8}$ and $.000\frac{1}{2}$
12. Of 216.86301, 48.1057, .029, 1.3, 1000.
13. Add 35 units, 35 tenths, 35 hundredths, 35 thousandths.
14. Add ten thousand and one millionths; four hundred-thousandths; 96 hundredths; forty-seven million sixty thousand and eight billionths.

SUBTRACTION OF DECIMALS.

161. Subtraction of Decimals is finding the difference between two decimals.

PROBLEM.—From 6.8 subtract 2.057

SOLUTION.—Write the numbers so that units of the same order stand in the same column; suppose ciphers to be annexed to the 8, and subtract as in whole numbers.

OPERATION.

```
 6.8
 2.057
 -----
 4.743, Ans.
```

PROBLEM.—From $13.256\frac{5}{9}$ subtract $6.77\frac{1}{3}$

EXPLANATION.—In this example, the complex decimals are carried out by division to the same place, and the common fractions treated by Art. 141.

OPERATION.

$$\begin{array}{r} 13.256\frac{5}{9} \\ 6.77\frac{1}{3} = \underline{6.773\frac{1}{3}} \\ 6.483\frac{2}{9}, \textit{ Ans.} \end{array}$$

Rule.—1. *Write the subtrahend beneath the minuend so that units of the same order stand in the same column.*

2. *Subtract as in simple numbers, and write the decimal point as in addition of decimals.*

NOTES.—1. If either or both of the given decimals be complex, proceed as directed in the second problem.

2. If the minuend has not as many decimal places as the subtrahend, annex decimal ciphers to it, or suppose them to be annexed, until the deficiency is supplied.

EXAMPLES FOR PRACTICE.

1. Subtract 8.00717 from 19.54
2. 3 thousandths from 3000.
3. 72.0001 from 72.01
4. Subtract $.93\frac{2}{35}$ from $1.169\frac{3}{7}$
5. How much is $19 - 8.999\frac{1}{9}$?
6. How much less is $.04\frac{1}{3}$ than .4?
7. How much is $.65007 - \frac{1}{2}$?
8. What is $2\frac{3}{4} - 1\frac{4}{5}$ in decimals?
9. Subtract 1 from 1.684
10. $\frac{5}{6}$ of a millionth from $.000\frac{4}{9}$
11. $1\frac{1}{2}$ hundredths from $49\frac{3}{8}$ tenths.
12. 10000 thousandths from 10 units.
13. $24\frac{1}{2}$ tenths from 3701 thousandths.
14. $1\frac{7}{8}$ units from 1875 thousandths.
15. $\frac{5}{9}$ of a hundredth from $\frac{1}{18}$ of a tenth
16. $64\frac{1}{6}$ hundredths from 100 units.

MULTIPLICATION OF DECIMALS.

162. Multiplication of Decimals is finding the product when either or when each of the factors is a decimal.

PRINCIPLE.—*The number of decimal places in the product equals the number of decimal places in both factors.*

PROBLEM.—Multiply 2.56 by .184

SOLUTION.—$2.56 = \frac{256}{100}$, and $.184 = \frac{184}{1000}$. Now, $\frac{256}{100} \times \frac{184}{1000} = \frac{47104}{100000}$; that is, the product of hundredths by thousandths is hundredth-thousandths. This requires five places of decimals, or as many as are found in both factors.

OPERATION
```
   2.56
   .184
   ----
   1024
  2048
  256
 ------
 .47104, Ans.
```

Rule.—1. *Multiply as in whole numbers.*

2. *Point off as many decimal places in the product as there are decimal places in the two factors.*

REMARKS.—1. If the product does not contain as many places as the factors, prefix ciphers till it does contain as many.

2. Ciphers to the right of the product are omitted after pointing.

EXAMPLES FOR PRACTICE.

1. $1 \times .1$
2. $16 \times .03\frac{1}{3}$
3. $.01 \times .1\frac{1}{2}$
4. $.080 \times 80.$
5. $37.5 \times 82\frac{1}{2}$
6. $64.01 \times .32$
7. $48000 \times 73.$
8. $64.66\frac{2}{3} \times 18.$
9. $.56\frac{1}{4} \times .03\frac{1}{16}$
10. 738×120.4

11. $.0001 \times 1.006$
12. 34 units $\times$ .193
13. 27 tenths $\times .4\frac{1}{5}$
14. $43.7004 \times .008$
15. $21.0375 \times 4.44\frac{4}{9}$
16. $9300.701 \times 251.$
17. $430.0126 \times 4000.$
18. $.059 \times .059 \times .059$
19. 42 units $\times$ 42 tenths.
20. $2\frac{1}{4}$ hundredths $\times$ 600.
21. $7100 \times \frac{1}{8}$ of a millionth.
22. 26 millions $\times$ 26 millionths.
23. 2700 hundredths $\times$ 60 tenths.
24. $6.3029 \times .03275$
25. 135.027×1.00327

163. Oughtred's Method for abbreviating multiplication, may be used when the product of two decimals is required for a definite number of decimal places less than is found in both factors.

PROBLEM.—Multiply 3.8640372 by 1.2479603, retaining only seven decimal places in the product.

OPERATION.

```
   3 0 6 9 7 4 2
 3.8 6 4 0 3 7 2
 1.2 4 7 9 6 0 3
 -----------------
 3 8 6 4 0 3 7 2 |
   7 7 2 8 0 7 4 | 4
   1 5 4 5 6 1 4 | 8
     2 7 0 4 8 2 | 6
       3 4 7 7 6 | 3
         2 3 1 8 | 4
             1 1 | 5
 -----------------
 4.8 2 2 1 6 5 0
```

EXPLANATION.—It is evident that we need regard only that portion of each partial product which affects the figures in and above the seventh decimal place.

Beginning with the highest figure of the multiplier, we obtain the first partial product. Taking the second figure of the multiplier, we carry each figure of the partial product one place to the right, so that figures of the same order shall be in the same column. This product is carried out one place further than is required, so as to secure accuracy in the seventh place, and we draw a perpendicular line to separate this portion. The product of .04 by the right-hand figure in the seventh place, would extend to the ninth place of decimals; so we may reject the last figure, and commence with the 7. With each succeeding figure of the multiplier, we commence to multiply at that figure of the multiplicand which will produce a product in the eighth place. It is also convenient to place each figure of the multiplier directly over the first figure of the multiplicand taken. In multiplying by .007, we have $7 \times 3 = 21$; but, if we had been expressing the complete work, we should have 5 to carry to this place; the corrected product is therefore $21 + 5 = 26$. The product from the last figure, 3, is carried two places to the right. In the total product, the eighth decimal is dropped; but the seventh decimal figure is corrected by the amount carried.

Rule.—1. *Multiply only such figures as shall produce one more than the required number of decimal places.*

2. *Begin with the highest order of the multiplier; under the right-hand figure of each partial product, place the right-hand figure of the succeeding one. In obtaining such right-hand figure, let that number be added which would be carried from multiplying the figure of the next lower order.*

3. *Add the partial products, and reject the right-hand figure.*

Remarks.—1. It will be found convenient to write the multiplier in a reverse order, with its *units'* figure under that *decimal* figure of the multiplicand whose order is next lower than the lowest required. Thus, in the fourth example, the 8 would be written under the 1.

2. In carrying the tens for what is left out on the right, carry also one ten for each 5 of units in the omitted part; thus, 1 ten for 5 or 14 units, 3 tens for 25 or 26 units, etc. Make the same correction for the final figure rejected in the product.

Examples for Practice.

1. Multiply 27.653 by 9.157, preserving three decimal places.

2. Multiply 43.2071 by 3.14159, preserving four decimal places.

3. Multiply 3.62741 by 1.6432, preserving four decimal places.

4. 9.012×48.75, preserving one place.
5. $4.804136 \times .010759$, preserving six places.
6. $814\frac{53}{719} \times 26\frac{35}{44}$, preserving three places.
7. $702.61 \times 1.258\frac{7}{36}$, preserving three places.
8. $849.93\frac{3}{4} \times .0424444$, preserving three places.
9. 880.695×131.72 true to units.
10. $.025381 \times .004907$, preserving five places.
11. $64.01082 \times .03537$, preserving six places.
12. $1380.37\frac{1}{2} \times .234\frac{5}{6}$, preserving two places.

DIVISION OF DECIMALS.

164. Division of Decimals is the process of finding the quotient when either or when each term is a decimal.

165. Since the dividend corresponds to the product in multiplication (Art. **73**), and the decimal places in the divi dend are as many as in both factors (Art. **162,** Prin.), we derive the following principles:

PRINCIPLES.—1. *The dividend must contain as many decimal places as the divisor; and when both have the same number, the quotient is an integer.*

2. *The quotient must contain as many decimal places as the number of those in the dividend exceeds the number of those in the divisor.*

PROBLEM.—Divide .50312 by .19

SOLUTION.—The division is performed as in integers. The quotient is pointed according to Principle 2. The quotient must have $5 - 2 = 3$ places.

OPERATION.

```
.19).50312(2.648, Ans.
     38
     ---
     123
     114
     ---
       91
       76
       ---
       152
       152
       ---
```

PROOF.—By expressing the decimals as common fractions we have:

$$\frac{50312}{100000} \div \frac{19}{100} = \frac{50312}{100000} \times \frac{100}{19} = \frac{2648}{1000} = 2.648$$

PROBLEM.—Divide .36 by .008

SOLUTION.—The dividend has a less number of decimal places than the divisor. Annex one cipher, making the number equal. The quotient is an integer.

OPERATION.

$$.008\overline{).360}$$
$$45, \textit{Ans.}$$

PROBLEM.—Divide $.002\frac{19}{40}$ by $.06\frac{3}{5}$

SOLUTION.—Reducing the mixed decimals to equivalent pure decimals, we have .002475 and .066. Dividing, we find one more decimal place necessary to make the division exact; and, pointing by Principle 2, we have .0375

OPERATION.

$.002\frac{19}{40} = .002475$
$.06\frac{3}{5} = .066$

```
.066).002475(.0375, Ans.
      198
       495
       462
        330
        330
```

Rule.—*Divide as in whole numbers, and point off as many decimal places in the quotient as those in the dividend exceed those in the divisor.*

NOTE.—When the division is not exact, annex ciphers to the dividend, and carry the work as far as may be necessary.

EXAMPLES FOR PRACTICE.

1. $63 \div 4000.$
2. $3.15 \div 375.$
3. $1.008 \div 18.$
4. $4096 \div .032$
5. $9.7 \div 97000.$
6. $.9 \div .00075$
7. $13 \div 78.12\frac{1}{2}$
8. $12.9 \div 8.256$
9. $81.2096 \div 1.28$
10. $1 \div 100.$
11. $10.1 \div 17.$
12. $.001 \div 100.$

13. 12755 ÷ 81632.
14. 2401 ÷ 21.4375
15. 21.13212 ÷ .916
16. 36.72672 ÷ .5025
17. 2483.25 ÷ 5.15625
18. 142.0281 ÷ 9.2376
19. .08⅓ ÷ .12½
20. .0001 ÷ .01
21. 95.3 ÷ .264
22. 1000 ÷ .001
23. Ten ÷ 1 tenth.
24. .000001 ÷ .01
25. .00001 ÷ 1000.
26. 16.275 ÷ .41664
27. 1 ten-millionth ÷ 1 hundredth.

166. Oughtred's Method.—If the quotient is not required to contain figures below a certain denomination, the work may sometimes be abridged.

PROBLEM.—Divide 84.27 by 1.27395607, securing a quotient true to four places of decimals.

OPERATION.

```
1.2̸7̸3̸9̸5̸6̸) 84.270000 (66.1483—
                7643736
                 783264
                 764374
                  18890
                  12740
                   6150
                   5096
                   1054
                   1019
                     35
                     38
```

SOLUTION.—Since the divisor is greater than 1 and less than 2, the quotient will contain six places,—two of integers and four of decimals. The highest denomination of the divisor, multiplied by the lowest denomination of the quotient, would obtain a figure in the fourth place. We take one place more as in multiplication (Art. **163**), and also cut off two figures of the divisor, since these can not affect the quotient above the fourth place. After obtaining the first figure of the quotient, we drop one right-hand figure of the divisor for each figure obtained. To prevent errors, we cancel the figure before each division.

Rule.—*Find the figure of the dividend that would result from multiplying a unit in the highest denomination of the divisor by a unit of the lowest denomination required in the quotient. Take one more figure of the dividend to secure accuracy. Cut off any figures of the divisor not needed for the abbreviated dividend.*

Divide as usual until the figures remaining in the dividend are all divided. At each subsequent division, drop a figure from the divisor, carrying the number necessary from the product of the figure omitted.

Continue until the divisor is reduced to two figures.

REMARK.—In the quotient, 5 units of an omitted order may be taken as 1 unit of the next higher order.

EXAMPLES FOR PRACTICE.

1. $1000 \div .98$, preserving two places.
2. $6215.75 \div .99\frac{1}{2}$, preserving three places.
3. $28012 \div .993$, preserving two places.
4. $52546.35 \div .99\frac{3}{4}$, preserving three places.
5. $4840 \div .9875$, preserving two places.
6. $2 \div 1.4142136$, preserving seven places.
7. $9.869604401 \div 3.14159265$, preserving eight places.

Topical Outline.

DECIMAL FRACTIONS.

Definitions.
Decimal Point.
Classes....... { Pure. / Mixed. / Complex. }
Principles.
Numeration, Rule.
Notation, Rule.

Reduction. { I. Dec. to a Com. Fraction / II. Com. Fraction to a Dec }
Addition, Rule.
Subtraction, Rule.
Multiplication, Rule.
  Abbreviated Multiplication.
Division, Rule.
  Abbreviated Division.

X. CIRCULATING DECIMALS.

167. In reducing common fractions to decimals, the process, in some cases, does not terminate. This gives rise to **Circulating Decimals.**

PRINCIPLE I.—*If any prime factors other than 2 and 5 are found in the denominator of a fraction in its lowest terms, the resulting decimal will be interminate.*

DEMONSTRATION.—If the fraction is in its lowest terms, the numerator and denominator are prime to each other (**131,** Rem. 2). In the process of reduction, the numerator is multiplied by 10. By this means the factors 2 and 5 may be introduced into the numerator as many times as necessary; but no others are introduced. Therefore, if any factors other than 2 and 5 are found in the denominator, the division can not be made complete, and the resulting decimal will be interminate.

Thus, $\frac{3}{320} = \frac{3}{2\times2\times2\times2\times2\times2\times5} = .009375$

and, $\frac{5}{16} = \frac{5}{2\times2\times2\times2} = .3125$

but, $\frac{7}{60} = \frac{7}{2\times2\times3\times5} = .116666+$

It is evident that the first will terminate if the numerator be multiplied six times by 10, carrying the decimal to the sixth place. In the same way we reduce $\frac{5}{16}$ to a decimal containing four places. But since the factor 3 is found in the denominator of $\frac{7}{60}$, the fraction can not be exactly reduced, though the numerator be multiplied by any power of 10.

PRINCIPLE II.—*Every interminate decimal arising from the reduction of a common fraction will, if the division be carried far enough, contain the same figure, or set of figures, repeated in the same order.*

DEMONSTRATION.—Each of the remainders must be less than the denominator which is used as the divisor (Art. 78, Note 2). If the division be carried far enough, some remainder must be found equal to some remainder already found, and the subsequent figures in the quotient must be similar to the figures found from the former remainder.

Thus, in reducing $\frac{1}{7}$, we find the decimal .142857, and then have the remainder 1, the number we started with; if we annex a cipher we shall get 1 for the next figure of the quotient, 4 for the next, etc.

168. 1. Interminate decimals, on this account, have received the name of *Circulating* or *Recurring Decimals.*

2. A **Circulate** or **Circulating Decimal** has one or more figures constantly repeated in the same order.

3. A **Repetend** is the figure or set of figures repeated, and it is expressed by placing a dot over the first and last figure; thus, $\frac{1}{7} = .\dot{1}4285\dot{7}$; if there be one figure repeated, the dot is placed over it, thus, $\frac{2}{3} = .6666 + = .\dot{6}$

4. A **Pure Circulate** has no figures but the repetend; as, $.\dot{5}$ and $.\dot{1}2\dot{4}$

5. A **Mixed Circulate** has other figures before the repetend; as, $.208\dot{3}$ and $.3\dot{1}24\dot{7}$

6. A **Simple Repetend** has one figure; as, $.\dot{4}$

7. A **Compound Repetend** has two or more figures; as, $.\dot{5}\dot{9}$

8. A **Perfect Repetend** is one which contains as many decimal places as there are units in the denominator, less 1; thus, $\frac{1}{7} = .\dot{1}4285\dot{7}$

9. **Similar Repetends** begin and end at the same decimal place; as, $.4\dot{2}\dot{7}$ and $.5\dot{3}\dot{6}$

10. **Dissimilar Repetends** begin or end at different decimal places; as, $.\dot{2}0\dot{5}$ and $.31\dot{2}46\dot{8}$

11. **Conterminous Repetends** end at the same place; as, $.50\dot{3}9\dot{7}$ and $.42\dot{6}1\dot{8}$

12. **Co-originous Repetends** begin at the same place; as, $.\dot{5}$ and $.\dot{1}2\dot{4}$

169. Any terminate decimal may be considered a circulate, its repetend being ciphers; as, $.35 = .35\dot{0} = .35\dot{0}00\dot{0}$ Any simple repetend may be made compound, and any compound repetend still more compound, by taking in one or more of the succeeding repetends; as, $.\dot{3} = .\dot{3}333\dot{3}$, and $.05\dot{6}\dot{2} = .05\dot{6}26\dot{2}$, and $.\dot{2}5\dot{7} = .\dot{2}5725725\dot{7}$

REMARKS.—1. When a repetend is thus enlarged, be careful to take in *no part* of a repetend without taking the whole of it; thus, if we take in 2 figures in the last example, the result, $.\dot{2}572\dot{5}$, would be incorrect, for the next figure understood being 7, shows that 25725 is not repeated.

2. A repetend may be made to begin at any lower place by carrying its dots forward, each the same distance; thus, $.\dot{5} = .55\dot{5}$, and $.29\dot{4}\dot{1} = .294\dot{1}\dot{4}$, and $5.1\dot{8}3\dot{6} = 5.183\dot{6}8\dot{3}$

3. Dissimilar repetends can be made similar, by carrying the dots forward till they all begin at the same place as the one farthest from the decimal point.

4. Co-originous repetends may be made conterminous by enlarging the repetends until they all contain the same number of figures. This number will be the least common multiple of the numbers of figures in the given repetends.

For, suppose one of the repetends to have 2, another 3, another 4, and the last 6 figures; in enlarging the first, figures must be taken in, 2 at a time, and in the others, 3, 4, and 6 at a time.

170. Circulating decimals originate, as has been already shown, in changing some common fractions to decimals. Then, having given a *circulate*, it can always be changed to an equivalent common fraction.

171. Circulating decimals may be added, subtracted, multiplied, and divided as other fractions.

CASE I.

172. To reduce a pure circulate to a common fraction.

PROBLEM.—Change $.\dot{5}\dot{3}$ to a common fraction.

OPERATION.

$$\begin{aligned} 100 \text{ times the repetend} &= 53.\dot{5}\dot{3} \\ \text{Once the repetend} &= .\dot{5}\dot{3} \\ \therefore 99 \text{ times the repetend} &= 53. \\ \text{Once the repetend} &= \tfrac{53}{99}, \textit{ Ans.} \end{aligned}$$

SOLUTION.—Removing the decimal point one place to the right, multiplies the repetend by 10; two places, by 100, and so on. Then, multiplying by 100, and subtracting the repetend from the product, removes all of that part to the right of the decimal point, and, dividing 53 by 99, we have the common fraction which produced the given repetend.

PROBLEM.—Change $.\dot{4}5\dot{6}$ to a common fraction.

OPERATION.

$$\begin{aligned} 1000 \text{ times the repetend} &= 456.\dot{4}5\dot{6} \\ \text{Once the repetend} &= .\dot{4}5\dot{6} \\ 999 \text{ times the repetend} &= 456. \\ \therefore \text{ once the repetend} &= \tfrac{456}{999}, \textit{ Ans.} \end{aligned}$$

PROBLEM.—Change $\dot{2}5.\dot{6}$ to a common fraction.

OPERATION.

$$\begin{aligned} \text{Carry the dot forward thus: } \dot{2}5.\dot{6} &= 25.\dot{6}2\dot{5} \\ 1000 \text{ times the repetend} &= 625.\dot{6}2\dot{5} \\ \text{Once the repetend} &= .\dot{6}2\dot{5} \\ 999 \text{ times the repetend} &= 625. \\ \therefore \text{ once the repetend} &= \tfrac{625}{999} \\ \text{Whence the } 25.\dot{6}2\dot{5} &= 25\tfrac{625}{999}, \textit{ Ans.} \end{aligned}$$

NOTE.—From these solutions the following rule is derived.

Rule.—*Write the repetend for the numerator, omitting the decimal point and the dots, and for the denominator write as many 9's as there are figures in the repetend, and reduce the fraction to its lowest terms.*

CASE II.

173. To reduce a mixed circulate to a common fraction.

PROBLEM.—Change $.821\dot{4}3\dot{7}$ to a common fraction.

OPERATION.

Omitting the decimal point, we have: $.821\dot{4}3\dot{7}=\frac{821\frac{437}{999}}{1000}=$

$$\frac{821\times999+437}{1000\times999}=\frac{821(1000-1)+437}{999000}=$$

$$\frac{821000-821+437}{999000}=\frac{820616}{999000}=\frac{102577}{124875}, \textit{ Ans.}$$

Or, briefly, $\frac{821437-821}{999000}=\frac{102577}{124875}$, *Ans.*

PROBLEM.—Change $.04\dot{8}$ to a common fraction.

OPERATION.

Omitting the decimal point, we have:

$$.04\dot{8}=\frac{4\frac{8}{9}}{100}=\frac{4}{100}+\frac{8}{900}=\frac{4(10-1)}{900}+\frac{8}{900}=$$

$$\frac{36}{900}+\frac{8}{900}=\frac{44}{900}=\frac{11}{225}, \textit{ Ans.}$$

Or, briefly, $\frac{48-4}{900}=\frac{44}{900}=\frac{11}{225}$, *Ans.*

NOTE.—The following rule is derived from the preceding solutions.

Rule. 1. *For the numerator, subtract the part which precedes the repetend from the whole expression, both quantities being considered as units.*

2. *For the denominator, write as many 9's as there are figures in the repetend, and annex as many ciphers as there are decimal figures before each repetend.*

Reduce to common fractions:

1. $.\dot{3}$
2. $.0\dot{5}$
3. $.\dot{1}2\dot{3}$
4. $2.\dot{6}\dot{3}$
5. $.3\dot{1}$
6. $.0\dot{2}1\dot{6}$
7. $\dot{4}8.\dot{1}$
8. $\dot{1}.00\dot{1}$
9. $.13\dot{8}$
10. $.208\dot{3}$
11. $\dot{8}5.714\dot{2}$
12. $.\dot{0}6349\dot{2}$
13. $.4\dot{4}7619\dot{0}$
14. $.0902\dot{7}$

ADDITION OF CIRCULATES.

174. Addition of Circulates is the process of finding the sum of two or more circulates. Similar circulates only can be added.

PROBLEM.—Add $.25\dot{6}$, $5.34\dot{7}\dot{2}$, $24.\dot{8}1\dot{5}$, and .9098

OPERATION.

$$
\begin{array}{r}
.25666\dot{6}66666\dot{6} \\
5.34727\dot{2}7272\dot{2} \\
24.81581\dot{1}5815\dot{8} \\
.90980\dot{0}0000\dot{0} \\
\hline
31.32955\dot{5}5209\dot{7}
\end{array}
$$

SOLUTION.—Make the circulates similar. The first column of figures which would appear, if the circulates were continued, is the same as the first figures of the repetends, 6, 7, 1, 0, whose sum, 14, gives 1 to be carried to the right-hand column. Since the last six figures in each number make a repetend, the last six figures of the sum also make a repetend.

Rule.—*Make the repetends similar, if they be not so; add, and point off as in ordinary decimals, increasing the right-hand column by the amount, if any, which would be carried to it if the circulates were continued; then make a repetend in the sum similar to those above.*

REMARK.—In finding the amount to be carried to the right-hand column, it may be necessary, sometimes, to use the *two* succeeding figures in each repetend.

EXAMPLES FOR PRACTICE.

1. Add $.45\dot{3}$, $.06\dot{8}$, $.32\dot{7}$, $.94\dot{6}$
2. Add $3.0\dot{4}$, $6.45\dot{6}$, $23.\dot{3}\dot{8}$, $.2\dot{4}\dot{8}$
3. Add $.\dot{2}\dot{5}$, $.10\dot{4}$, $.\dot{6}\dot{1}$, and $.56\dot{3}\dot{5}$
4. Add $\dot{1}.0\dot{3}$, $.25\dot{7}$, $5.\dot{0}\dot{4}$, $28.0445\dot{2}4\dot{5}$
5. Add $.\dot{6}$, $.13\dot{8}$, $.0\dot{5}$, $.097\dot{2}$, $.041\dot{6}$
6. Add $9.21\dot{1}0\dot{7}$, $.\dot{6}\dot{5}$, $5.00\dot{4}$, $3.5\dot{6}2\dot{2}$
7. Add $.20\dot{4}\dot{5}$, $.\dot{0}\dot{9}$, and $.25$
8. Add $5.0\dot{7}7\dot{0}$, $.\dot{2}\dot{4}$, and $7.\dot{1}2494\dot{3}$
9. Add $3.\dot{4}88\dot{4}$, $1.6\dot{3}\dot{7}$, $130.8\dot{1}$, $.06\dot{6}$

SUBTRACTION OF CIRCULATES.

175. Subtraction of Circulates is the process of finding the difference between two circulates. The two circulates must be similar.

PROBLEM.—Subtract $9.3\dot{1}5\dot{6}$ from $12.90\dot{2}\dot{1}$

OPERATION.

$$\begin{array}{r} 12.90\dot{2}1212\dot{1} \\ 9.31\dot{5}6156\dot{1} \\ \hline 3.58\dot{6}5055\dot{9} \end{array}$$

SOLUTION.—Prepare the numbers for subtraction. If the circulates were continued, the next figure in the subtrahend (5) would be larger than the one above it (2); therefore, carry 1 to the right-hand figure of the subtrahend.

Rule.—*Make the repetends similar, if they be not so; subtract and point off as in ordinary decimals, carrying* 1, *however, to the right-hand figure of the subtrahend, if on continuing the circulates it be found necessary; then make a repetend in the remainder, similar to those above.*

REMARK.—It may be necessary to observe more than one of the succeeding figures in the circulates, to ascertain whether 1 is to be carried to the right-hand figure of the subtrahend or not.

EXAMPLES FOR PRACTICE.

1. Subtract $.\dot{0}7\dot{4}$ from $.2\dot{6}$
2. Subtract $\dot{9}.0\dot{9}$ from $15.35\dot{4}6\dot{5}$
3. Subtract $\dot{4}.5\dot{1}$ from $18.23\dot{6}7\dot{3}$
4. Subtract $37.01\dot{2}\dot{8}$ from 100.73
5. Subtract $8.\dot{2}\dot{7}$ from $10.0\dot{5}6\dot{3}$
6. Subtract $\dot{1}90.47\dot{6}$ from $199.6\dot{4}2857\dot{1}$
7. Subtract $13.\dot{6}3\dot{7}$ from $10\dot{4}.\dot{1}$

MULTIPLICATION OF CIRCULATES.

176. Multiplication of Circulates is the process of finding the product when either or when each of the factors is a circulate.

PROBLEM.—Multiply $.375\dot{4}$ by $17.4\dot{3}$

OPERATION.

$$
\begin{array}{r}
.375\dot{4} \\
17.4\dot{3} = 17.4\tfrac{1}{3} \\
\hline
.1501\dot{7}7\dot{7} \\
2.6281\dot{1}1\dot{1} \\
3.7544\dot{4}4\dot{4} \\
125\dot{1}4\dot{8} \\
\hline
6.5452\dot{4}8\dot{1}
\end{array}
\quad \textit{Ans.}
$$

SOLUTION.—In forming the partial products, carry to the right-hand figures of each respectively, the numbers 1, 3, 0, arising from the multiplication of the figures that do not appear. The repetend of the multiplier being equal to $\frac{1}{3}$, $\frac{1}{3}$ of the multiplicand is $125\dot{1}4\dot{8}$, whose figures are set down under those of the multiplicand from which they were obtained. Point the several products, carry them forward until their repetends are similar, and add for answer.

Rule.—1. *If the multiplier contain a repetend, change it to a common fraction.*

2. *Then multiply as in multiplication of decimals, and add to the right-hand figure of each partial product the amount necessary if the repetend were repeated.*

3. *Make the partial products similar, and find their sum.*

Examples for Practice.

1. $4.7\dot{3}\dot{5} \times 7.349$
2. $.0\dot{7}06\dot{7} \times .9\dot{4}3\dot{2}$
3. $714.3\dot{2} \times 3.45\dot{6}$
4. $16.\dot{2}0\dot{4} \times 32.\dot{7}\dot{5}$
5. $19.0\dot{7}\dot{2} \times .208\dot{3}$
6. $3.\dot{7}54\dot{3} \times 4.7157$
7. $1.\dot{2}5678\dot{4} \times 6.420\dot{8}\dot{1}$

DIVISION OF CIRCULATES.

177. Division of Circulates is the process of finding the quotient when either or when each of the terms is a circulate.

Problem.—Divide $.\dot{1}5\dot{4}$ by $.\dot{2}$

OPERATION.

$$.\dot{1}5\dot{4} = \tfrac{154}{999},\ .\dot{2} = \tfrac{2}{9}.$$

$$\therefore\ \tfrac{154}{999} \div \tfrac{2}{9} = \tfrac{154}{999} \times \tfrac{9}{2} = \tfrac{77}{111} = .\dot{6}9\dot{3},\ \textit{Ans.}$$

Rule.—*Change the terms to common fractions; then divide as in division of fractions, and reduce the result to a repetend.*

Remark.—This is the easiest method of solving problems in division of circulates. The terms may be made similar, however, and the division performed without changing the circulates to common fractions.

Examples for Practice.

1. $.\dot{7}\dot{5} \div .\dot{1}$
2. $51.49\dot{1} \div 17.$
3. $681.55\dot{9}887\dot{9} \div 94.$
4. $90.5\dot{2}0374\dot{9} \div 6.7\dot{5}\dot{4}$
5. $11.\dot{0}6873540\dot{2} \div .\dot{2}4\dot{5}$
6. $9.5\dot{3}3066399\dot{7} \div 6.\dot{2}1\dot{7}$
7. $3.500\dot{6}9135802\dot{4} \div 7.68\dot{4}$

Topical Outline.

Circulating Decimals.

1. Principles.
2. Definitions......
   1. Circulate.
   2. Repetend.
   3. Pure Circulate.
   4. Mixed Circulate.
   5. Simple Repetend.
   6. Compound Repetend.
   7. Perfect Repetend.
   8. Similar Repetends.
   9. Dissimilar Repetends.
   10. Conterminous Repetends.
   11. Co-originous Repetends.
3. Reduction.......
   - Case I.
   - Case II.
4. Addition ....
   - Definition.
   - Rule.
   - Applications.
5. Subtraction.....
   - Definition.
   - Rule.
   - Applications.
6. Multiplication
   - Definition.
   - Rule.
   - Applications.
7. Division ....
   - Definition.
   - Rule.
   - Applications.

XI. COMPOUND DENOMINATE NUMBERS.

178. 1. A **Measure** is a standard unit used in estimating quantity. Standard units are fixed by law or custom.

2. A quantity is measured by finding how many times it contains the unit.

3. **Denomination** is the name of a unit of measure of a concrete number.

4. A **Denominate Number** is a concrete number which expresses a particular kind of quantity; as, 3 feet, 7 pounds.

5. A **Compound Denominate Number** is one expression of a quantity by *different denominations* under *one kind* of measure; as, 5 yards, 2 feet, and 8 inches.

6. All measures of denominate numbers may be embraced under the following divisions: *Value*, *Weight*, *Extension*, and *Time*.

MEASURES OF VALUE.

179. 1. **Value** is the worth of one thing as compared with another.

2. Value is of three kinds: *Intrinsic*, *Commercial*, and *Nominal*.

3. The **Intrinsic Value** of any thing is measured by the amount of labor and skill required to make it useful.

4. The **Commercial Value** of any thing is its purchasing power, exchangeability, or its worth in market.

5. The **Nominal Value** is the name value of any thing.

6. Value is estimated among civilized people by its price in *money*.

7. **Money** is a standard of value, and is the medium of exchange; it is usually stamped metal, called coins, and printed bills or notes, called paper money.

8. The money of a country is its **Currency**. Currency is national or foreign.

United States Money.

180. United States Money is the legal currency of the United States. It is based upon the decimal system; that is, ten units of a lower order make one of the next higher.

The **Dollar** is the unit. The same unit is the standard of Canada, the Sandwich Islands, and Liberia.

TABLE.

10 mills, marked m.,	make	1 cent,	marked	ct.
10 cents	"	1 dime,	"	d.
10 dimes	"	1 dollar,	"	$.
10 dollars	"	1 eagle,	"	E.

NOTE.—The *cent* and *mill*, which are $\frac{1}{100}$ and $\frac{1}{1000}$ of a dollar, derive their names from the Latin *centum* and *mille*, meaning *a hundred* and *a thousand;* the *dime*, which is $\frac{1}{10}$ of a dollar, is from the French word *disme*, meaning *ten*.

REMARKS.—1. United States money was established, by act of Congress, in 1786. The first money coined, by the authority of the United States, was in 1793. The coins first made were copper cents. In 1794 silver dollars were made. Gold eagles were made in 1795; gold dollars, in 1849. Gold and silver are now both legally standard. The *trade dollar* was minted for Asiatic commerce

2. The coins of the United States are classed as *bronze, nickel, silver,*

and *gold.* The name, value, composition, and weight of each coin are shown in the following table:

TABLE.

COIN.	VALUE.	COMPOSITION.	WEIGHT.
BRONZE.			
One cent.	1 cent.	95 parts copper, 5 parts tin & zinc.	48 grains Troy.
NICKEL.			
3-cent piece.	3 cents.	75 parts copper, 25 parts nickel.	30 grains Troy.
5-cent piece.	5 cents.	75 " " 25 " "	77.16 " "
SILVER.			
Dime.	10 cents.	90 parts silver, 10 parts copper.	2.5 grams.
Quarter dollar.	25 cents.	90 " " 10 " "	6.25 "
Half dollar.	50 cents.	90 " " 10 " "	12.5 "
Dollar.	100 cents.	90 " " 10 " "	412.5 grains Troy.
GOLD.			
Dollar.	100 cents.	90 parts gold, 10 parts copper.	25.8 grains Troy.
Quarter eagle.	2½ dollars.	90 " " 10 " "	64.5 " "
Three dollar.	3 dollars.	90 " " 10 " "	77.4 " "
Half eagle.	5 dollars.	90 " " 10 " "	129 " "
Eagle.	10 dollars.	90 " " 10 " "	258 " "
Double eagle.	20 dollars.	90 " " 10 " "	516 " "

3. A deviation in weight of ½ a grain to each piece, is allowed by law in the coinage of Double Eagles and Eagles; of ¼ of a grain in the other gold pieces; of 1½ grains in all silver pieces; of 3 grains in the five-cent piece; and of 2 grains in the smaller pieces.

4. The mill is not coined. It is used only in calculations.

181. In reading U. S. Money, *name the dollars and all higher denominations together as dollars, the dimes and cents as cents, and the next figure, if there be one, as mills;*

Or, *name the whole number as dollars, and the rest as a decimal of a dollar.*

Thus, $9.124 is read 9 dollars 12 ct. 4 mills, or 9 dollars 124 thousandths of a dollar.

English or Sterling Money.

182. English or **Sterling Money** is the currency of the British Empire.

The **pound sterling** (worth $4.8665 in U. S. money) is the unit, and is represented by the sovereign and the £1 bank-note.

TABLE.

4 farthings, marked qr.,	make	1 penny, marked	d.
12 pence	"	1 shilling, "	s.
20 shillings	"	1 pound, "	£.

EQUIVALENT TABLE.

£	s.	d.	qr.
1 =	20 =	240 =	960.
	1 =	12 =	48.
		1 =	4.

REMARKS.—1. The abbreviations, £, s., d., q., are the initials of the Latin words *libra*, *solidarius*, *denarius*, *quadrans*, signifying, respectively, pound, shilling, penny, and quarter.

2. The coins are gold, silver, and copper. The *gold* coins are the sovereign (=£1), half sovereign (=10s.), guinea (=21s.), and half-guinea (=10s. 6d.) The *silver* coins are the crown (=5s.), the half crown (=2s. 6d.), the florin (=2s.), the shilling, and the six-penny, four-penny, and three-penny pieces. The penny, half-penny, and farthing are the *copper* coins. The guinea, half-guinea, crown, and half-crown are no longer coined, but some of them are in circulation.

3. The standard for gold coins is $\frac{11}{12}$ pure gold and $\frac{1}{12}$ alloy; the standard for silver is $\frac{37}{40}$ pure silver and $\frac{3}{40}$ copper; pence and half-pence are pure copper.

French Money.

183. French Money is the legal currency of France. Its denominations are decimal.

The **franc** (worth 19.3 cents in U. S. money) is the unit. The franc is also used in Switzerland and Belgium, and, under other names, in Italy, Spain, and Greece.

TABLE.

10 millimes, marked m.,	make	1 centime,	marked c
10 centimes	"	1 decime,	" d.
10 decimes	"	1 franc,	" fr.

EQUIVALENT TABLE.

1 fr. = 10 d. = 100 c. = 1000 m.
1 d. = 10 c. = 100 m.
1 c. = 10 m.

REMARK.—The coins are of gold, silver, and bronze. The *gold* coins are 100, 50, 20, 10, and 5-franc pieces; they are .9 pure gold. The 20-franc piece weighs 99.55 gr. The *silver* coins are 5, 2, and 1-franc pieces, and 50 and 20-centime pieces; they are now .835 pure silver. The *bronze* coins are 10, 5, 2, and 1-centime pieces. The decime is not used in practice. All sums are given in francs and centimes, or hundredths.

German Money.

184. German Money is the legal currency of the German Empire.

The **mark, or reichsmark** (worth 23.8 cents in U. S. money), is the unit. The only other denomination is the pfennig (penny).

TABLE.

100 pfennige, marked Pf., make 1 mark, marked RM.

REMARK.—The coins are of gold, silver, and copper. The *gold* coins are of the value of 40, 20, 10, and 5 marks; the *silver* coins, 3, 2, and 1-mark pieces, and 50, 20, and 10 pfennige. The *copper*, 2 and 1-pfennig pieces. Gold and silver are .9 fine.

MEASURES OF WEIGHT.

185. Weight is the measure of the force called gravity, which draws bodies toward the center of the earth.

The standard unit of weight in the United States is the **Troy pound** of the Mint.

186. Three kinds of weight are in use,—*Troy Weight, Apothecaries' Weight*, and *Avoirdupois Weight.*

Troy Weight.

187. Troy Weight is used in weighing gold, silver, platinum, and jewels. Formerly it was used in philosophical and chemical works.

TABLE.

24 grains, marked gr.,	make	1 pennyweight,	marked	pwt.
20 pwt.	"	1 ounce,	"	oz.
12 oz.	"	1 pound,		℔.

EQUIVALENT TABLE.

℔.		oz.		pwt.		gr.
1	=	12	=	240	=	5760.
		1	=	20	=	480.
				1	=	24.

REMARK.—The Troy pound is equal to the weight of 22.7944 cubic inches of pure water at its maximum density, the barometer being at 30 inches. The standard pound weight is identical with the Troy pound of Great Britain.

Apothecaries' Weight.

188. Apothecaries' Weight is used by physicians and apothecaries in prescribing and mixing dry medicines. Medicines are bought and sold by Avoirdupois Weight.

TABLE.

20 grains, marked gr.,	make	1 scruple,	marked	℈.
3 ℈	"	1 dram,	"	ʒ.
8 ʒ	"	1 ounce,	"	℥.
12 ℥	"	1 pound,	"	℔.

EQUIVALENT TABLE.

℔.		℥.		ʒ.		℈.		gr.
1	=	12	=	96	=	288	=	5760.
		1	=	8	=	24	=	480.
				1	=	3	=	60.
						1	=	20.

REMARK.—The pound, ounce, and grain of this weight are the same as those of Troy weight; the pound in each contains 12 oz. = 5760 gr.

Avoirdupois or Commercial Weight.

189. Avoirdupois or **Commercial Weight** is used for weighing all ordinary articles.

TABLE.

16 ounces, marked oz.,	make	1 pound,	marked	lb.
25 lb.	"	1 quarter,	"	qr.
4 qr.	"	1 hundred-weight,	"	cwt.
20 cwt.	"	1 ton,	"	T.

EQUIVALENT TABLE.

T.		cwt.		qr.		lb.		oz.
1	=	20	=	80	=	2000	=	32000.
		1	=	4	=	100	=	1600.
				1	=	25	=	400.
						1	=	16.

REMARKS.—1. In Great Britain, the qr. = 28 lb., the cwt. = 112 lb., the ton = 2240 lb. These values are used at the United States custom-houses in invoices of English goods, and are still used in some lines of trade, such as coal and iron.

2. Among other weights sometimes mentioned in books, are: 1 stone, horseman's weight, = 14 lb.; 1 stone of butcher's meat = 8 lb.: 1 clove of wool = 7 lb.

3. The lb. avoirdupois is equal to the weight of 27.7274 cu. in. of distilled water at 62° (Fahr.); or 27.7015 cu. in. at its maximum density, the barometer at 30 inches. For ordinary purposes, 1 cubic foot of water can be taken $62\frac{1}{4}$ lb. avoirdupois.

4. The terms *gross* and *net* are used in this weight. *Gross weight* is the weight of the goods, together with the box, cask, or whatever contains them. *Net weight* is the weight of the goods alone.

5. The word *avoirdupois* is from the French *avoirs, du, pois,* signifying *goods of weight.*

6. The ounce is often divided into halves and quarters in weighing. The sixteenth of an ounce is called a *dram.*

COMPARISON OF WEIGHTS.

190. The pound Avoirdupois weighs 7,000 grains Troy, and the Troy pound weighs 5,760 grains, hence there are 1,240 grains more in the Avoirdupois pound than in the Troy pound.

The following table exhibits the relation between certain denominations of Avoirdupois, Troy, and Apothecaries' Weight.

Avoirdupois.	Troy.	Apothecaries'.
1 lb. =	$1\frac{31}{144}$ ℔. =	$1\frac{31}{144}$ ℔.
1 oz. =	$\frac{175}{192}$ oz. =	$\frac{175}{192}$ ℥.
	1 ℔. =	1 ℔.
	1 oz. =	1 ℥.
	1 gr. =	1 gr.
	1 pwt. =	$\frac{2}{5}$ ʒ.
	1 pwt. =	$1\frac{1}{5}$ ℈.

Remark.—In addition to the foregoing, the following, called *Diamond Weight,* is used in weighing diamonds and other precious stones.

Table.

16 parts	make	1 carat grain	= .792	Troy grains.
4 carat grains	"	1 carat	= 3.168	" "

Note.—This carat is entirely different from the *assay carat,* which has reference to the *fineness* of gold. The mass of gold is considered as divided into twenty-four parts, called carats, and is said to be so many carats fine, according to the number of twenty-fourths of pure gold which it contains.

MEASURES OF EXTENSION.

191. 1. **Extension** is that property of matter by which it occupies space. It may have one or more of the three dimensions,—length, breadth, and thickness.

2. A **line** has only one dimension,—length.

3. A **surface** has two dimensions, length and breadth.

4. A **solid** or **volume** has three dimensions,—length, breadth, and thickness.

192. Measures of Extension embrace:

1. Linear Measure.
   - Long Measure.
   - Chain Measure.
   - Mariners' Measure.
   - Cloth Measure.
2. Superficial Measure.
   - Square Measure.
   - Surveyors' Measure.
3. Solid Measure.
4. Measures of Capacity.
   - Liquid Measure.
   - Apothecaries' Measure.
   - Dry Measure.
5. Angular Measure.

Long or Linear Measure.

193. Linear Measure is used in measuring distances, or length, in any direction.

The standard unit for all measures of extension is the **yard**, which is identical with the Imperial yard of Great Britain.

TABLE.

12	inches, marked in.,	make	1 foot,	marked	ft.
3	ft.	"	1 yard,	"	yd.
$5\frac{1}{2}$	yd. or $16\frac{1}{2}$ ft.	"	1 rod,	"	rd.
320	rd.	"	1 mile,	"	mi.

EQUIVALENT TABLE.

mi.	rd.	yd.	ft.	in.
1 =	320 =	1760 =	5280 =	63360.
	1 =	$5\frac{1}{2}$ =	$16\frac{1}{2}$ =	198.
		1 =	3 =	36.
			1 =	12.

REMARKS.—1. The *standard yard* of the United States was obtained from England in 1856. It is of bronze, and of due length at 59.8° Fahr. A copy of the former standard is deposited at each state capital: this was about $\frac{1}{1000}$ of an inch too long.

2. The *rod* is sometimes called *perch* or *pole.* The *furlong*, equal to 40 rods, is seldom used.

3. The *inch* may be divided into halves, fourths, eighths, etc., or into tenths, hundredths, etc.

4. The following measures are sometimes used:

12 lines	make	1 inch.
3 barleycorns	"	1 "
3 inches	"	1 palm.
4 inches	"	1 hand.
9 inches	"	1 span.
18 inches	"	1 cubit.
3 feet	"	1 pace.

194. Chain Measure is used by surveyors in measuring land, laying out roads, establishing boundaries, etc.

TABLE.

7.92 inches, marked in.,	make	1 link,	marked li.
100 li.	"	1 chain,	" ch.
80 ch.	"	1 mile,	" mi.

EQUIVALENT TABLE.

mi.		ch.		li.		in.
1	=	80	=	8000	=	63360.
		1	=	100	=	792.
				1	=	7.92.

REMARKS.—1. The surveyors' chain, or Gunter's chain, is 4 rods, or 66 feet in length. Since it consists of 100 links, the chains and links may be written as integers and hundredths; thus, 2 chains 56 links are written 2.56 ch.

2. The engineers' chain is 100 feet long, and consists of 100 links.

3. The engineers' leveling rod is used for measuring *vertical* distances. It is divided into feet, tenths, and hundredths, and, by means of a vernier, may be read to thousandths.

195. Mariners' Measure is used in measuring the depth of the sea, and also distances on its surface.

TABLE.

6 feet	make	1 fathom.
720 feet	"	1 cable-length.

REMARKS.—1. A nautical mile is one minute of longitude, measured on the equator at the level of the sea. It is equal to $1.152\frac{2}{3}$ statute miles. 60 nautical miles = 1 degree on the equator, or 69.16 statute miles. A league is equal to 3 nautical miles, or 3.458 statute miles.

2. Depths at sea are measured in fathoms; distances are usually measured in nautical miles.

196. Cloth Measure is used in measuring dry-goods. The standard yard is the same as in Linear Measure, but is divided into *halves*, *quarters*, *eighths*, *sixteenths*, etc., in place of feet and inches.

REMARKS.—1. There was formerly a recognized table for Cloth Measure, but it is now obsolete. The denominations were as follows:

$2\frac{1}{4}$ inches, marked in.,	make	1 nail,	marked	na.
4 na. or 9 in.	"	1 quarter,	"	qr.
4 qr.	"	1 yard,	"	yd.

2. At the custom-house, the yard is divided decimally.

Superficial or Surface Measure.

197. 1. **Superficial Measure** is used in estimating the numerical value of surfaces; such as, land, weather-boarding, plastering, paving, etc.

2. A **surface** has length and breadth, but not thickness

3. The **area** of a surface is its numerical value; or the number of times it contains the *measuring unit.*

4. A **superficial unit** is an assumed unit of measure for surfaces.

Usually the square, whose side is the linear unit, is the **unit of measure;** as, the *square inch, square foot, square yard.*

5. A **Rectangle** may be defined as a surface bounded by four straight lines forming four square corners; as either of the figures A, B.

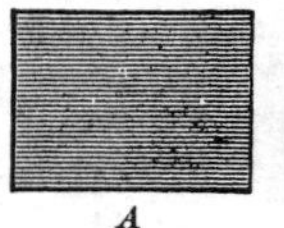
A

B

6. When the four sides are equal the rectangle is called a **square;** as the figure C.

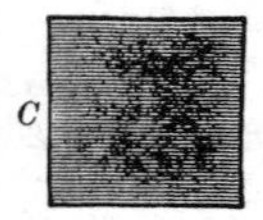
C

7. *The area of a rectangle is equal to its length multiplied by its breadth.*

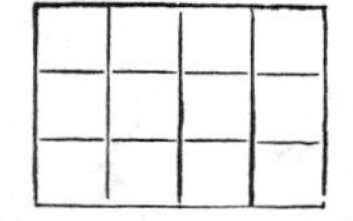

EXPLANATION.—Take a rectangle 4 inches long by 3 inches wide. If upon each of the inches in the length, a square inch be conceived to stand, there will be a row of 4 square inches, extending the whole length of the rectangle, and reaching 1 inch of its width. As the rectangle contains as many such rows as there are inches in its width, its area must be equal to the number of square inches in a row (4) multiplied by the number of rows (3), =12 square inches. This statement (7), as commonly understood, can present no exception to Prin. 2, Art. 60.

TABLE.

144 square inches (sq. in.)	make	1 square foot,	marked	sq. ft.
9 sq. ft.	"	1 square yard,	"	sq. yd.
$30\frac{1}{4}$ sq. yd.	"	1 square rod,	"	sq. rd.
160 sq. rd.	"	1 acre,	"	A.

EQUIVALENT TABLE.

A.		sq. rd.		sq. yd.		sq. ft.		sq. in.
1	=	160	=	4840	=	43560	=	6272640.
		1	=	$30\frac{1}{4}$	=	$272\frac{1}{4}$	=	39204.
				1	=	9	=	1296.
						1	=	144.

NOTE.—The following, though now seldom used, are often found in records of calculations:

40 perches (P.), or sq. rds.,	make	1 rood,	marked	R.
4 roods	"	1 acre,	"	A.

198. Surveyors' Measure is a kind of superficial measure, which is used chiefly in *government* surveys.

TABLE.

625 square links (sq. li.)	make	1 square rod,	sq. rd.
16 sq. rd.	"	1 square chain,	sq. ch.
10 sq. ch.	"	1 acre,	A.
640 A.	"	1 square mile,	sq. mi
36 sq. mi. (6 miles square)	"	1 township,	Tp.

EQUIVALENT TABLE.

Tp.	sq. mi.	A.	sq. ch.	sq. rd.	sq. li.
1 =	36 =	23040 =	230400 =	3686400 =	2304000000.
	1 =	640 =	6400 =	102400 =	64000000.
		1 =	10 =	160 =	100000.
			1 =	16 =	10000.
				1 =	625.

Solid Measure.

199. **A Solid** has length, breadth, and thickness.

Solid Measure is used in estimating the *contents* or *volume* of solids.

A **Cube** is a solid, bounded by six equal squares, called *faces*. Its length, breadth, and thickness are all equal.

REMARK.—The size or name of any cube, like that of a square, depends upon its side, as cubic inch, cubic foot, cubic yard.

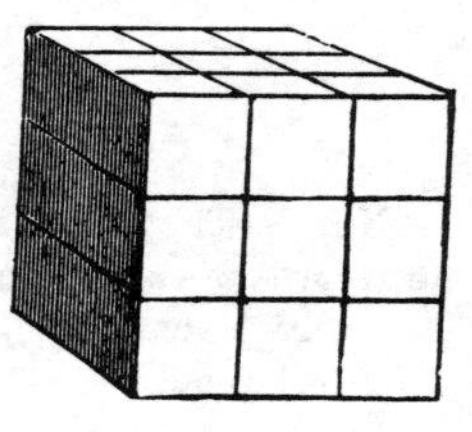

EXPLANATION.—If each side of a cube is 1 inch long, it is called a cubic inch; if each side is 3 feet (1 yard) long, as represented in the figure, it is a cubic or solid yard.

When the base of a cube is 1 square yard, it contains $3 \times 3 = 9$ square feet; and 1 foot high on this base, contains 9 solid feet; 2 feet high contains $9 \times 2 = 18$ solid feet; 3 feet high contains $9 \times 3 = 27$ solid feet. Also it may be shown that 1 solid or *cubic* foot contains $12 \times 12 \times 12 = 1728$ solid or *cubic* inches.

The unit by which all solids are measured is a cube, whose side is a linear inch, foot, etc., and their size or solidity will be the number of times they contain this unit.

REMARK.—The simplest solid is the *rectangular* solid, which is bounded by six rectangles, called its *faces*, each opposite pair being equal, and perpendicular to the other four; as, for example, the ordinary form of a brick or a box of soap. If the length, breadth, and thickness are the same, the faces are squares, and the solid is a cube.

TABLE.

1728 cubic inches (cu. in.) make 1 cubic foot, cu. ft.
27 cu. ft. " 1 cubic yard, cu. yd.

EQUIVALENT TABLE.

cu. yd.		cu. ft.		cu. in.
1	=	27	=	46656.
		1	=	1728.

REMARKS.—1. A *perch* of stone is a mass $16\frac{1}{2}$ ft. long, $1\frac{1}{2}$ ft. wide, and 1 ft. high, and contains $24\frac{3}{4}$ cu. ft.

2. Earth, rock-excavations, and embankments are estimated by the cubic yard.

3. Round timber will lose $\frac{1}{5}$ in being sawed, hence 50 cubic ft. of round timber is said to be equal to 40 cubic ft. of hewn timber, which is a *ton.*

4. Fire-wood is usually measured by the *cord.* A pile of wood 4 ft. high, 4 ft. wide, and 8 ft. long, contains 128 cubic feet or one cord. One foot in length of this pile, or 16 cu. ft., is called a *cord foot.*

5. Planks and scantling are estimated by *board measure.* In this measure, 1 *reduced foot,* 1 ft. long, 1 ft. wide, and 1 in. thick, contains $12 \times 12 \times 1 = 144$ cu. in. All planks and scantling less than an inch thick, are reckoned at that thickness; but, if more than an inch thick, allowance must be made for the excess.

Measures of Capacity.

200. Capacity means *room* for things.

Measures of Capacity are divided into *Measures of Liquids* and *Measures of Dry Substances.*

201. Liquid Measure is used in measuring liquids, and in estimating the capacities of cisterns, reservoirs, etc.

The **gallon,** which contains 231 cu. in., is the unit of measure in liquids.

NOTE.—This gallon of 231 cubic inches was the standard in England at the time of Queen Anne. The present *imperial gallon* of England contains 10 lb. of water at 62° Fahr., or 277.274 cubic inches.

TABLE.

4 gills, marked gi.,	make	1 pint,	marked pt.
2 pt.	"	1 quart,	" qt.
4 qt.	"	1 gallon,	" gal.

EQUIVALENT TABLE.

gal.	qt.	pt.	gi.
1 =	4 =	8 =	32.
	1 =	2 =	8.
		1 =	4.

NOTE.—Sometimes the barrel is estimated at 31½ gal., and the hogshead at 63 gal.; but usually each package of this description is gauged separately.

202. Apothecaries' Fluid Measure is used for measuring all liquids that enter into the composition of medical prescriptions.

TABLE.

60 minims, marked ♏.,	make	1 fluid drachm,	marked fʒ.
8 fʒ	"	1 fluid ounce,	" f℥.
16 f℥	'	1 pint,	" O.
8 O	"	1 gallon,	" cong.

EQUIVALENT TABLE.

cong.	O.	f℥.	fʒ.	♏.
1 =	8 =	128 =	1024 =	61440.
	1 =	16 =	128 =	7680.
		1 =	8 =	480.
			1 =	60.

NOTES.—1. Cong. is an abbreviation for *congiarium*, the Latin for gallon; O. is the initial of *octans*, the Latin for *one eighth*, the pint being one eighth of a gallon.

2. For ordinary purposes, 1 tea-cup = 2 wine-glasses = 8 table-spoons = 32 tea-spoons = 4 f℥.

203. Dry Measure is used for measuring grain, fruit vegetables, coal, salt, etc.

The **Winchester bushel** is the unit; it was formerly used in England, and so called from the town where the standard was kept. It is 8 in. deep, and 18½ in. in diameter, and contains 2150.42 cu. in., or 77.6274 lb. av. of distilled water at maximum density, the barometer at 30 inches.

NOTE.—This bushel was discarded by Great Britain in 1826, and the *imperial bushel* substituted; the latter contains 2218.192 cu. in., or eighty pounds avoirdupois of distilled water.

TABLE.

2 pints, marked pt.,	make	1 quart,	marked	qt.
8 qt.	"	1 peck,	"	pk.
4 pk.	"	1 bushel,	"	bu.

EQUIVALENT TABLE.

bu.		pk.		qt.		pt.
1	=	4	=	32	=	64.
		1	=	8	=	16.
				1	=	2.

REMARKS.—1. 4 qt. or ½ peck = 1 dry gal. = 268.8 cu. in. nearly.

2. The *quarter* is still used in England for measuring *wheat*, of which it holds eight bushels, or 480 pounds avoirdupois.

3. When articles usually measured by the above table are sold by weight, the *bushel* is taken as the unit. The following table gives the legal weight of a bushel of various articles in avoirdupois pounds:

TABLE.

ARTICLES.	LB.	EXCEPTIONS.
Beans.	60	Me., 64; N. Y., 62.
Coal.	80	Ohio, 70 of cannel; Ky. 76 of anthracite.
Corn (Indian).	56	N. Y., 58; Cal., 52; Arizona, 54.
Flax Seed.	56	N. Y. and N. J., 55; Kan., 54.
Oats.	32	Md., 26; Me., N. H., N. J., Pa., 30; Neb., 34; Montana, 35; Oregon and Wash., 36.
Potatoes (Irish).	60	Ohio, 58; Wash., 50.
Rye.	56	La., 32; Cal., 54.
Salt.	50	Mass., 70; Pa., coarse, 85; ground, 70; fine, 62; Ky. and Ill., fine, 55; Mich., 56; Col. and Dak., 80.
Wheat.	60	

COMPARATIVE TABLE OF MEASURES.

	cu. in. gal.	cu. in. qt.	cu. in. pt.	cu. in. gi.
Liquid Measure,	231	$57\frac{3}{4}$	$28\frac{7}{8}$	$7\frac{7}{32}$
Dry Measure ($\frac{1}{2}$ pk.),	$268\frac{4}{5}$	$67\frac{1}{5}$	$33\frac{3}{5}$	$8\frac{2}{5}$

Angular or Circular Measure.

204. A **plane angle** is the difference of direction of two straight lines which meet at a point.

EXPLANATION.—Thus, the two lines AB and AC meet at the point A, called the *apex*. The lines AB and AC are the *sides* of the angle, and the difference in direction, or the *opening* of the lines, is the *angle itself*.

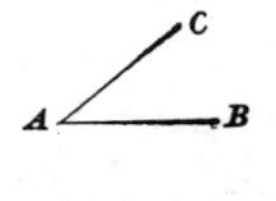

Angular Measure is used to measure angles, directions, latitude, and longitude, in navigation, astronomy, etc.

A **circle** is a plane surface bounded by a line, all the points of which are equally distant from a point within.

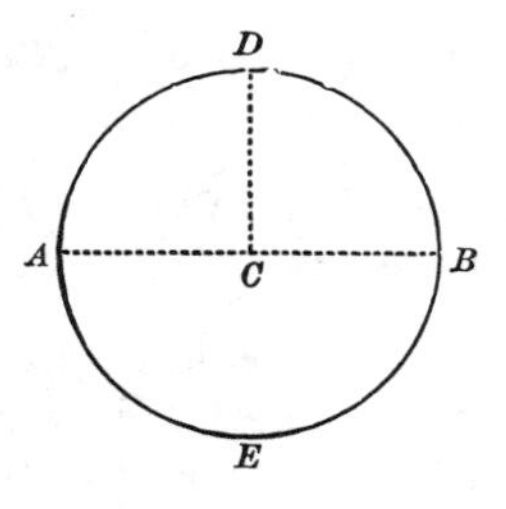

Explanations.—The bounding line ADBEA is a *circumference.* Every point of this line is at the same distance from the point C, which is called the *center.* The *circle* is the area included within the circumference. Any straight line drawn from the center to the circumference is called a *radius;* thus, CD and CB are radii. Any part of the circumference, as AEB or AD, is an *arc.* A straight line, like AB, drawn through the center, and having its ends in the circumference, is a *diameter;* it divides the circle into two equal parts.

Notes.—1. Every circumference contains 360 degrees; and, the apex of an angle being taken as the center of a circle, the angle is measured by the number of degrees in the arc included by the sides of the angle.

2. The angle formed by two lines perpendicular to each other, as the radii AC and DC in the above figure, is a *right angle,* and is measured by the fourth part of a circumference, 90°, called a *quadrant.*

Table.

60 seconds, marked ″,	make	1 minute,	marked,	′.
60′	"	1 degree,	"	°.
360°	"	1 circumference,	"	c.

Equivalent Table.

c.		°		′		″
1	=	360	=	21600	=	1296000.
		1	=	60	=	3600.
				1	=	60.

Note.—The twelfth part of a circumference, or 30°, is called a *sign.*

MEASURE OF TIME.

205. 1. **Time** is a measured portion of duration.

2. A **Year** is the time of the revolution of the earth

around the sun; a **Day** is the time of the revolution of the earth on its axis.

3. The **Solar Day** is the interval of time between two successive passages of the sun over the same meridian.

4. The **Mean Solar Day** is the mean, or average, length of all the solar days in the year. Its duration is 24 hours, and it is the unit of Time Measure.

5. The **Civil Day,** used for ordinary purposes, commences at midnight and closes at the next midnight.

6. The **Astronomical Day** commences at noon and closes at the next noon.

TABLE.

60	seconds, marked sec.,	make	1 minute,	marked	min.
60	min.	"	1 hour,	"	hr.
24	hr.	"	1 day,	"	da.
7	da.	"	1 week,	"	wk.
4	wk.	"	1 month,	"	mon.
12	calendar mon.	"	1 year,	"	yr.
365	da.	"	1 common year.		
366	da.	"	1 leap year.		
100	yr.	"	1 century, marked cen.		

NOTE.—1 Solar year = 365 da. 5 hr. 48 min. 46.05 sec. = 365¼ da., nearly.

EQUIVALENT TABLE.

yr.	mo.	wk.	da.	hr.	min.	sec.
1 =	12 =	52 =	{ 365 =	8760 =	525600 =	31536000.
			{ 366 =	8784 =	527040 =	31622400.
		1 =	7 =	168 =	10080 =	604800.
			1 =	24 =	1440 =	86400.
				1 =	60 =	3600.
					1 =	60.

NOTE.—The ancients were unable to find accurately the number of days in a year. They had 10, afterward 12, calendar months, corresponding to the revolutions of the moon around the earth. In the time of Julius Cæsar the year contained 365¼ days; instead of taking account of the ¼ of a day every year, the common or civil year was reckoned 365 days, and every 4th year a day was inserted (called the *intercalary* day), making the year then have 366 days. The extra day was introduced by repeating the 24th of February, which, with the Romans, was called the *sixth day before the kalends of March.* The years containing this day twice, were on this account called *bissextile,* which means *having two sixths.* By us they are generally called *leap* years.

But 365¼ days (365 days and 6 hours) are a little longer than the true year, which is 365 days 5 hours 48 minutes 46.05 seconds. The difference, 11 minutes 13.95 seconds, though small, produced, in a long course of years, a sensible error, which was corrected by Gregory XIII., who, in 1582, suppressed the 10 days that had been gained, by decreeing that the 5th of October should be the 15th.

206. To prevent difficulty in future, it has been decided to adopt the following rule.

Rule for Leap Years.—*Every year that is divisible by* 4 *is a leap year, unless it ends with two ciphers; in which case it must be divisible by* 400 *to be a leap year.*

ILLUSTRATION.—Thus, 1832, 1648, 1600, and 2000 are leap years; but 1857, 1700, 1800, 1918, are not.

NOTES.—1. The *Gregorian* calendar was adopted in England in 1752. The error then being 11 days, Parliament declared the 3d of September to be the 14th, and at the same time made the year begin January 1st, instead of March 25th. Russia, and all other countries of the Greek Church, still use the Julian calendar; consequently their dates (*Old Style*) are now 12 days later than ours (*New Style*). The error in the Gregorian calendar is small, amounting to a day in 3600 years.

2. The year formerly began with March instead of January; consequently, September, October, November, and December were the 7th, 8th, 9th, and 10th months, as their names indicate; being derived from the Latin numerals Septem (7), Octo (8), Novem (9), Decem (10).

COMPARISON OF TIME AND LONGITUDE.

207. The **longitude** of a place is its distance in degrees, minutes, and seconds, east or west of an established meridian.

NOTE.—The *difference of longitude* of two places on the same side of the established meridian, is found by subtracting the less longitude from the greater; but, of two places on opposite sides of the meridian, the difference of longitude is found by adding the longitude of one to the longitude of the other.

The circumference of the earth, like other circles, is divided into 360 equal parts, called *degrees of longitude.*

The sun appears to pass entirely round the earth, 360°, once in 24 hours, *one day;* and in 1 hour it passes over 15°. $(360° \div 24 = 15°.)$

As 15° equal 900′, and 1 hour equals 60 minutes of *time*, therefore, the sun in 1 minute of *time* passes over 15′ of a *degree.* $(900' \div 60 = 15'.)$

As 15′ equal 900″, and 1 minute of *time* equals 60 seconds of *time*, therefore, in 1 second of *time* the sum passes over 15″ of a *degree.* $(900'' \div 60 = 15''.)$

TABLE FOR COMPARING LONGITUDE AND TIME.

15° of longitude = 1 hour of time.
15′ of longitude = 1 min. of time.
15″ of longitude = 1 sec. of time.

NOTE.—If one place has greater east or less west longitude than another, its time must be later; and, conversely, if one place has later time than another, it must have greater east or less west longitude.

MISCELLANEOUS TABLES.

208. The words folio, quarto, octavo, etc., used in speaking of books, show how many leaves a sheet of paper makes.

A sheet folded into

2 leaves, called	a	folio,	makes	4	pages.
4 " "	a	quarto or 4to,	"	8	"
8 " "	an	octavo or 8vo,	"	16	"
12 " "	a	duodecimo or 12mo,	"	24	"
16 " "	a	16mo,	"	32	"
32 " "	a	32mo,	"	64	"

Also,	24 sheets of paper	make	1 quire.
	20 quires	"	1 ream.
	2 reams	"	1 bundle.
	5 bundles	"	1 bale.

12 things	make	1 dozen.
12 dozens or 144 things	"	1 gross.
12 gross or 144 dozens	"	1 great gross.
20 things	"	1 score.
56 lb.	"	1 firkin of butter.
100 lb.	"	1 quintal of fish.
196 lb.	"	1 bbl. of flour.
200 lb.	"	1 bbl. of pork.

THE METRIC SYSTEM.

Historical.

The Metric System is an outgrowth of the French Revolution of 1789. At that time there was a general disposition to break away from old customs; and the revolutionists contended that every thing needed remodeling. A commission was appointed to determine an invariable standard for all measures of length, area, solidity, capacity, and weight. After due deliberation, an accurate survey was made of that portion of the terrestrial meridian through Paris, between Dunkirk, France, and Barcelona, Spain; and from this, the distance on that meridian from the equator to the pole was computed. The quadrant thus obtained was divided into ten million equal parts; one part was called a *meter*, and is the *base* of the system. From it all measures are derived.

In 1795 the Metric System was adopted in France. It is now used in nearly all civilized countries. It was authorized by an act of Congress in the United States in 1866.

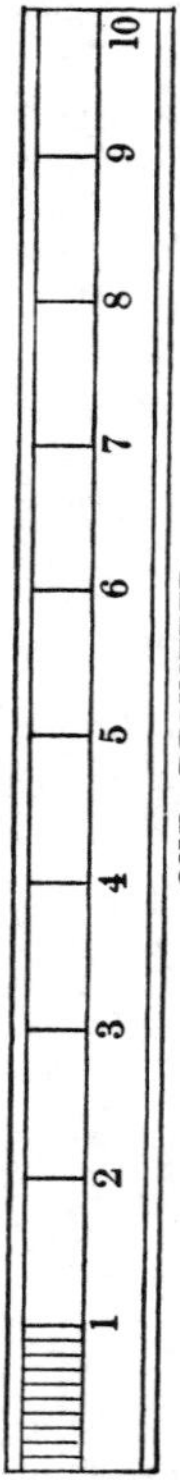

ONE DECIMETER

1 2 3 4 5 6 7 8 9 10

209. The **Metric System** is a decimal system of weights and measures.

The **meter** is the primary unit upon which the system is based, and is also the unit of length. It is 39.37043 inches long, which is very nearly one ten-millionth part of the distance on the earth's surface from the equator to the pole, as measured on the meridian through Paris.

REMARK.—The *standard meter*, a bar of platinum, is kept among the archives in Paris. Duplicates of this bar have been furnished to the United States.

210. The **names** of the *lower* denominations in each measure of the Metric System are formed by prefixing the *Latin* numerals, *deci* (.1), *centi* (.01), and *milli* (.001) to the *unit* of that measure; those of the *higher* denominations, by prefixing the *Greek* numerals, *deka* (10), *hekto* (100), *kilo* (1000), and *myria* (10000), to the same unit.

These prefixes may be grouped about the unit of measure, showing the decimal arrangement of the system, as follows:

Lower Denominations.	milli =	.001
	centi =	.01
	deci =	.1
	Unit of Measure =	1.
Higher Denominations.	deka =	10.
	hekto =	100.
	kilo =	1000.
	myria =	10000.

211. The units of the various measures, to which these prefixes are attached, are as follows:

The Meter,	which	is	the	unit	of	Length.
The Ar,	"	"	"	"	"	Surface.
The Liter,	"	"	"	"	"	Capacity.
The Gram,	"	"	"	"	"	Weight.

REMARK.—The name of each denomination thus derived, immediately shows its relation to the unit of measure. Thus, a *centimeter* is one one-hundredth of a meter; a *kilogram* is a thousand grams; a *hektoliter* is one hundred liters, etc.

Measure of Length.

212. The **Meter** is the unit of Length, and is the denomination used in all ordinary measurements.

TABLE.

10 millimeters, marked mm.,	make	1 centimeter,	marked	cm.
10 centimeters	"	1 decimeter,	"	dm.
10 decimeters	"	1 meter,	"	m.
10 meters	"	1 dekameter,	"	Dm.
10 dekameters	"	1 hektometer,	"	Hm.
10 hektometers	"	1 kilometer,	"	Km.
10 kilometers	"	1 myriameter,	"	Mm.

REMARKS.—1. The figure on page 148 shows the exact length of the decimeter, and its subdivisions the centimeter and millimeter.

2. The centimeter and millimeter are most often used in measuring very short distances; and the kilometer, in measuring roads and long distances.

Measure of Surface.

213. The **Ar** (*pro.* är) is the unit of Land Measure; it is a square, each side of which is 10 meters (1 dekameter) in length, and hence its area is one square dekameter.

TABLE.

100 centars, marked ca.,	make	1 ar,	marked	a.
100 ars	"	1 hektar,	"	Ha.

REMARK.—The *square meter* (marked m^2) and its subdivisions are used for measuring small surfaces.

Measure of Capacity.

214. The **Liter** (*pro.* lē'ter) is the unit of Capacity. It is equal in volume to a cube whose edge is a decimeter; that is, one tenth of a meter.

TABLE.

10 milliliters, marked ml.,	make	1 centiliter,	marked	cl.
10 centiliters	"	1 deciliter,	"	dl.
10 deciliters	"	1 liter,	"	l.
10 liters	"	1 dekaliter,	"	Dl.
10 dekaliters	"	1 hektoliter,	"	Hl.

REMARKS.—1. This measure is used for liquids and for dry substances. The denominations most used are the liter and hektoliter; the former in measuring milk, wine, etc., in moderate quantities, and the latter in measuring grain, fruit, etc., in large quantities.

2. Instead of the milliliter and the kiloliter, it is customary to use the cubic centimeter and the cubic meter (marked m^3), which are their equivalents.

3. For measuring wood the *ster* (pro. stêr) is used. It is a cubic meter in volume.

Measure of Weight.

215. The **Gram** (*pro.* grăm) is the unit of Weight. It was determined by the weight of a cubic centimeter of distilled water, at the temperature of maximum density (39.2° F.)

TABLE.

10 milligrams, marked mg.,	make	1 centigram,	marked cg.
10 centigrams	"	1 decigram,	" dg.
10 decigrams	"	1 gram,	" g.
10 grams	"	1 dekagram,	" Dg.
10 dekagrams	"	1 hektogram,	" Hg.
10 hektograms	"	1 kilogram,	" Kg.
10 kilograms	"	1 myriagram,	" Mg.
10 myriagrams, or 100 kilograms	"	1 quintal,	" Q.
10 quintals, or 1000 "	"	1 metric ton,	" M.T.

REMARKS.—1. The gram, kilogram (*pro.* kĭl′o-gram), and metric ton are the weights commonly used.

2. The gram is used in all cases where great exactness is required; such as, mixing medicines, weighing the precious metals, jewels, letters, etc.

3. The kilogram, or, as it is commonly abbreviated, the "kilo," is used in weighing coarse articles, such as groceries, etc.

4. The metric ton is used in weighing hay and heavy articles generally.

216. Since, in the Metric System, 10, 100, 1000, etc., units of a lower denomination make a unit of the higher denomination, the following principles are derived:

PRINCIPLES.—1. *A number is reduced to a* LOWER *denomination by removing the decimal point as many places to the* RIGHT *as there are ciphers in the multiplier.*

2. *A number is reduced to a* HIGHER *denomination by removing the decimal point as many places to the* LEFT *as there are ciphers in the divisor.*

ILLUSTRATIONS.—Thus, 15.03 m. is read 15 meters and 3 centimeters; or, 15 and 3 hundredths meters. Again, 15.03 meters = 1.503 dekameters = .1503 hektometer = 150.3 decimeters = 1503 centimeters. As will be seen, the reduction is effected by changing the decimal point in precisely the same manner as in United States Money

217. The following table presents the legal and approximate values of those denominations of the Metric System which are in common use.

TABLE.

DENOMINATION.	LEGAL VALUE.	APPROX. VALUE.
Meter.	39.37 inches.	3 ft. $3\frac{3}{8}$ inches.
Centimeter.	.3937 inch.	$\frac{2}{5}$ inch.
Millimeter.	.03937 inch.	$\frac{1}{25}$ inch.
Kilometer.	.62137 mile.	$\frac{5}{8}$ mile.
Ar.	119.6 sq. yards.	4 sq. rods.
Hektar.	2.471 acres.	$2\frac{1}{2}$ acres.
Square Meter.	1.196 sq. yards.	$10\frac{3}{4}$ sq. feet.
Liter.	1.0567 quarts.	1 quart.
Hektoliter.	2.8375 bushels.	2 bu. $3\frac{1}{3}$ pecks.
Cubic Centimeter.	.061 cu. inch.	$\frac{1}{16}$ cu. inch.
Cubic Meter.	1.308 cu. yards.	$35\frac{1}{3}$ cu. feet.
Ster.	.2759 cord.	$\frac{1}{4}$ cord.
Gram.	15.432 grains troy.	$15\frac{1}{2}$ grains.
Kilogram.	2.2046 pounds av.	$2\frac{1}{5}$ pounds.
Metric Ton.	2204.6 pounds av.	1 T. 204 pounds.

Topical Outline.

COMPOUND NUMBERS.

1. Preliminary Definitions.

2. Value........
- 1. Definitions.
- 2. United States and Canadian Money.
- 3. English Money.
- 4. French Money.
- 5. German Money.

3. Weight......
- 1. Troy Weight.
- 2. Apothecaries' Weight.
- 3. Avoirdupois Weight.

Compound Numbers.—(*Concluded.*)

- **4. Extension**
  - 1. Linear...
    - 1. Long Measure.
    - 2. Chain Measure.
    - 3. Mariners' Measure.
    - 4. Cloth Measure.
  - 2. Superficial Measure....
    - 1. Surface Measure.
    - 2. Surveyors' Measure.
  - 3. Solid Measure.
  - 4. Measures of Capacity.
    - 1. Liquid.
    - 2. Dry.
  - 5. Angular Measure.
- **5. Time.**
- **6. Comparison of Time and Longitude.**
- **7. Miscellaneous Tables.**

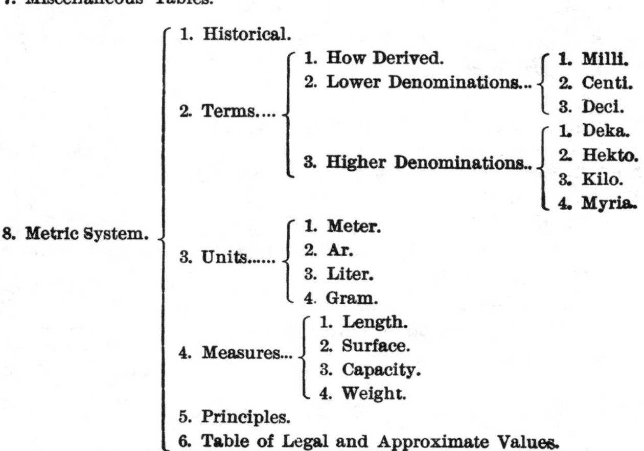

- **8. Metric System.**
  - 1. Historical.
  - 2. Terms....
    - 1. How Derived.
    - 2. Lower Denominations...
      - 1. Milli.
      - 2. Centi.
      - 3. Deci.
    - 3. Higher Denominations..
      - 1. Deka.
      - 2. Hekto.
      - 3. Kilo.
      - 4. Myria.
  - 3. Units......
    - 1. Meter.
    - 2. Ar.
    - 3. Liter.
    - 4. Gram.
  - 4. Measures...
    - 1. Length.
    - 2. Surface.
    - 3. Capacity.
    - 4. Weight.
  - 5. Principles.
  - 6. Table of Legal and Approximate Values.

REDUCTION OF COMPOUND NUMBERS.

218. Reduction of Compound Numbers is the process of changing them to equivalent numbers of a different denomination.

Reduction takes place in two ways:

From a higher denomination to a lower.

From a lower denomination to a higher.

PRINCIPLES.—1. *Reduction from a higher denomination to a lower, is performed by multiplication.*

2. *Reduction from a lower denomination to a higher, is performed by division.*

PROBLEM.—Reduce 18 bushels to pints.

SOLUTION.—Since 1 bu. = 4 pk., 18 bu. = 18 times 4 pk. = 72 pk., and since 1 pk. = 8 qt., 72 pk. = 72 times 8 qt. = 576 qt.; and since 1 qt. = 2 pt., 576 qt. = 576 times 2 pt. = 1152 pt. Or, since 1 bu. = 64 pt., multiply 64 pt. by 18, which gives 1152 pt. as before. This is sometimes called *Reduction Descending.*

OPERATION.

```
   18 bu.
    4
   ---
   72 pk
    8
  ---
  576 qt.
    2
 ----
 1152 pt.
18 bu. = 1152 pt.
```

PROBLEM.—Reduce 236 inches to yards.

SOLUTION.—Since 12 inches = 1 ft., 236 inches will be as many feet as 12 in. is contained times in 236 in., which is $19\frac{2}{3}$ ft., and since 3 ft. = 1 yd., $19\frac{2}{3}$ ft will be as many yd. as 3 ft. is contained times in $19\frac{2}{3}$ ft., which is $6\frac{5}{9}$ yd. Or, since 1 yd. = 36 in., divide 236 in. by 36 in., which gives $6\frac{5}{9}$ yd., as before. This is sometimes called *Reduction Ascending.*

OPERATION.

$$12)\ 236 \text{ in.}$$
$$3)\ 19\tfrac{2}{3} \text{ ft.}$$
$$6\tfrac{5}{9} \text{ yd.}$$
$$236 \text{ in.} = 6\tfrac{5}{9} \text{ yd.}$$

NOTE.—In the last example, instead of dividing 236 in. by 36 in. the unit of value of yards, since 1 inch is equal to $\frac{1}{36}$ yards, 236 inches $= 236 \times \frac{1}{36} = \frac{236}{36}$ yd. $= 6\frac{5}{9}$ yd. The operation by division is generally more convenient.

REMARK.—Reduction Descending diminishes the *size*, and, therefore, increases the *number* of units given; while Reduction Ascending increases the *size*, and, therefore, diminishes the *number* of units given. This is further evident from the fact, that the multipliers in Reduction Descending are *larger* than 1; but in Reduction Ascending *smaller* than 1.

PROBLEM.—Reduce $\frac{3}{8}$ gallons to pints.

SOLUTION.—Multiply by 4 to reduce gal. to qt.; then by 2 to reduce qt. to pt. Indicate the operation, and cancel.

OPERATION.

$$\frac{3}{\not 8} \times \not 4 \times \not 2 = 3 \text{ pt.}$$

$$\tfrac{3}{8} \text{ gal.} = 3 \text{ pt.}$$

PROBLEM.—Reduce $5\frac{5}{7}$ gr. to ℥.

SOLUTION. — Although this is Reduction Ascending, we use Prin. 1, in multiplying by the successive unit values, $\frac{1}{20}$, $\frac{1}{3}$, and $\frac{1}{8}$.

OPERATION.

$$5\tfrac{5}{7} \text{ gr.} = \frac{\overset{2}{\not 4 \not 0}}{7} \times \frac{1}{\not 2 \not 0} \times \frac{1}{3} \times \frac{1}{\underset{4}{\not 8}} = \frac{1}{84} \text{ ℥.}$$

$$5\tfrac{5}{7} \text{ gr.} = \tfrac{1}{84} \text{ ℥.}$$

PROBLEM.—Reduce 9.375 acres to square rods.

OPERATION.

```
     9.375
       160
   -------
    562500
    9375
   -------
  1500.000 sq. rd.
```

9.375 A. = 1500 sq. rd.

PROBLEM.—Reduce 2000 seconds to hours.

OPERATION.

$$20\not 0\not 0 \times \frac{1}{6\not 0} \times \frac{1}{6\not 0} = \frac{20}{36} = \frac{5}{9} \text{ hr.}$$

$$2000 \text{ sec.} = \tfrac{5}{9} \text{ hr.}$$

PROBLEM.—Reduce 1238.73 hektograms to grams.

OPERATION.

$$1238.73 \times 100 = 123873 \text{ grams.}$$

PROBLEM.—How many yards in 880 meters?

OPERATION.

$$\frac{39.37 \text{ in.} \times 880}{12 \times 3} = 962.377 + \text{yd.}$$

REMARK.—Abstract factors can not produce a concrete result *sometimes*, however, in the steps of an indicated solution, where the change of denomination is *very* obvious, the abbreviations may be omitted until the result is written.

From the preceding exercises, the following rules are derived:

219. For reducing from higher to lower denominations.

Rule.—1. *Multiply the highest denomination given, by that number of the next lower which makes a unit of the higher.*

2. *Add to the product the number, if any, of the lower denomination.*

3. *Proceed in like manner with the result thus obtained, till the whole is reduced to the required denomination.*

220. For reducing from lower to higher denominations.

Rule.—1. *Divide the given quantity by that number of its own denomination which makes a unit of the next higher.*

2. *Proceed in like manner with the quotient thus obtained, till the whole is reduced to the required denomination.*

3. *The last quotient, with the several remainders, if any, annexed, will be the answer.*

NOTE.—In the Metric System the operations are performed by removing the point to the right or to the left.

EXAMPLES FOR PRACTICE.

1. How many square rods in a rectangular field 18.22 chains long by 4.76 ch. wide?

2. Reduce 16.02 chains to miles.

3. How many bushels of wheat would it take to fill 750 hektoliters?

4. Reduce 35.781 sq. yd. to sq. in.

5. Reduce 10240 sq. rd. to sq. ch.

6. How many perches of masonry in a rectangular solid wall 40 ft. long by $7\frac{1}{2}$ ft. high, and $2\frac{2}{3}$ ft. average thickness?

7. How many ounces troy in the Brazilian Emperor's diamond, which weighs 1680 carats?

8. Reduce 75 pwt. to ʒ.

9. Reduce $\frac{4}{7}$ gr. to ℥.

10. Reduce $18\frac{3}{4}$ ʒ to oz. av.

11. Reduce 96 oz. av. to oz. troy.

12. How many gal. in a tank 3 ft. long by $2\frac{1}{4}$ ft. wide and $1\frac{1}{2}$ ft. deep?

13. How many bushels in a bin 9.3 ft. long by $3\frac{5}{8}$ ft. wide and $2\frac{1}{4}$ ft. deep?

14. How many sters in 75 cords of wood?

15. Reduce $2\frac{1}{4}$ years to seconds.

16. Forty-nine hours is what part of a week?

17. Reduce 90.12 kiloliters to liters.

18. Reduce 25″ to the decimal of a degree.

19. Reduce 192 sq. in. to sq. yd.

20. Reduce $6\frac{2}{3}$ cu. yd. to cu. in.

21. Reduce $117.14 to mills.

22. Reduce 6.19 cents to dollars.

23. Reduce 1600 mills to dollars.

24. Reduce $$5\frac{3}{8}$ to mills.

25. Reduce 12 lb. av. to ℔. troy.

26. How many grams in 6.45 quintals?

27. Reduce .216 gr. to oz. troy.

28. Reduce 47.3084 sq. mi. to sq. rd.

29. Reduce $4\frac{1}{2}$ ℈ to ℔.

30. Reduce $7\frac{1}{9}$ oz. av. to cwt.

31. Reduce 99 yd. to miles.

32. How many acres in a rectangle $24\frac{1}{2}$ rd. long by 16.02 rd. wide?

33. How many cubic yards in a box $6\frac{1}{4}$ ft. long by $2\frac{1}{2}$ ft. wide and 3 ft. high?

34. Reduce 169 ars to square meters.

35. Reduce $2\frac{1}{2}$ f℥ to ♏.

36. If a piece of gold is $\frac{6}{7}$ pure, how many carats fine is it?

37. In $18\frac{3}{4}$ carat gold, what part is pure and what part alloy?

38. How many square meters of matting are required to cover a floor, the dimensions of which are 6 m., $1\frac{1}{2}$ dm by 5 m., 3 cm.?

39. How many cords of wood in a pile 120 ft. long, $6\frac{1}{2}$ ft. wide, and $8\frac{3}{4}$ ft. high?

40. How many sq. ft. in the four sides of a room $21\frac{1}{2}$ ft. long, $16\frac{1}{2}$ ft. wide, and 13 ft. high?

41. What will be the cost of 27 T. 18 cwt. 3 qr. 15 lb. 12 oz. of potash, at $48.20 a ton?

42. What is the value of a pile of wood 16 m., 1 dm., 5 cm. long, 1 m., 2 dm., 2 cm. wide, and 1 m., 6 dm., 8 cm. high, at $2.30 a ster?

43. What is the cost of a field 173 rods long and 84 rods wide, at $25.60 an acre?

44. If an open court contain 160 sq. rd. 85 sq. in.; how many stones, each 5 inches square, will be required to pave it?

45. A lady had a grass-plot 20 meters long and 15 meters wide; after reserving two plots, one 2 meters square and the other 3 meters square, she paid 51 cents a square meter to have it paved with stones: what did the paving cost?

46. A cubic yard of lead weighs 19,128 lb.: what is the weight of a block 5 ft. $3\frac{1}{4}$ in. long, 3 ft. 2 in. wide, and 1 ft. 8 in. thick?

47. A lady bought a dozen silver spoons, weighing 3 oz. 4 pwt. 9 gr., at $2.20 an oz., and a gold chain weighing 13 pwt., at $$1\frac{1}{4}$ a pwt.: required the total cost of the spoons and chain.

48. A wagon-bed is $10\frac{1}{2}$ ft. long, $3\frac{1}{2}$ ft. wide, and $1\frac{1}{2}$ ft. deep, inside measure: how many bushels of corn will it hold, deducting one half for cobs?

49. If a man weigh 160 lb. avoirdupois, what will he weigh by troy weight?

50. The fore-wheel of a wagon is 13 ft. 6 in. in circumference, and the hind wheel 18 ft. 4 in.: how many more revolutions will the fore-wheel make than the hind one in 50 miles?

51. An apothecary bought 5 ℔. 10 ℥. of quinine, at $2.20 an ounce, and sold it in doses of 9 gr., at 10 cents a dose: how much did he gain?

52. How many steps must a man take in walking from Kansas City to St. Louis, if the distance be 275 miles, and each step, 2 ft. 9 in.?

53. The area of Missouri is 65350 sq. mi.: how many hektars does it contain?

54. A school-room is 36 ft. long, 24 ft. wide, and 14 ft. high; required the number of gallons of air it will contain?

55. Allowing 8 shingles to the square foot, how many shingles will be required to cover the roof of a barn which is 60 feet long, and 15 feet from the comb to the eaves?

56. A boy goes to bed 30 minutes later, and gets up 40 minutes earlier than his room-mate: how much time does he gain over his room-mate for work and study in the two years 1884 and 1885, deducting Sundays only?

ADDITION OF COMPOUND NUMBERS.

221. Compound Numbers may be added, subtracted, multiplied, and divided. The principles upon which these operations are performed are the same as in Simple Numbers, with this variation; namely, that in Simple Numbers *ten* units of a lower denomination make one of the next higher, while in Compound Numbers the *scales vary.*

Addition of Compound Numbers is the process of finding the sum of two or more similar Compound Numbers.

PROBLEM.—Add 3 bu. $2\frac{1}{4}$ pk.; 1 pk. $1\frac{1}{2}$ pt.; 5 qt. 1 pt.; 2 bu. $1\frac{1}{3}$ qt.; and .125 pt.

OPERATION.

bu.	pk.	qt.	pt.	
3	2	2	0	= 3 bu. $2\frac{1}{4}$ pk.
	1	0	$1\frac{1}{2}$	= 1 pk. $1\frac{1}{2}$ pt.
		5	1	= 5 qt. 1 pt.
2	0	1	$\frac{2}{3}$	= 2 bu. $1\frac{1}{3}$ qt.
			$\frac{1}{8}$	= .125 pt.
6	0	1	$1\frac{7}{24}$	= *Ans.*

SOLUTION.—Reduce the fraction in each number to lower denominations, and write units of the same kind in the same column. The right-hand column, when added, gives $3\frac{7}{24}$ pt. = 1 qt. $1\frac{7}{24}$ pt.; write the $1\frac{7}{24}$ and add the 1 qt. with the next column, making 9 qt. = 1 pk. 1 qt.; write the 1 qt. and carry the 1 pk. to the next column, making 4 pk. = 1 bu.; as there are no pk. left, write down a cipher and carry 1 bu. to the next column, making 6 bu.

PROBLEM.—Add 2 rd. 9 ft. $7\frac{1}{4}$ in.; 13 ft. 5.78 in.; 4 rd. 11 ft. 6 in.; 1 rd. $10\frac{2}{3}$ ft.; 6 rd. 14 ft. $6\frac{5}{8}$ in.

OPERATION.

rd.	ft.	in.
2	9	7.25
	13	5.78
4	11	6
1	10	8
6	14	6.625
16	$9\frac{1}{2}$	9.655
	but $\frac{1}{2}$ ft.	= 6.
16	10	3.655

SOLUTION.—The numbers are prepared, written, and added, as in the last example; the answer is 16 rd. $9\frac{1}{2}$ ft. 9.655 in. The $\frac{1}{2}$ foot is then reduced to 6 inches, and added to the 9.655 in., making 15.655 in. = 1 ft. 3.655 in. Write the 3.655 in., and carry the 1 ft., which gives 16 rd. 10 ft. 3.655 in. for the final answer.

Rule.—1. *Write the numbers to be added, placing units of the same denomination in the same column.*

2. *Begin with the lowest denomination, add the numbers, and divide their sum by the number of units of this denomination which make a unit of the next higher.*

3. *Write the remainder under the column added, and carry the quotient to the next column.*

4. *Proceed in the same manner with all the columns to the last, under which write its entire sum.*

REMARK.—The proof of each fundamental operation in Compound Numbers is the same as in Simple Numbers.

EXAMPLES FOR PRACTICE.

1. Add $\frac{2}{7}$ mi.; $146\frac{1}{3}$ rd.; 10 mi. 14 rd. 7 ft. 6 in.; 209.6 rd.; 37 rd. 16 ft. $2\frac{1}{4}$ in.; 1 mi. 12 ft. 8.726 in.

2. Add 6.19 yd.; 2 yd. 2 ft. $9\frac{3}{4}$ in.; 1 ft. 4.54 in.; 10 yd. 2.376 ft.; $\frac{3}{4}$ yd.; $1\frac{5}{6}$ ft.; $\frac{7}{8}$ in.

3. Add 3 yd. 2 qr. 3 na. $1\frac{1}{2}$ in.; 1 qr. $2\frac{2}{5}$ na.; 6 yd. 1 na. 2.175 in.; 1.63 yd.; $\frac{2}{3}$ qr.; $\frac{5}{9}$ na.

4. If the volume of the earth is 1; Mercury, .06; Venus, .957; Mars, .14; Jupiter, 1414.2; Saturn, 734.8; Uranus, 82; Neptune, 110.6; the Sun, 1407124; and the Moon, .018, what is the volume of all?

5. James bought a balloon for 9 francs and 76 centimes, a ball for 68 centimes, a hoop for one franc and 37 centimes, and gave to the poor 2 francs and 65 centimes, and had 3 francs and 4 centimes left. How much money did he have at first?

6. Add 15 sq. yd. 5 sq. ft. 87 sq. in.; $16\frac{1}{2}$ sq. yd.; 10 sq. yd. 7.22 sq. ft.; 4 sq. ft. 121.6 sq. in.; $\frac{11}{32}$ sq. yd.

7. Add 101 A. 98.35 sq. rd.; 66 A. $74\frac{1}{2}$ sq. rd.; 20 A.; 12 A. 113 sq. rd.; 5 A. $13.33\frac{1}{3}$ sq. rd.

8. Add 23 cu yd. 14 cu. ft. 1216 cu. in.; 41 cu. yd. 6 cu. ft. 642.132 cu. in.; 9 cu. yd. 25.065 cu. ft.; $\frac{7}{15}$ cu. yd.

9. Add $\frac{5}{6}$ C.; $\frac{3}{8}$ cu. ft.; 1000 cu. in.

10. Add 2 ℔. troy, $6\frac{2}{3}$ oz.; $1\frac{3}{4}$ ℔.; 12.68 pwt.; 11 oz. 13 pwt. $19\frac{1}{5}$ gr.; $\frac{3}{8}$ ℔. $\frac{15}{16}$ oz.; $\frac{5}{9}$ pwt.

11. Add 8 ℥ 14.6 gr.; 4.18 ℥; $7\frac{5}{12}$ ʒ; 2 ʒ 2 ℈ 18 gr.; 1 ℥ 12 gr.; $\frac{1}{5}$ ℈.

12. Add $\frac{5}{16}$ T.; 9 cwt. 1 qr. 22 lb.; 3.06 qr.; 4 T. 8.764 cwt.; 3 qr. 6 lb.; $\frac{7}{12}$ cwt.

13. Add .3 lb. av.; $\frac{5}{9}$ oz.

14. Add 6 gal. $3\frac{1}{3}$ qt.; 2 gal. 1 qt. .83 pt.; 1 gal. 2 qt. $\frac{1}{2}$ pt.; $\frac{3}{5}$ gal.; $\frac{2}{9}$ qt.; $\frac{7}{8}$ pt.

15. Add 4 gal. .75 pt.; 10 gal. 3 qt. $1\frac{1}{2}$ pt.; 8 gal. $\frac{2}{3}$ pt.; 5.64 gal.; 2.3 qt.; 1.27 pt.; $\frac{3}{20}$ pt.

16. Add 1 bu. $\frac{1}{2}$ pk.; $\frac{6}{11}$ bu.; 3 pk. 5 qt. $1\frac{1}{4}$ pt.; 9 bu. 3.28 pk.; 7 qt. 1.16 pt.; $\frac{5}{12}$ pk.

17. Add $\frac{3}{8}$ bu.; $\frac{2}{5}$ pk.; $\frac{7}{9}$ qt.; $\frac{2}{3}$ pt.

18. Add 6 f℥ 2 fʒ 25 ♏; $2\frac{1}{2}$ f℥; 7 fʒ 42 ♏; 1 f℥ $2\frac{2}{3}$ fʒ; 3 f℥ 6 fʒ 51 ♏.

19. Add $\frac{1}{2}$ wk.; $\frac{1}{2}$ da.; $\frac{1}{2}$ hr.; $\frac{1}{2}$ min.; $\frac{1}{2}$ sec.

20. Add 3.26 yr. (365 da. each); 118 da. 5 hr. 42 min. $37\frac{1}{2}$ sec.; 63.4 da.; $7\frac{4}{5}$ hr.; 1 yr. 62 da. 19 hr. $24\frac{4}{5}$ min.; $\frac{73}{192}$ da.

21. Add 27° 14′ 55.24″; 9° $18\frac{1}{4}$″; 1° $15\frac{1}{3}$′; 116° 44′ 23.8″

22. Add $\$\frac{1}{2}$; $\frac{1}{4}$ ct.; $\frac{1}{8}$ m.

23. Add 3 dollars 7 m.; 5 dollars 20 ct.; 100 dollars 2 ct. 6 m.; 19 dollars $\frac{1}{8}$ ct.

24. Add £21 6s. $3\frac{1}{2}$d.; £5 $17\frac{3}{4}$s.; £9.085; 16s. $7\frac{1}{4}$d.; £$\frac{5}{12}$.

25. Add $\frac{2}{3}$ A.; $\frac{3}{4}$ sq. rd.; $\frac{4}{5}$ sq. ft.

SUBTRACTION OF COMPOUND NUMBERS.

222. Subtraction of Compound Numbers is the process of finding the difference between two similar Compound Numbers.

PROBLEM.—From 9 yd. 1 ft. $6\frac{1}{2}$ in. take 1 yd. 2.45 ft.

SOLUTION.—Change the $\frac{1}{2}$ in. to a decimal, making the minuend 9 yd. 1 ft. 6.5 in.; reduce .45 ft. to inches, making the subtrahend 1 yd. 2 ft. 5.4 in. The first term of the difference is 1.1 in. To subtract the 2 ft., increase the minuend term by 3 feet, and the next term of the subtrahend by the equivalent, 1 yard. Taking 2 ft. from 4 ft. we have a remainder 2 ft., and 2 yd. from 9 yd. leaves 7 yd., making the answer 7 yd. 2 ft. 1.1 in.

OPERATION.

yd.	ft.	in.	
9	1	6.5	
1	2	5.4	
7	2	1.1	*Ans.*

PROBLEM.—From 2 sq. rd. 1 sq. ft. take 1 sq. rd. 30 sq. yd. 2 sq. ft.

SOLUTION.—Write the numbers as before. If the 1 sq. ft. be increased by a *whole* sq. yd. and the next higher part of the subtrahend by the same amount, we shall have to make an inconvenient reduction of the first remainder as itself a minuend. Hence, we make the convenient addition of $\frac{1}{4}$ sq. yd., and, subtracting 2 sq. ft. from $3\frac{1}{4}$ sq. ft., we have $1\frac{1}{4}$ sq. ft. Then, giving the same increase to the 30 sq. yd., we proceed as in the former case, increasing the upper by *one* of the next higher, and, having no remainder higher than *feet*, the answer is, simply, $1\frac{1}{4}$ sq. ft. $=$ 1 sq. ft. 36 sq. in.

OPERATION.

sq.rd.	sq.yd.	sq. ft.
2	0	1
1	30	2
	Ans.	$1\frac{1}{4}$ sq. ft

Rule.—1. *Place the subtrahend under the minuend, so that numbers of the same denomination stand in the same column. Begin at the lowest denomination, and, subtracting the parts successively from right to left, write the remainders beneath.*

2. *If any number in the subtrahend be greater than that of the same denomination in the minuend, increase the upper by a unit, or such other quantity of the next higher denomina-*

tion as will render the subtraction possible, and give an equal increase to the next higher term of the subtrahend.

REMARK.—The increase required at any part of the minuend is, commonly, a unit of the next higher denomination. In a few instances it will be convenient to use more, and, if less be required, the tables will show what fraction is most convenient. Sometimes it is an advantage to alter the form of one of the given quantities before subtracting.

EXAMPLES FOR PRACTICE.

1. Subtract $\frac{2}{5}$ mi. from 144.86 rd.
2. Subtract 1.35 yd. from 4 yd. 2 qr. 1 na. $1\frac{3}{4}$ in.

3. Subtract 2 sq. rd. 24 sq. yd. 91 sq. in. from 5 sq. rd 16 sq. yd. $6\frac{2}{3}$ sq. ft.
4. Subtract 384 A. 43.92 sq. rd. from 1.305 sq. mi.

5. Subtract 13 cu. yd. 25 cu. ft. 1204.9 cu. in. from 20 cu. yd. 4 cu. ft. 1000 cu. in.
6. Subtract 9.362 oz. troy from 1 ℔. 15 pwt. 4 gr.

7. Subtract $\frac{10}{13}$ ʒ from $\frac{6}{11}$ ℥.
8. Subtract 56 T. 9 cwt. 1 qr. 23 lb. from 75.004 T.

9. Subtract $\frac{5}{24}$ ℔. troy from $\frac{6}{35}$ lb. avoirdupois.
10. Subtract 12 gal. 1 qt. 3 gills from 31 gal. $1\frac{1}{2}$ pt.

11. Subtract .0625 bu. from 3 pk. 5 qt. 1 pt.

12. Subtract 1 f℥ 4 fʒ 38 ♏ from 4 f℥ 2 fʒ.

13. Subtract 275 da. 9 hr. 12 min. 59 sec. from 2.4816 yr. (allowing $365\frac{1}{4}$ days to the year.)

14. Find the difference of time between Sept. 22d, 1855, and July 1st, 1856.

15. Find the difference of time between December 31st, 1814, and April 1st, 1822.

16. Subtract 43° 18′ 57.18″ from a quadrant.

17. Subtract 161° 34′ 11.8″ from 180°.

18. Subtract $\frac{37}{40}$ ct. from $\$\frac{3}{32}$.

19. Subtract 5 dollars 43 ct. $2\frac{1}{2}$ m; from 12 dollars 6 ct. $8\frac{1}{3}$ m.

20. Subtract £9 18s. $6\frac{1}{2}$d. from £20.

21. From $\frac{1}{2}$ A. 10 sq. in. take 79 sq. rd. 30 sq. yd. 2 sq. ft. 30 sq. in.

22. From 3 sq. rd. 1 sq. ft. 1 sq. in. take 1 sq. rd. 30 sq. yd. 1 sq. ft. 140 sq. in.

23. From 3 rd. 2 in. take 2 rd. 5 yd. 1 ft. 4 in.

24. From 7 mi. 1 in. take 4 mi. 319 rd. 16 ft. 3 in.

25. From 13 A. 3 sq. rd. 5 sq. ft. take 11 A. 30 sq. yd. 8 sq ft. 40 sq. in.

26. From 18 A. 3 sq. ft. 3 sq. in. take 15 A. 3 sq. rd. 30 sq. yd. 1 sq. ft. 142 sq. in.

MULTIPLICATION OF COMPOUND NUMBERS.

223. Compound Multiplication is the process of multiplying a Compound Number by an Abstract Number.

Problem.—Multiply 9 hr. 14 min. 8.17 sec. by 10.

OPERATION.

da.	hr.	min.	sec.
	9	14	8.17
			10
3	20	21	21.7

Solution.—Ten times 8.17 sec. = 81.7 sec. = 1 min. 21.7 sec. Write 21.7 sec. and carry 1 min. to the 140 min. obtained by the next multiplication. This gives 141 min. = 2 hr. 21 min. Write 21 min. and carry 2 hr. This gives 92 hr. = 3 da. 20 hr.

PROBLEM.—Multiply 12 A. 148 sq. rd. 28$\frac{3}{8}$ sq. yd. by 84.

SOLUTION.—Since $84 = 7 \times 12$, multiply by one of these factors, and this product by the other; the last product is the one required. The same result can be obtained by multiplying by 84 at once; performing the work separately, at one side, and transferring the results.

OPERATION.

A.	sq. rd.	sq. yd.
12	148	28$\frac{3}{8}$
		7
90	82	17$\frac{1}{8}$
		12
1086	30	24

Rule.—1. *Write the multiplier under the lowest denomination of the multiplicand.*

2. *Multiply the lowest denomination first, and divide the product by the number of units of this denomination which make a unit of the next higher; write the remainder under the denomination multiplied, and carry the quotient to the product of the next higher denomination.*

3. *Proceed in like manner with all the denominations, writing the entire product at the last.*

EXAMPLES FOR PRACTICE.

1. Multiply 7 rd. 10 ft. 5 in. by 6.
2. Multiply 1 mi. 14 rd. 8$\frac{1}{4}$ ft. by 97.

3. Multiply 5 sq. yd. 8 sq. ft. 106 sq. in. by 13.

4. Multiply 41 A. 146.1087 sq. rd. by 9.046

5. Multiply 10 cu. yd. 3 cu. ft. 428.15 cu. in. by 67.

6. Multiply 7 oz. 16 pwt. 5$\frac{3}{4}$ gr. by 174.

7. Multiply 2 ℥ 1 ℈ 13 gr. by 20.
8. Multiply 16 cwt. 1 qr. 7.88 lb. by 11.

9. Multiply 5 gal. 3 qt. 1 pt. 2 gills by 35.108

10. Multiply 26 bu. 2 pk. 7 qt. .37 pt. by 10.

11. Multiply 3 f℥ 48 ♏ by 12.
12. Multiply 18 da. 9 hr. 42 min. 29.3 sec. by $16\frac{7}{11}$.

13. Multiply £215 16s. $2\frac{1}{4}$d. by 75.
14. Multiply 10° 28′ $42\frac{1}{2}$″ by 2.754

DIVISION OF COMPOUND NUMBERS.

224. Division of Compound Numbers is the process of dividing when the dividend is a Compound Number. The divisor may be Simple or Compound, hence there are two cases:

1. *To divide a Compound Number into a number of equal parts.*
2. *To divide one Compound Number by another of the same kind.*

NOTE.—Problems under the second case are solved by reducing both Compound Numbers to the same denomination, and then dividing as in simple division.

PROBLEM.—Divide 5 cwt. 3 qr. 24 lb. $14\frac{3}{4}$ oz. of sugar equally among 4 men.

OPERATION.

	cwt.	qr.	lb.	oz.
4)	5	3	24	$14\frac{3}{4}$
	1	1	24	$15\frac{11}{16}$

SOLUTION.—4 into 5 cwt. gives a quotient 1 cwt., with a remainder 1 cwt., = 4 qr., to be carried to 3 qr., making 7 qr.; 4 into 7 qr. gives 1 qr., with 3 qr., = 75 lb., to be carried to 24 lb., = 99 lb.; 4 into 99 lb. gives 24 lb., with 3 lb., = 48 oz., to be carried to $14\frac{3}{4}$ oz., making $62\frac{3}{4}$ oz.; 4 into $62\frac{3}{4}$ oz. gives $15\frac{11}{16}$ oz., and the operation is complete.

PROBLEM.—If $42 purchase 67 bu. 2 pk. 5 qt. $1\frac{3}{4}$ pt. of meal, how much will $1 purchase?

SOLUTION.—Since $42 = 6 \times 7$, divide first by one of these factors, and the resulting quotient by the other; the last quotient will be the one required.

OPERATION.

	bu.	pk.	qt.	pt.
6)	67	2	5	$1\frac{3}{4}$
7)	11	1	0	$1\frac{23}{24}$
	1	2	3	$1\frac{23}{168}$

Rule.—1. *Write the quantity to be divided in the order of its denominations, beginning with the highest; place the divisor on the left.*

2. *Begin with the highest denomination, divide each number separately, and write the quotient beneath.*

3. *If a remainder occur after any division, reduce it to the next lower denomination, and, before dividing, add to it the number of its denomination.*

EXAMPLES FOR PRACTICE.

1. Divide 16 mi. 109 rd. by 7.
2. Divide 37 rd. 14 ft. 11.28 in. by 18.

3. Divide 675 C. 114.66 cu. ft. by 83.

4. Divide 10 sq. rd. 29 sq. yd. 5 sq. ft. 94 sq. in. by 17.

5. Divide 6 sq. mi. 35 sq. rd. by $22\frac{1}{2}$.
6. Divide 1245 cu. yd. 24 cu. ft. 1627 cu. in. by 11.303

7. Divide 3 ℥ 7 ʒ 18 gr. by 12.
8. Divide 600 T. 7 cwt. 86 lb. by 29.06

9. Divide 312 gal. 2 qt. 1 pt. 3.36 gills by $72\frac{5}{8}$.

10. Divide 19302 bu. by 6.215

11. Divide 76 yr. 108 da. 2 hr. 38 min. 26.18 see. by 45.

12. Divide 152° 46′ 2″ by 9.

225. Longitude and Time give rise to two cases:

1. *To find the difference of longitude between two places when the difference of time is given.*

2. *To find the difference of time when their longitudes are given.*

PROBLEM.—The difference of time between two places is 4 hr. 18 min. 26 sec.: what is their difference of longitude?

SOLUTION.—Every hour of time corresponds to 15° of longitude; every minute of time to 15′ of longitude; every second of time to 15″ of longitude (Art. 207). Hence, multiplying the hours in the difference of time by 15 will give the degrees in the difference of longitude, multiplying the minutes of time by 15 will give minutes (′) of longitude, and multiplying the seconds of time by 15 will give seconds (″) of longitude.

OPERATION.

hr.	min.	sec.
4	18	26
		15
64°	36′	30″

PROBLEM.—The difference of longitude between two places is 81° 39′ 22″: what is their difference of time?

SOLUTION.—15 into 81° gives 5 (marked *hr.*), and 6° to be carried. Instead of multiplying 6 by 60, adding the 39′, and then dividing, proceed thus: 15 into 6° is the same as 15 into $6 \times 60' = \frac{6 \times 60\ 4}{15} = 24'$, and as 15 into 39′ gives 2′ for a quotient and 9′ remainder, the whole quotient is 26′ (marked *min.*), and remainder $9' = 9 \times 60''$, which, divided by 15, gives $\frac{9 \times 60\ 4}{15} = 36''$, which with $1\frac{7}{15}$, obtained by dividing 22″ by 15, gives $37\frac{7}{15}''$ (marked *sec.*). The ordinary mode of dividing will give the same result, and may be used if preferred.

OPERATION.

15)81°	39′	22″
5 hr.	26 min.	$37\frac{7}{15}$ sec.

226. From these solutions we may obtain the following rules:

CASE I.

Rule.—*Multiply the difference of time by* 15, *according to the rule for Multiplication of Compound Numbers, and mark the product* ° ′ ″ *instead of* hr. min. sec.

CASE II.

Rule.—*Divide the difference of longitude by* 15, *according to the rule for Division of Compound Numbers, and mark the quotient* hr. min. sec., *instead of* ° ′ ″.

NOTE.—The following table of longitudes, as given in the records of the U. S. Coast Survey, is to be used for reference in the solution of exercises. "W." indicates longitude *West*, and "E." longitude *East* of the meridian of Greenwich, England.

TABLE OF LONGITUDES.

PLACE.	LONGITUDE.				PLACE.	LONGITUDE.			
	°	′	″			°	′	″	
Portland, Me., ...	70	15	18	W.	Des Moines, Iowa, . . .	93	37	16	W.
Boston, Mass., ...	71	3	50	"	Omaha, Neb., ...	95	56	14	"
New Haven, Conn., .	72	55	45	"	Austin, Tex., ...	97	44	12	"
New York City, . . .	74	0	24	"	Denver, Col..	104	59	33	"
Philadelphia, Pa., .	75	9	3	"	Salt Lake City, Utah, .	111	53	47	"
Baltimore, Md., . . .	76	36	59	"	San Francisco, Cal., . .	122	27	49	"
Washington, D. C., .	77	0	36	"	Sitka, Alaska, ...	135	19	42	"
Richmond, Va., . . .	77	26	4	"	St. Helena Island, . . .	5	42	0	"
Charleston, S. C., . .	79	55	49	"	Reykjavik, Iceland, . .	22	0	0	"
Pittsburgh, Pa., . . .	80	2	0	"	Rio Janeiro, Brazil, . .	43	20	0	"
Savannah, Ga., . . .	81	5	26	"	St. Johns, N. F., ...	52	43	0	"
Detroit, Mich., . . .	83	3	0	"	Honolulu, Sandwich Is.	157	52	0	"
Cincinnati, O., . . .	84	29	45	"	Greenwich, Eng., ...	0	0	0	
Louisville, Ky., . . .	85	25	0	"	Paris, France, ...	2	20	0	E.
Indianapolis, Ind., .	86	6	0	"	Rome, Italy.	12	28	0	"
Nashville, Tenn., . .	86	49	0	"	Berlin, German Emp., .	13	23	0	"
Chicago, Ill., ...	87	35	0	"	Vienna, Austria.	16	20	0	"
Mobile, Ala., . . .	88	2	28	"	Constantinople, Turkey,	28	59	0	"
Madison, Wis., . . .	89	24	3	"	St. Petersburg, Russia, .	30	16	0	"
New Orleans, La., . .	90	3	28	"	Bombay, India, ...	72	48	0	"
St. Louis, Mo., ...	90	12	14	"	Pekin, China.	116	26	0	"
Minneapolis, Minn.,	93	14	8	"	Sydney, Australia, . . .	151	11	0	"

Examples for Practice.

1. It is six o'clock A. M. at New York; what is the time at Cincinnati?

2. The difference of time between Springfield, Ill., and Philadelphia being 58 min. $1\frac{2}{15}$ sec., what is the longitude of Springfield?

3. At what hour must a man start, and how fast would he have to travel, at the equator, so that it would be noon for him for twenty-four hours?

4. What is the relative time between Mobile and Chicago?

5. A man travels from Halifax to St. Louis; on arriving, his watch shows 9 A. M. Halifax time. The time in St. Louis being 13 min. $32\frac{4}{15}$ sec. after 7 o'clock A. M., what is the longitude of Halifax?

6. Noon occurs 46 min. 58 sec. sooner at Detroit than at Galveston, Texas: what is the longitude of the latter place?

7. When it is five minutes after four o'clock on Sunday morning at Honolulu, what is the hour and day of the week at Sydney, Australia?

8. What is the difference in time between St. Petersburg and New Orleans?

9. When it is one o'clock P. M. at Rome, it is 54 min. 34 sec. after 6 o'clock A. M. at Buffalo, N. Y.: what is the longitude of the latter?

10. When it is six o'clock P. M. at St. Helena, what is the time at San Francisco?

11. A ship's chronometer, set at Greenwich, points to 4 hr. 43 min. 12 sec. P. M.: the sun being on the meridian, what is the ship's longitude?

ALIQUOT PARTS.

227. An **aliquot part** is an exact divisor of a number.

Aliquot parts may be used to advantage in finding a product when either or when each of the factors is a Compound Number.

PROBLEM.—Find the cost of 28 A. 145 sq. rd. $15\frac{1}{8}$ sq. yd. of land, at \$16 per acre.

SOLUTION.—Multiply \$16, the price of 1 A., by 28; the product, \$448, is the price of 28 A. 145 sq. rd. is made up of 120 sq. rd., 20 sq. rd., and 5 sq. rd. 120 sq. rd. $=\frac{3}{4}$ of an acre; hence take $\frac{3}{4}$ of \$16, the price per acre, to find the cost of 120 sq. rd. 20 sq. rd. $=\frac{1}{6}$ of 120 sq. rd., and cost $\frac{1}{6}$ as much. 5 sq. rd. $=\frac{1}{4}$ of 20 sq. rd., and cost $\frac{1}{4}$ as much. $15\frac{1}{8}$ sq. yd. $=\frac{1}{2}$ of 1 sq. rd., or $\frac{1}{10}$ of 5 sq. rd., and cost $\frac{1}{10}$ as much as the latter. Add the cost of the several aliquot parts to the cost of 28 acres. The result is the cost of the entire tract of land.

OPERATION.

	\$16
	28
	128
	32
	\$448
120 sq. rd. $=\frac{3}{4}$	12.
20 sq. rd. $=\frac{1}{6}$	2.
5 sq. rd. $=\frac{1}{4}$	.50
$15\frac{1}{8}$ sq. yd. $=\frac{1}{10}$	.05
	\$462.55

PROBLEM.—A man travels 3 mi. 20 rd. 5 yd. in 1 hr.; how far will he go in 6 da. 9 hr. 18 min. 45 sec. (12 hr. to a day)?

SOLUTION.—This example is solved like the preceding, except that here the multiplications and divisions are performed on a Compound instead of a Simple Number.

OPERATION.

	mi.	rd.	yd.
	3	20	5
			9
	27	188	1
6 da. $=8\times9$ hr.	220	225	$2\frac{1}{2}$
15 min. $=\frac{1}{4}$ of 1 "		245	$1\frac{1}{4}$
3 " $=\frac{1}{5}$ of 15 min.		49	$\frac{1}{4}$
45 sec. $=\frac{1}{4}$ of 3 min.		12	$1\frac{7}{16}$
	249	80	$\frac{15}{16}$

NOTE.—In all questions in aliquot parts, one of the numbers indicates a *rate*, and the other is a Compound Number whose *value* at this rate is to be found.

Rule for Aliquot Parts.—*Multiply the number indicating the rate by the number of that denomination for whose unit the rate is given, and separate the numbers of the other denominations into parts whose values can be obtained directly by a simple division or multiplication of one of the preceding values. Add these different values; the result will be the entire value required.*

NOTES.—1. Sometimes one of the values may be obtained by *adding* or *subtracting two preceding* values instead of by multiplying or dividing.

2. Aliquot parts are generally used in examples involving U. S. money, and the following table should be memorized for future use.

ALIQUOT PARTS OF 100.

$5 = \frac{1}{20}$	$12\frac{1}{2} = \frac{1}{8}$	$25 = \frac{1}{4}$
$6\frac{1}{4} = \frac{1}{16}$	$16\frac{2}{3} = \frac{1}{6}$	$33\frac{1}{3} = \frac{1}{3}$
$10 = \frac{1}{10}$	$20 = \frac{1}{5}$	$50 = \frac{1}{2}$

REMARK.—The following *multiples* of aliquot parts of 100, are often used: $18\frac{3}{4} = \frac{3}{16}$, $37\frac{1}{2} = \frac{3}{8}$, $40 = \frac{2}{5}$, $60 = \frac{3}{5}$, $62\frac{1}{2} = \frac{5}{8}$, $75 = \frac{3}{4}$, $87\frac{1}{2} = \frac{7}{8}$.

EXAMPLES FOR PRACTICE.

1. If a man travel 2 mi. 105 rd. $6\frac{1}{6}$ ft. in 1 hour, how far can he travel in 30 hr. 29 min. 52 sec.?

2. An old record says that 694 A. 1 R. 22 P. of land, at $11.52 per acre, brought $8009.344; what is the error in the calculation?

3. At $15.46 an oz., what will be the cost of 7 ℔. 8 oz. 16 pwt. 11 gr. of gold?

4. What is the cost of 88 gal. 3 qt. 1 pt. of vinegar, at $37\frac{1}{2}$ ct. a gallon?

5. If the heart should beat 97920 times in each day, how many times would it beat in 8 da. 5 hr. 25 min. 30 sec.?

6. At a cost of $8190.50 per mile over the plain, and at a rate of $84480 per mile of tunnel, what is the cost of a railway 17 mi. 150 rd. plain, and 70 rd. tunnel?

7. What is the value of 20 T. 1 cwt. 13 lb. of sugar, at $\$3\frac{3}{4}$ per cwt.?

8. If £3 6s. silver weigh 1 ℔. troy, how much will 17 ℔. 11 oz. 16 pwt. 9 gr. be worth?

9. If a steam-ship could make 16 mi. $67\frac{3}{8}$ rd. in 1 hr., how far could it go in 24 da. 22 hr. 56 min. 12 sec.?

Topical Outline.

OPERATIONS WITH COMPOUND NUMBERS.

1. Reduction
   1. Definition.
   2. Cases
      1. From Higher to Lower.
      2. From Lower to Higher.
   3. Rules.
2. Addition
   1. Definition.
   2. Rule.
3. Subtraction
   1. Definition.
   2. Rule.
4. Multiplication
   1. Definition.
   2. Rule.
5. Division
   1. Definition.
   2. Cases.
   3. Rule.
6. Longitude and Time.
   - Case I.—Rule.
   - Case II.—Rule.
   - Table of Longitudes.
7. Aliquot Parts
   1. Definition.
   2. Rule.
   3. Table.

XII. RATIO.

DEFINITIONS.

228. 1. **Ratio** is a Latin word, signifying *relation* or *connection*; in Arithmetic, it is *the measure of the relation of one number to another of the same kind, expressed by their quotient.*

2. A Ratio is found by dividing the first number by the second; as, the ratio of 8 to 4 is 2. The ratio is abstract.

3. The **Sign of Ratio** is the colon (:), which is the sign of division, with the horizontal line omitted; thus, 6 : 4 signifies the ratio of 6 to $4 = \frac{6}{4}$.

4. Each number is called a term of the ratio, and both together a **couplet** or **ratio.** The first term of a ratio is the **antecedent,** which means *going before;* the second term is the **consequent,** which means *following.*

5. A **Simple Ratio** is a single ratio consisting of two terms; as, $3 : 4 = \frac{3}{4}$.

6. A **Compound Ratio** is the product of two or more simple ratios; as, $\left\{\begin{matrix} 3 : 7 \\ 5 : 8 \end{matrix}\right\} = \frac{3 \times 5}{7 \times 8}$.

7. The **Reciprocal of a Ratio** is 1 divided by the ratio, or the ratio inverted; thus, the reciprocal of 2 : 3, or $\frac{2}{3}$, is $1 \div \frac{2}{3} = \frac{3}{2}$.

8. **Inverse Ratio** is the quotient of the consequent divided by the antecedent; thus, $\frac{5}{4}$ is the inverse ratio of 4 to 5.

9. The **Value of the Ratio** depends upon the relative size of the terms.

229. From the preceding definitions the following principles are derived:

PRINCIPLES.—1. $Ratio = \frac{Antecedent}{Consequent}$.

2. $Antecedent = Consequent \times Ratio$.

3. $Consequent = \frac{Antecedent}{Ratio}$.

Hence, by Art. **87**:

1. *The Ratio is multiplied by multiplying the Antecedent or dividing the Consequent.*
2. *The Ratio is divided by dividing the Antecedent or multiplying the Consequent.*
3. *The Ratio is not changed by multiplying or dividing both terms by the same number.*

General Law.—*Any change in the Antecedent produces a like change in the Ratio, but any change in the Consequent produces an opposite change in the Ratio.*

PROBLEM.—What is the ratio of 15 to 36?

OPERATION.

$$15 : 36 = \frac{15}{36} = \frac{5}{12}.$$

Rule.—*Divide the Antecedent by the Consequent.*

EXAMPLES FOR PRACTICE.

1. What is the ratio of 2 ft. 6 in. to 3 yd. 1 ft. 10 in.?
2. What is the ratio of 4 mi. 260 rd. to 1 mi. 96 rd.?
3. What is the ratio of 13 A. 145 sq. rd. : 6 A. 90 sq. rd.?
4. What is the ratio of 3 lb. 10 oz. 6 pwt. $10\frac{1}{2}$ gr. : 2 lb. $14\frac{3}{4}$ pwt.?

5. What is the ratio of 10 gal. 1.54 pt. : 7 gal. 2 qt. .98 pt.?

6. What is the ratio of 56 bu. 2 pk. 1 qt. : 35 bu. 3 pk. 6.055 qt.?

7. If the antecedent is 7 and the ratio $1\frac{1}{2}$, what is the consequent?

8. If the consequent is $\frac{3}{7}$ and the ratio $\frac{2}{9}$, what is the antecedent?

9. What is the ratio of a yard to a meter, and of a meter to a yard?

10. What is the ratio of a pound avoirdupois to a pound troy?

11. Find the difference between the compound ratios $\left\{\begin{matrix} 3 : 4 \\ 5 : 9 \end{matrix}\right\}$ and $\left\{\begin{matrix} 1 : 6 \\ 2 : 7 \end{matrix}\right\}$.

12. Find the difference between the ratio $4\frac{2}{3} : 7\frac{1}{7}$ and the inverse ratio.

13. If the consequent is $6\frac{1}{2}$, and the ratio is $2\frac{1}{4}$, what is the antecedent, and what is the inverse ratio of the two numbers?

XIII. PROPORTION.

DEFINITIONS.

230. 1. **Proportion** is an equality of ratios.

Thus, 4 : 6 : : 8 : 12 is a proportion, and is read 4 *is to* 6 *as* 8 *is to* 12.

2. The **Sign of Proportion** is the double colon (: :).

NOTE.—It is the same in effect as the sign of equality, which is sometimes used in its place.

3. The two ratios compared are called **couplets.** The first couplet is composed of the first and second terms, and the second couplet of the third and fourth terms.

4. Since each ratio has an antecedent and consequent, every proportion has *two* antecedents and *two* consequents, the 1st and 3d terms being the antecedents, and the 2d and 4th the consequents.

5. The first and last terms of a proportion are called the **extremes**; the middle terms, the **means.** All the terms are called **proportionals,** and the last term is said to be a *fourth proportional* to the other three in their order.

6. When three numbers are proportional, the second number is a **mean proportional** between the other two.

Thus, 4 : 6 : : 6 : 9; six is a mean proportional between 4 and 9.

7. Proportion is either **Simple** or **Compound**: Simple when both ratios are simple; Compound when one or both ratios are compound.

PRINCIPLES.—1. *In every proportion the product of the means is equal to the product of the extremes.*

2. *The product of the extremes divided by either mean, will give the other mean.*

3. *The product of the means divided by either extreme, will give the other extreme.*

SIMPLE PROPORTION.

231. 1. **Simple Proportion** is an expression of equality between *two simple ratios.*

2. It is employed when three terms are given and we wish to find the fourth. Two of the three terms are alike, and the other is of the same kind as the fourth which is to be found.

3. All proportions must be true according to Principle 1, which is the test. Principles 2 and 3 indicate methods of finding the wanting term.

4. The **Statement** is the proper arrangement of the terms of the proportion.

PROBLEM.—If 6 horses cost \$300, what will 15 horses cost?

STATEMENT.

6 horses : 15 horses :: \$300 : (\$).

OPERATION.

$$\frac{\$300 \times 15}{6} = \$750. \quad \text{Or, } (\$300 \times 15) \div 6 = \$750, \textit{Ans.}$$

SOLUTION.—Since 6 horses and 15 horses may be compared, they form the first couplet; also, \$300 and \$ — may be compared, as they are of the same unit of value.

NOTES.—1. To find the missing extreme, we use Prin. 3.

2. To prove the proportion, we use Prin. 1. Thus, $\$750 \times 6 = \$300 \times 15$.

PROBLEM.—If 15 men do a piece of work in $9\frac{3}{5}$ da., how long will 36 men be in doing the same?

SOLUTION.—Since 36 men will require *less* time than 15 men to do the same work, the answer should be *less* than $9\frac{3}{5}$ da.; make a decreasing ratio, $\frac{15}{36}$, and multiply the remaining quantity by it.

STATEMENT.

men men da. da.
36 : 15 :: $9\frac{3}{5}$: ( )

OPERATION.

$9\frac{3}{5} = \frac{48}{5}$ da.

$\frac{48}{5} \times \frac{15}{36} = 4$ da., *Ans.*

Rule.—1. *For the third term, write that number which is of the same denomination as the number required.*

2. *For the second term, write the* GREATER *of the two remaining numbers, when the fourth term is to be greater than the third; and the* LESS, *when the fourth term is to be less than the third.*

3. *Divide the product of the second and third terms by the first; the quotient will be the fourth term, or number required.*

Examples for Practice.

Note.—Problems marked with an asterisk are to be solved mentally.

1.* If I walk $10\frac{1}{2}$ mi. in 3 hr., how far will I go in 10 hr., at the same rate?

2. If the fore-wheel of a carriage is 8 ft. 2 in. in circumference, and turns round 670 times, how often will the hind-wheel, which is 11 ft. 8 in. in circumference, turn round in going the same distance?

3. If a horse trot 3 mi. in 8 min. 15 sec., how far can he trot in an hour, at the same rate?

4. What is a servant's wages for 3 wk. 5 da., at $1.75 per week?

5. What should be paid for a barrel of powder, containing 132 lb., if 15 lb. are sold for $\$5.43\frac{3}{4}$?

6. A body of soldiers are 42 in rank when they are 24 in file: if they were 36 in rank, how many in file would there be?

7. If a pulse beats 28 times in 16 sec., how many times does it beat in a minute?

8. If a cane 3 ft. 4 in. long, held upright, casts a shadow 2 ft. 1 in. long, how high is a tree whose shadow at the same time is 25 ft. 9 in.?

9. If a farm of 160 A. rents for $450, how much should be charged for one of 840 A?

10. A grocer has a false gallon, containing 3 qt. $1\frac{1}{2}$ pt.: what is the worth of the vanilla that he sells for $240, and what is his gain by the cheat?

11. If he uses $14\frac{3}{4}$ oz. for a pound, how much does he cheat by selling sugar for $27.52?

12. An equatorial degree is 365000 ft.: how many ft. in $80° \; 24' \; 37''$ of the same?

13. If a pendulum beats 5000 times a day, how often does it beat in 2 hr. 20 min. 5 sec.?

14.* If it takes 108 days, of $8\frac{1}{2}$ hr., to do a piece of work, how many days of $6\frac{3}{4}$ hr. would it take?

15. A man borrows $1750, and keeps it 1 yr. 8 mon.: how long should he lend $1200 to compensate for the favor?

16. A garrison has food to last 9 mon., giving each man 1 lb. 2 oz. a day: what should be a man's daily allowance, to make the same food last 1 yr. 8 mon.?

17. A garrison of 560 men have provisions to last during a siege, at the rate of 1 lb. 4 oz. a day per man; if the daily allowance is reduced to 14 oz. per man, how large a reinforcement could be received?

18. A shadow of a cloud moves 400 ft. in $18\frac{3}{4}$ sec.: what was the wind's velocity per hour?

19. If 1 ℔. troy of English standard silver is worth £3 6s., what is 1 lb. av. worth?

20. If I go a journey in $12\frac{3}{4}$ days, at 40 mi. a day, how long would it take me at $29\frac{3}{4}$ mi. a day?

21.* If $\frac{5}{9}$ of a ship is worth $6000, what is the whole of it worth?

22. If A, worth $5840, is taxed $78.14, what is B worth, who is taxed $256.01?

23.* What are 4 lb. 6 oz. of butter worth, at 28 ct. a lb.?

24. If I gain $160.29 in 2 yr. 3 mon., what would I gain in 5 yr. 6 mon., at that rate?

25. If I gain $92.54 on $1156.75 worth of sugar, how much must I sell to gain $67.32?

26. If coffee costing $255 is now worth $318.75, what did $1285.20 worth cost?

27. A has cloth at $3.25 a yd., and B has flour at $5.50 a barrel. If, in trading, A puts his cloth at $3.62\frac{1}{2}$, what should B charge for his flour?

28.* If a boat is rowed at the rate of 6 miles an hour, and is driven 44 feet in 9 strokes of the oar, how many strokes are made in a minute?

29. If I gain \$7.75 by trading with \$100, how much ought I to gain on \$847.56?

30. What is a pile of wood, 15 ft. long, $10\frac{1}{2}$ ft. high, and 12 ft. wide, worth, at \$4.25 a cord?

REMARK.—In Fährenheit's thermometer, the freezing point of water is marked 32°, and the boiling point 212°: in the Centigrade, the freezing point is 0°, and the boiling point 100°: in Reaumer's, the freezing point is 0°, and the boiling point 80°.

31. From the above data, find the value of a degree of each thermometer in the degrees of the other two.

1° F. $= \frac{4}{9}^\circ$ R. $= \frac{5}{9}^\circ$ C.; 1° C. $= 1\frac{4}{5}^\circ$ F. $= \frac{4}{5}^\circ$ R.; 1° R. $= 1\frac{1}{4}^\circ$ C. $= 2\frac{1}{4}^\circ$ F.

32. Convert 108° F. to degrees of the other two thermometers.

33. Convert 25° R. to degrees of the other two thermometers.

34. Convert 46° C. to degrees of the other two thermometers.

REMARKS.—1. In the working of machinery, it is ascertained that *the available power is to the weight overcome, inversely as the distances they pass over in the same time.*

2. *Inverse variation* exists between two numbers when one increases as the other decreases.

3. The *available* power is taken $\frac{2}{3}$ of the whole power, $\frac{1}{3}$ being allowed for friction and other impediments.

35. If the whole power applied is 180 lb. and moves 4 ft., how far will it lift a weight of 960 lb.?

36. If 512 lb. be lifted 1 ft. 3 in. by a power moving 6 ft. 8 in., what is the power?

37. A lifts a weight of 1440 lb. by a wheel and axle; for every 3 ft. of rope that passes through his hands the weight rises $4\frac{1}{2}$ in.: what power does he exert?

38. A man weighing 198 lb. lets himself down 54 ft. with a uniform motion, by a wheel and axle: if the weight at the hook rises 12 ft., how much is it?

39. Two bodies free to move, attract each other with forces that vary inversely as their weights. If the weights are 9 lb. and 4 lb., and the smaller is attracted 10 ft., how far will the larger be attracted?

40. Suppose the earth and moon to approach each other in obedience to this law, their weights being 49147 and 123 respectively, how many miles would the moon move while the earth moved 250 miles?

Can the three following questions be solved by proportion?

41. If 3 men mow 5 A. of grass in a day, how many men will mow $13\frac{1}{3}$ A. in a day?

42.* If 6 men build a wall in 7 da., how long would 10 men be in doing the same?

43.* If I gain 15 cents each, by selling books at $4.80 a doz., what is my gain on each at $5.40 a doz.?

44. A clock which loses 5 minutes a day, was set right at 6 in the morning of January 1st: what will be the right time when that clock points to 11 on the 15th?

45. If water begin and continue running at the rate of 80 gal. an hour, into a cellar 12 ft. long, 8 ft. wide, and 6 ft. deep, while it soaks away at the rate of 35 gal. an hour, in what time will the cellar be full?

46. Take the proportion of 4 : 9 :: 252 : a fourth term. If the third and fourth terms each be increased by 7, while the first remains unchanged, what multiplier is needed by the second to make a proportion?

47. Prove that there is no number which can be added to each term of 6 : 3 :: 18 : 9 so that the resulting numbers shall stand in proportion.

48. A certain number has been divided by one more than itself, giving a quotient $\frac{1}{5}$: what is the number?

49. If 48 lb. of sea-water contain $1\frac{1}{2}$ lb. of salt, how much fresh water must be added to these 48 lb. so that 40 lb. of the mixture shall contain $\frac{1}{2}$ lb. of salt?

COMPOUND PROPORTION.

232. **Compound Proportion** is an expression of equality between two ratios when either or when each ratio is Compound.

PROBLEM.—If 3 men mow 8 A. of grass in 4 da., how long would 10 men be in mowing 36 A.?

STATEMENT.

$$\left.\begin{array}{l}10 \text{ men} : 3 \text{ men} \\ 8 \text{ A.} : 36 \text{ A.}\end{array}\right\} :: 4 \text{ da.} : (\ ) \text{ da.}$$

OPERATION.

$$\frac{3 \times \overset{\overset{9}{18}}{36} \times 4}{\underset{5}{10} \times \underset{2}{8}} = \frac{27}{5} = 5\tfrac{2}{5} \text{ da. } Ans.$$

SOLUTION.—Since the denomination of the required term is days, make the third term 4 da. In forming the first and second terms, consider each denomination separately; 10 men can do the same amount of work in less time than 3 men; hence, the first ratio is, 10 men : 3 men, the *less* number being the second term. Since it takes 4 da. to mow 8 A., it will take a greater number of days to mow 36 A., and the second ratio is, 8 A. : 36 A., the *greater* number being the second term. Then dividing the continued product of the means by that of the extremes (Art. **230**, Prin. 3), after cancellation, we have $5\frac{2}{5}$ da., the required term.

Rule.—1. *For the third term, write that number which is of the same denomination as the number required.*

2. *Arrange each pair of numbers having the same denomination in the compound ratio, as if, with the third term, they formed a simple proportion.*

3. *Divide the product of the numbers in the second and third terms by the product of the numbers in the first term: the quotient will be the required term.*

233. Problems in Compound Proportion are readily solved by separating all the quantities involved into *two causes* and *two effects.*

Problem.—If 6 men, in 10 days of 9 hr. each, build 25 rd. of fence, how many hours a day must 8 men work to build 48 rd. in 12 days?

Solution.—6 men 10 da. and 9 hr. constitute the first cause, whose effect is 25 rd.; 8 men 12 da. and ( ) hr. constitute the second cause, whose effect is 48 rd. Hence,

STATEMENT.

$$\left.\begin{matrix}\text{6 men.}\\\text{10 da.}\\\text{9 hr.}\end{matrix}\right\} : \left.\begin{matrix}\text{8 men.}\\\text{12 da.}\\\text{( ) hr.}\end{matrix}\right\} :: \text{25 rd.} : \text{48 rd.}$$

OPERATION.

$$\frac{6 \times \overset{5}{\not{1}\not{0}} \times 9 \times \overset{4}{\not{4}\not{8}}}{\underset{2}{\not{8}} \times \not{1}\not{2} \times \underset{5}{\not{2}\not{5}}} = \frac{54}{5} = 10\tfrac{4}{5} \text{ hr. } \textit{Ans.}$$

Rule of Cause and Effect.—1. *Separate all the quantities contained in the question into two causes and their effects.*

2. *Write, for the first term of a proportion, all the quantities that constitute the first cause; for the second term, all that constitute the second cause; for the third, all that constitute the effect of the first cause; and for the fourth, all that constitute the effect of the second cause.*

3. *The required quantity may be indicated by a bracket, and found by* Art. **230,** Principles.

Note.—The two causes must be exactly alike in the *number* and *kind* of their terms; and so must the two effects.

Examples for Practice.

1. If 18 pipes, each delivering 6 gal. per minute, fill a cistern in 2 hr. 16 min., how many pipes, each delivering 20 gal. per minute, will fill a cistern $7\frac{1}{2}$ times as large as the first, in 3 hr. 24 min.?

2. The use of $100 for 1 year is worth $8: what is the use of $4500 for 2 yr. 8 mon. worth?

3. If 12 men mow 25 A. of grass in 2 da. of $10\frac{1}{2}$ hr., how many hours a day must 14 men work to mow an 80 A. field in 6 days?

4. If 4 horses draw a railroad car 9 miles an hour, how many miles an hour can a steam engine of 150 horse-power drive a train of 12 such cars, the locomotive and tender being counted 3 cars?

5. If 12 men, working 20 days 10 hours a day, mow 247.114 hektars of timothy, how many men in 30 days, working 8 hours a day, will mow 1976.912 centars of timothy of the same quality?

6. If the use of $3750 for 8 mon. is worth $68.75, what sum is that whose use for 2 yr. 4 mon. is worth $250?

7. If the use of $1500 for 3 yr. 8 mon. 25 da. is worth $336.25, what is the use of $100 for 1 yr. worth?

8. A garrison of 1800 men has provisions to last $4\frac{1}{2}$ months, at the rate of 1 lb. 4 oz. a day to each: how long will 5 times as much last 3500 men, at the rate of 12 oz. per day to each man?

9. What sum of money is that whose use for 3 yr., at the rate of $$4\frac{1}{2}$ for every hundred, is worth as much as the use of $540 for 1 yr. 8 mon., at the rate of $7 for every hundred?

10. A man has a bin 7 ft. long by $2\frac{1}{2}$ ft. wide, and 2 ft. deep, which contains 28 bu. of corn: how deep must he make another, which is to be 18 ft. long by $1\frac{7}{8}$ ft. wide, in order to contain 120 bu.?

11. If it require 4500 bricks, 8 in. long by 4 in wide, to pave a court-yard 40 ft. long by 25 ft. wide, how many tiles, 10 in. square, will be needed to pave a hall 75 ft. long by 16 ft. wide?

12. If 150000 bricks are used for a house whose walls average $1\frac{1}{2}$ ft. thick, 30 ft. high, and 216 ft. long, how

many will build one with walls 2 ft. thick, 24 ft. high, and 324 ft. long?

13. If 240 panes of glass 18 in. long, 10 in. wide, glaze a house, how many panes 16 in. long by 12 in. wide will glaze a row of 6 such houses?

14. If it require 800 reams of paper to publish 5000 volumes of a duodecimo book containing 320 pages, how many reams will be needed to publish 24000 copies of a book, octavo size, of 550 pages?

15. If 15 men cut 480 sters of wood in 10 days, of 8 hours each: how many boys will it take to cut 1152 sters of wood, only $\frac{2}{5}$ as hard, in 16 days, of 6 hours each, provided that while working a boy can do only $\frac{3}{4}$ as much as a man, and that $\frac{1}{3}$ of the boys are idle at a time throughout the work?

Topical Outline.

Ratio and Proportion.

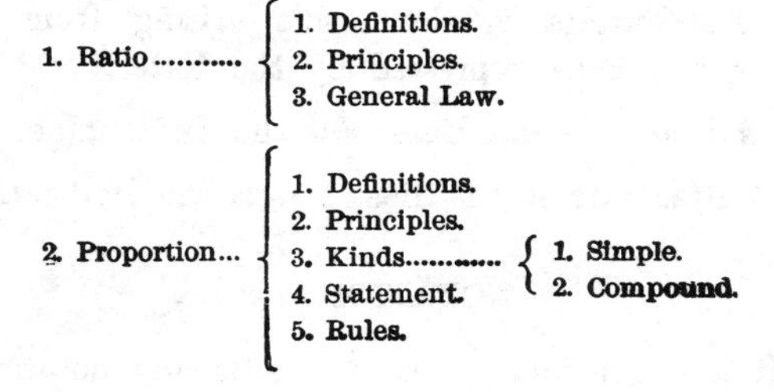

XIV. PERCENTAGE.

DEFINITIONS.

234. 1. **Percentage** is a term applied to all calculations in which 100 is the basis of comparison; it is also used to denote the result arising from taking so many hundredths of a given number.

2. **Per Cent** is derived from the Latin phrase *per centum*, which means *by* or *on the hundred.*

3. The **Sign of Per Cent** is %; the expression $4\% = \frac{4}{100}$, is read "4 per cent" equals $\frac{4}{100}$; or, decimally, .04

4. The elements in Percentage are the *Base*, the *Rate*, the *Percentage*, and the *Amount* or *Difference.*

5. The **Base** is the number on which the Percentage is estimated.

6. The **Rate** is the number of hundredths to be taken.

7. The **Percentage** is the result arising from taking that part of the Base expressed by the Rate.

8. The **Amount** is the Base *plus* the Percentage.

9. The **Difference** is the Base *minus* the Percentage.

NOTATION.

235. It is convenient to use the following notation:

1. Base = B.
2. Rate = R.
3. Percentage = P.
4. { Amount = A. / Difference = D. }

REMARK.—All problems in Percentage refer to two or more of the above terms. Owing to the relations existing among these terms, any two of them being given the others can be found. These relations give rise to the following cases :

CASE I.

236. Given the base and the rate, to find the percentage.

PRINCIPLE.—*The percentage of any number is the same part of that number as the given rate is of* 100%.

PROBLEM.—If I have 160 sheep, and sell 35% of them: how many do I sell?

OPERATIONS.

1. 160 sheep $\times .35 = 56$ sheep, *Ans.*

2. $100\% = 160$ sheep.
   $1\% = 1.60$ sheep.
   $\therefore\ 35\% = 1.60 \times 35 = 56$ sheep, *Ans.*

3. $160 \times \frac{7}{20} = 56$ sheep, *Ans.* (Art. **218**, Rem.)

SUGGESTION.—$35\% = \frac{35}{100} = \frac{7}{20}$.

SOLUTION.—Take .35 of the base; the result is the percentage.

FORMULA.—$B \times R = P$.

Rule 1.—*Multiply the base by the rate % expressed decimally; the product is the percentage.*

Rule 2.—*Find that part of the base which the rate % is of* 100.

EXAMPLES FOR PRACTICE.

1. Find $62\frac{1}{2}\%$ of 1664 men.
2. Find 35% of $\frac{2}{7}$.
3. Find $9\frac{3}{8}\%$ of 48 mi. 256 rd.

4. Find $11\frac{1}{9}\%$ of \$3283.47
5. Find $33\frac{1}{3}\%$ of 127 gal. 3 qt. 1 pt.
6. Find 98% of 14 cwt. 2 qr. 20 lb.

7. Find 40% of 6 hr. 28 min. 15 sec.

8. Find 104% of 75 A. 75 sq. rd.
9. Find $15\frac{5}{8}\%$ of a book of 576 pages.
10. Find $56\frac{1}{4}\%$ of 144 cattle.
11. Find $16\frac{2}{3}\%$ of 1932 hogs.
12. Find 1000% of \$5.43$\frac{3}{4}$
13. Find $2\frac{1}{2}\%$ of $\frac{4}{5}$.
14. What part is 25% of a farm?
15. What part of a quantity is $18\frac{3}{4}\%$ of it; $31\frac{1}{4}\%$; $37\frac{1}{2}\%$; $43\frac{3}{4}\%$; $56\frac{1}{4}\%$; $62\frac{1}{2}\%$; $68\frac{3}{4}\%$; $81\frac{1}{4}\%$; $83\frac{1}{3}\%$; $87\frac{1}{2}\%$; $93\frac{3}{4}\%$?
16. How much is 100% of a quantity; 125% of it; 250%; 675%; 1000%; $9437\frac{1}{2}\%$?

17.* A man owning $\frac{3}{8}$ of a ship, sold 40% of his share: what part of the ship did he sell, and what part did he still own?

18.* A owed B a sum of money; at one time he paid him 40% of it; afterward he paid him 25% of what he owed; and finally he paid him 20% of what he then owed: how much does he still owe?

19. Out of a cask containing 47 gal. 2 qt. 1 pt., leaked $6\frac{2}{3}\%$: how much was that?

20. A has an income of \$1200 a year; he pays 23% of it for board; $10\frac{2}{5}\%$ for clothing; $6\frac{3}{4}\%$ for books; $\frac{7}{12}\%$ for newspapers; $12\frac{7}{8}\%$ for other expenses: how much does he pay for each item, and how much does he save at the end of the year?

21. Find 10% of 20% of \$13.50
22. Find 40% of 15% of 75% of \$133.33$\frac{1}{3}$

23. A man contracts to supply dressed stone for a court-house for \$119449, if the rough stone costs him 16 ct. a cu. ft.; but if he can get it for 15 ct. a cu. ft., he will deduct 3% from his bill; how many cu. ft. would be needed, and what does he charge for dressing a cu. ft.?

24. 48% of brandy is alcohol; how much alcohol does a man swallow in 40 years, if he drinks a gill of brandy 3 times a day?

25. A had \$1200; he gave 30% to a son, 20% of the remainder to his daughter, and so divided the rest among four brothers that each after the first had \$12 less than the preceding: how much did the last receive?

26. What number increased by 20% of 3.5, diminished by $12\frac{1}{2}$% of 9.6, gives $3\frac{1}{2}$?

CASE II.

237. Given the base and the percentage, to find the rate.

PRINCIPLE.—*The rate equals the number of hundredths that the percentage is of the base.*

PROBLEM.—What per cent of 45 is 9?

SOLUTION.—9 is $\frac{1}{5}$ of 45; but $\frac{1}{5}$ of any number is equal to 20% of that number; hence, 9 is 20% of 45.

OPERATION.

$\frac{9}{45} = \frac{1}{5} = 20\%$, *Ans.*

FORMULA.—$\frac{P}{B} = R$.

Rule 1.—*Divide the percentage by the base; the quotient is the rate;*

Rule 2.—*Find that part of 100% that the percentage is of the base.*

EXAMPLES FOR PRACTICE.

1. 15 ct. is how many % of \$2?
2. 2 yd. 2 ft. 3 in. is how many % of 4 rd.?
3. 3 gal. 3 qt. is what % of $31\frac{1}{2}$ gal.?
4. $\frac{2}{3}$ is how many % of $\frac{4}{5}$?
5. $\frac{1}{2}$ of $\frac{3}{5}$ of $\frac{4}{7}$ is what % of $1\frac{7}{10}$?
6. $\frac{2\frac{1}{2}}{3}$ is how many % of $\frac{3\frac{3}{4}}{10}$?
7. \$5.12 is what % of \$640?
8. \$3.20 is what % of \$2000?
9. 750 men is what % of 12000 men?
10. 3 qt. $1\frac{1}{2}$ pt. is what % of 5 gal. $2\frac{1}{2}$ qt.?
11. A's money is 50% more than B's; then B's money is how many % less than A's?
12. What % of a number is 8% of 35% of it?
13. What % of a number is $2\frac{1}{2}$% of $2\frac{1}{2}$% of it?
14. What % of a number is 40% of $62\frac{1}{2}$% of it?
15. 12% of \$75 is what % of \$108?
16. U. S. standard gold and silver are 9 parts pure to 1 part alloy: what % of alloy is that?
17. What % of a meter is a yard?
18. How many % of a township 6 miles square, does a man own who has 9000 acres?
19. How many % of a quantity is 40% of 25% of it? also, 16% of $37\frac{1}{2}$% of it? also, $4\frac{1}{6}$% of 120% of it? also, 2% of 80% of $66\frac{2}{3}$% of it? also, $\frac{3}{5}$% of 36% of 75% of it? also, $6\frac{7}{8}$% of $22\frac{1}{2}$% of 96% of it?

20. 30% of the whole of an article is how many % of $\frac{2}{3}$ of it?
21. 25% of $\frac{2}{5}$ of an article is how many % of $\frac{3}{4}$ of it?

22. How many % of his time does a man rest, who sleeps 7 hr. out of every 24?

CASE III.

238. Given the rate and the percentage, to find the base.

PRINCIPLE.—*The base bears the same ratio to the percentage that* 100% *does to the rate.*

PROBLEM.—95 is 5% of what number?

OPERATION.

$$5\% = \tfrac{1}{20} = 95;$$
$$1\% = \tfrac{1}{100} = 19;$$
$$100\% = \tfrac{100}{100} = 19 \times 100 = 1900, \textit{ Ans.}$$

Or, $95 \div .05 = 1900$, *Ans.*

FORMULA.—$\dfrac{P}{R} = B.$

Rule 1.—*Divide the percentage by the rate, and then multiply the quotient by* 100; *the product is the base.*

Rule 2.—*Divide the percentage by the rate expressed decimally; the quotient is the base.*

EXAMPLES FOR PRACTICE.

1. \$3.80 is 5% of what sum?
2. $\frac{2}{11}$ is 80% of what number?
3. 16 is $1\frac{1}{2}\%$ of what number?
4. $31\frac{1}{4}$ ct. is $15\frac{5}{8}\%$ of what?
5. \$10.75 is $3\frac{1}{3}\%$ of what?
6. 162 men is $4\frac{4}{5}\%$ of how many men?
7. \$19.20 is $\frac{6}{10}\%$ of what?
8. \$189.80 is 104% of what?
9. 16 gal. 1 pt. is $6\frac{1}{7}\%$ of what?
10. 10 mi. 316 rd. is 75% of what?

11. Thirty-six men of a ship's crew die, which is $42\frac{6}{7}\%$ of the whole: what was her crew?

12. A stock-farmer sells 144 sheep, which is $12\frac{4}{5}\%$ of his flock: how many sheep had he?

13. A merchant sells 35% of his stock for \$6000: what is it all worth at that rate?

14. I shot 12 pigeons, which was $2\frac{2}{3}\%$ of the flock: how many pigeons escaped?

15. A, owing B, hands him a \$10 bill, and says, "there is $6\frac{1}{4}\%$ of your money:" what was the debt?

16. \$25 is $62\frac{1}{2}\%$ of A's money, and $41\frac{2}{3}\%$ of B's: how much has each?

17. A found \$5, which was $13\frac{1}{3}\%$ of what he had before: how much had he then?

18. I drew 48% of my funds in bank, to pay a note of \$150: how much had I left?

19. A farmer gave his daughter at her marriage 65 A. 106 sq. rd. of land, which was 3% of his farm: how much land did he own?

20. A pays \$13 a month for board, which is 20% of his salary: what is his salary?

21. Paid 40 ct. for putting in 25 bu. of coal, which was $11\frac{3}{7}\%$ of its cost: what did it cost a bu.?

22. 81 men is 5% of 60% of what?

23. A, owning 60% of a ship, sells $7\frac{1}{2}\%$ of his share for \$2500: what is the ship worth?

24. A father, having a basket of apples, took out $33\frac{1}{3}\%$ of them; of these, he gave $37\frac{1}{2}\%$ to his son, who gave 20% of his share to his sister, who thus got 2 apples: how many apples were in the basket at first?

25. B lost three dollars, which was $31\frac{1}{4}\%$ of what he had left: how much had he at first?

26. Bought 8000 bu. of wheat, which was $57\frac{1}{7}\%$ of my whole stock: how much had I before?

27. If 32% of 75% of 800% of a number is 1539, what is that number?

CASE IV.

239. Given the rate and the amount or the difference, to find the base.

PRINCIPLE.—*The base is equal to the amount divided by 1 plus the rate, or the difference divided by 1 minus the rate.*

PROBLEM.—A rents a house for \$377, which is an advance of 16% on the rent of last year: what amount did he pay last year?

OPERATION.

$\$377 \div 1.16 = \$325.00$, *Ans.*

Or, 100% = rental last year;
16% = increase this year;
hence, 100% + 16% = 116% = \$377;
1% = \$3.25;
∴ 100% = 3.25 × 100 = \$325.00, *Ans.*

PROBLEM.—John has \$136, which is 20% less than Joseph's money: how many dollars has Joseph?

OPERATION.

$\$136 \div (1 - .2) = \$170.00$, *Ans.*

Or, 100% = Joseph's money;
80% = John's money = \$136;
1% = \$1.7;
∴ 100% = \$1.7 × 100 = \$170.00, *Ans.*

Or, $80\% = \frac{4}{5} = \$136$; $\frac{1}{5} = \$34$; and $\frac{5}{5} = \$34 \times 5 = \$170.00$, *Ans.*

FORMULA.—$B = \left\{ \begin{matrix} A \div (1 + R) \\ D \div (1 - R) \end{matrix} \right\}.$

Rule.—*Divide the sum by 1 plus the rate, or divide the difference by 1 minus the rate; the quotient will be the base.*

EXAMPLES FOR PRACTICE.

1. \$4.80 is $33\frac{1}{3}\%$ more than what?
2. $\frac{5}{6}$ is 50% more than what?
3. 96 da. is 100% more than what?
4. 2576 bu. is 60% less than what?
5. $87\frac{1}{2}$ ct. is $87\frac{1}{2}\%$ less than what?
6. 42 mi. 60 rd. is 55% less than what?
7. 2 lb. $9\frac{29}{96}$ oz. is 50% less than what number of pounds?
8. $\frac{7}{12}$ is $99\frac{5}{8}\%$ less than what?
9. \$$920.93\frac{3}{4}$ is $337\frac{1}{2}\%$ more than what?
10. \$$4358.06\frac{1}{4}$ is $233\frac{1}{3}\%$ more than what?
11. In $64\frac{1}{2}$ gal. of alcohol, the water is $7\frac{1}{2}\%$ of the spirit: how many gal. of each?
12.* A coat cost \$32; the trimmings cost 70% less, and the making 50% less, than the cloth: what did each cost?

13.* If a bushel of wheat make $39\frac{1}{5}$ lb. of flour, and the cost of grinding be 4%, how many barrels of flour can a farmer get for 80 bu. of wheat?
14.* How many eagles, each containing 9 pwt. 16.2 gr. of pure gold, can I get for 455.6538 oz. pure gold at the mint, allowing $1\frac{1}{2}\%$ for expense of coinage?
15. 2047 is 10% of 110% less than what number?
16. $4246\frac{1}{2}$ is 6% of 50% of $466\frac{2}{3}\%$ more than what number?
17. A drew out of bank 40% of 50% of 60% of 70% of his money, and had left \$1557.20: how much had he at first?
18. I gave away $42\frac{6}{7}\%$ of my money, and had left \$2: what had I at first?
19. In a school, 5% of the pupils are always absent, and the attendance is 570: how many on the roll, and how many absent?

20. A man dying, left $33\frac{1}{3}\%$ of his property to his wife, 60% of the remainder to his son, 75% of the remainder to his daughter, and the balance, \$500, to a servant: what was the whole property, and each share?

21. In a company of 87, the children are $37\frac{1}{2}\%$ of the women, who are $44\frac{4}{9}\%$ of the men: how many of each?

22.* Our stock decreased $33\frac{1}{3}\%$, and again 20%; then it rose 20%, and again $33\frac{1}{3}\%$; we have thus lost \$66: what was the stock worth at first?

23.* A brewery is worth 4% less than a tannery, and the tannery 16% more than a boat; the owner of the boat has traded it for 75% of the brewery, losing thus \$103: what is the tannery worth?

ADDITIONAL FORMULAS.

240. The following **additional formulas,** derived from preceding data, may also be employed to advantage:

1. By definition, $A = B + P$; also, $D = B - P$.

2. From Case IV. $\begin{cases} \dfrac{A}{B} = 1 + R. \\ \dfrac{D}{B} = 1 - R. \\ A = B + (B \times R) = B + P. \end{cases}$

APPLICATIONS OF PERCENTAGE.

241. The **Applications of Percentage** may be divided into two classes, those in which time is *not* an essential element, and those in which it *is* an essential element, as follows:

Those in which Time is not an Essential Element......
1. Profit and Loss.
2. Stocks and Dividends.
3. Premium and Discount.
4. Commission and Brokerage.
5. Stock Investments.
6. Insurance.
7. Taxes.
8. United States Revenue.

Those in which Time is an Essential Element......
1. Simple Interest.
2. Partial Payments.
3. True Discount.
4. Bank Discount.
5. Exchange.
6. Equation of Payments.
7. Settlement of Accounts.
8. Compound Interest.
9. Annuities.

NOTE.—These topics will be presented in the order in which they stand.

Topical Outline.

PERCENTAGE.

1. Definitions.
2. Notation.
3. Cases......
   - I.
     1. Principle.
     2. Formula.
     3. Rules.
   - II.
     1. Principle.
     2. Formula.
     3. Rules.
   - III.
     1. Principle.
     2. Formula.
     3. Rules.
   - IV.
     1. Principle.
     2. Formula.
     3. Rule.
4. Additional Formulas.
5. Applications of Percentage.
   - Time not an Element.
   - Time an Element.

XV. PERCENTAGE.—APPLICATIONS.

I. PROFIT AND LOSS.

DEFINITIONS.

242. 1. **Profit** and **Loss** are commercial terms, and presuppose a cost price.

2. The **Cost** is the price paid for any thing.

3. The **Selling Price** is the price received for whatever is sold.

4. **Profit** is the excess of the Selling Price above the Cost.

5. **Loss** is the excess of the Cost above the Selling Price.

243. There are **four cases** of Profit or Loss, solved like the four corresponding cases of Percentage.

The cost corresponds to the *Base;* the per cent of profit or loss, to the *Rate;* the profit or loss, to the *Percentage;* the cost plus the profit, or the selling price, to the *Amount;* and the cost minus the loss, or the selling price, to the *Difference.*

CASE I.

244. Given the cost and the rate, to find the profit or loss.

PROBLEM.—Having invested $4800, my rate of profit is 13%: what is my profit?

SOLUTION.—Since the cost is $4800, and the rate of profit is 13%; the profit is 13% of $4800, which is $624.

OPERATION.

```
 $4800
   .13
 -----
 14400
 4800
 -------
$624.00 Profit.
```

EXAMPLES FOR PRACTICE.

1. If a man invests $1450 so as to gain $14\frac{1}{2}$%, what is his profit?

2. I bought $1760 worth of grain, and sold it so as to make $26\frac{1}{4}$% profit: what did I receive for it?

3. If a man invests $42540, and loses $11\frac{2}{3}$% of his capital: to what does his loss amount, and how much money has he left?

4. A man buys 576 sheep, at $10 a head. If his flock increases $21\frac{19}{36}$ per cent, and he sells it at the same rate per head, how much money does he receive?

5. The cost of publishing a book is 50 ct. a copy; if the expense of sale be 10% of this, and the profit 25%: what does it sell for by the copy?

6. A began business with $5000: the 1st year he gained $14\frac{3}{4}$%, which he added to his capital; the 2d year he gained 8%, which he added to his capital; the 3d year he lost 12%, and quit: how much better off was he than when he started?

7. A bought a farm of government land, at $1.25 an acre; it cost him 160% to fence it, 160% to break it up, 80% for seed, 100% to plant it, 100% to harvest it, 112% for threshing, 100% for transportation; each acre produced 35 bu. of wheat, which he sold at 70 ct. a bushel: how much did he gain on every acre above all expenses the first year?

8. For what must I sell a horse, that cost me $150, to gain 35%?

9. Bought hams at 8 ct. a lb.; the wastage is 10%: how must I sell them to gain 30%?

10. I started in business with $10000, and gained 20% the first year, and added it to what I had; the 2d year I gained 20%, and added it to my capital; the 3d year I gained 20%: what had I then?

11. I bought a cask of brandy, containing 46 gal., at \$2.50 per gal.; if 6 gal. leak out, how must I sell the rest, so as to gain 25% ?

CASE II.

245. Given the cost and the profit or loss, to find the rate.

PROBLEM.—A man bought part of a mine for \$45000, and sold it for \$165000: how many per cent profit did he make?

OPERATION.

$$\$165000 - \$45000 = \$120000 \text{ profit.}$$
$$100\% = \$45000;$$
$$1\% = \$450; \text{ and } 120000 \div 450 = 266\tfrac{2}{3}\%, \textit{ Ans.}$$

SOLUTION.—Here the profit is \$120000, which, compared with \$45000, the cost, is $\frac{120000}{45000} = \frac{8}{3}$, that is, $\frac{8}{3}$ of $100\% = 266\frac{2}{3}\%$.

EXAMPLES FOR PRACTICE.

1. If I buy at \$1 and sell at \$4, how many per cent do I gain?

2. If I buy at \$4 and sell at \$1, how many per cent do I lose?

3. If I sell $\frac{5}{9}$ of an article for what the whole cost me, how many per cent do I gain?

4.* Paid \$125 for a horse, and traded him for another, giving 60% additional money. For the second horse I received a third and \$25; I then sold the third horse for \$150: what was my per cent of profit or loss?

5.* A man bought a farm for \$1635, which depreciated in value 25%. Selling out, he invested the proceeds so as to make $33\frac{1}{3}\%$ profit: what was his per cent of profit or loss on the entire transaction?

6. It cost me $1536 to raise my wheat crop: if I sell it for $1728, what per cent profit is that per bushel?

7. If I pay for a lb. of sugar, and get a ℔. troy, what % do I lose, and what % does the grocer gain by the cheat?

8. A, having failed, pays B $1750 instead of $2500, which he owed him: what % does B lose?

9. An article has lost 20% by wastage, and is sold for 40% above cost: what is the gain per cent?

10. If my retail profit is $33\frac{1}{3}$%, and I sell at wholesale for 10% less than at retail, what is my wholesale profit?

11. Bought a lot of glass; lost 15% by breakage: at what % above cost must I sell the remainder, to clear 20% on the whole?

12. If a bushel of corn is worth 35 ct. and makes $2\frac{1}{2}$ gal. of whisky, which sells at $1.14 a gal., what is the profit of the distiller who pays a tax of 90 ct. a gallon?

13. A horse cost me $80, and I sold him for $90. What per cent did I gain?

CASE III.

246. Given the profit or loss and the rate, to find the cost.

PROBLEM.—By selling a lot for $34\frac{3}{8}$% more than I gave, my profit is $423.50: what did it cost me?

SOLUTION.—Since $34\frac{3}{8}$% = $423.50, 1% = $423.50 ÷ $34\frac{3}{8}$, = $12.32; and 100%, or the whole cost, = 100 times $12.32 = $1232, as in Case III of Percentage.

OPERATION.

$34\frac{3}{8}$% = $423.50;
1% = $12.32
100% = $1232, *Ans.*

REMARK.—After the cost is known, the profit or loss may be added to it, or subtracted from it, to get the selling price (amount or difference).

EXAMPLES FOR PRACTICE.

1. How large sales must I make in a year, at a profit of 8%, to clear $2000?

2. I lost $50 by selling sugar at $22\frac{1}{2}$% below cost: what was the cost?

3. If I sell tea at $13\frac{1}{3}$% profit, I make 10 ct. a lb.: how much a pound did I give?

4. I lost a $2\frac{1}{2}$ dollar gold coin, which was $7\frac{1}{7}$% of all I had: how much had I?

5. A and B each lost $5, which was $2\frac{7}{9}$% of A's and $3\frac{1}{3}$% of B's money: which had the most money, and how much?

6. I gained this year $2400, which is 120% of my gain last year, and that is $44\frac{4}{9}$% of my gain the year before: what were my profits the two previous years?

7. The dogs killed 40 of my sheep, which was $4\frac{1}{6}$% of my flock: how many had I left?

CASE IV.

247. Given the selling price (amount or difference) and the rate, to find the cost.

PROBLEM.—Sold goods for $25.80, by which I gained $7\frac{1}{2}$%: what was the cost?

SOLUTION.—The cost is 100%; the $25.80 being $7\frac{1}{2}$% more, is $107\frac{1}{2}$%; then, 1% = $25.80 ÷ $107\frac{1}{2}$ = 24 ct., and 100%, or the cost, = 100 times 24 ct. = $24; as in Case IV of Percentage.

OPERATION.

100% = cost price;
$107\frac{1}{2}$% = $25.80;
∴ 1% = 24 ct.
100% = $24, *Ans.*

REMARK.—After the cost is found, the difference between it and the selling price (amount or difference) will be the profit or loss.

EXAMPLES FOR PRACTICE.

1. Sold cloth at \$3.85 a yard; my profit was 10%: how much a yard did I pay?

2. Gold pens, sold at \$5 apiece, yield a profit of $33\frac{1}{3}$%: what did they cost apiece?

3. Sold out for \$952.82 and lost 12%: what was the cost, and what would I have received if I had sold out at a profit of 12%?

4. Sold my horse at 40% profit; with the proceeds I bought another, and sold him for \$238, losing 20%: what did each horse cost me?

5. Sold flour at an advance of $13\frac{1}{3}$%; invested the proceeds in flour again, and sold this lot at a profit of 24%, receiving \$3952.50: how much did each lot cost me?

6. An invoice of goods purchased in New York, cost me 8% for transportation, and I sold them at a gain of $16\frac{2}{3}$% on their total cost on delivery, realizing \$1260: at what were they invoiced?

7. For 6 years my property increased each year, on the previous, 100%, and became worth \$100000: what was it worth at first?

II. STOCKS AND BONDS.

DEFINITIONS.

248. 1. A **Company** is an association of persons united for the transaction of business.

2. A company is called a **Corporation** when authorized by law to transact business as one person.

Corporations are regulated by general laws or special acts, called *Charters*.

3. A **Charter** is the law which defines the powers, rights, and legal obligations of a corporation.

4. **Stock** is the capital of the corporation invested in business. Those owning the stock are **Stockholders.**

5. Stock is divided into **Shares,** usually of $50 or $100 each.

6. **Scrip** or **Certificates of Stock** are the papers issued by a corporation to the stockholders. Each stockholder is entitled to certificates showing the number of shares that he holds.

7. **Stocks** is a general term applied to bonds, state and national, and to the certificates of stock belonging to corporations.

8. A **Bond** is a written or printed obligation, under seal, securing the payment of a certain sum of money at or before a specified time.

Bonds bear a fixed rate of interest, which is usually payable either annually or semi-annually.

The principal classes of bonds are government, state, city, county, and railroad.

9. An **Assessment** is a sum of money required of the stockholders in proportion to their amounts of stock.

REMARK.—Usually, in the formation of a company, the stock subscribed is not all paid for at once; but *assessments* are made as the needs of the business require. The stock is then said to be paid for in *installments.* Other assessments may be made to meet losses or to extend the business.

10. A **Dividend** is a sum of money to be paid to the stockholders in proportion to their amounts of stock.

REMARK.—The *gross earnings* of a company are its total receipts in the transaction of the business; the *net earnings* are what is left of the receipts after deducting all expenses. The dividends are paid out of the net earnings.

249. Problems involving dividends and assessments give rise to **four cases,** solved like the four corresponding cases of Percentage.

The **quantities involved** are, the *Stock,* the *Rate,* and the *Dividend* or *Assessment.*

The stock corresponds to the *Base;* the dividend or assessment, to the *Percentage;* the stock plus the dividend, to the *Amount;* and the stock minus the assessment, to the *Difference.*

CASE I.

250. Given the stock and the rate, to find the dividend or assessment.

FORMULA.—*Stock* $\times$ *Rate* = *Dividend or Assessment.*

EXAMPLES FOR PRACTICE.

1. I own 18 shares, of $50 each, in the City Insurance Co., which has declared a dividend of $7\frac{1}{2}\%$: what do I receive?

2. I own 147 shares of railroad stock ($50 each), on which I am entitled to a dividend of 5%, payable in stock: how many additional shares do I receive?

3. The Western Stage Co. declares a dividend of $4\frac{1}{2}$ per cent: if their whole stock is $150000, how much is distributed to the stockholders?

4. The Cincinnati Gas Co. declares a dividend of 18%: what do I get on 50 shares ($100 each)?

5. A railroad company, whose stock account is $4256000, declared a dividend of $3\frac{1}{2}\%$: what sum was distributed among the stockholders?

6. A telegraph company, with a capital of $75000, declares a dividend of 7%, and has $6500 surplus: what has it earned?

7. I own 24 shares of stock ($25 each) in a fuel company, which declares a dividend of 6%; I take my dividend in coal, at 8 ct. a bu.: how much do I get?

CASE II.

251. Given the stock and dividend or assessment, to find the rate.

FORMULA.—$\frac{\textit{Dividend or Assessment}}{\textit{Stock}} = \textit{Rate}.$

EXAMPLES FOR PRACTICE.

1. My dividend on 72 shares of bank stock ($50 each) is $324: what was the rate of dividend?

2. A turnpike company, whose stock is $225000, earns during the year $16384.50: what rate of dividend can it declare?

3. The receipts of a certain canal company, whose stock is $3650000, amount in one year to $256484; the outlay is $79383: what rate of dividend can the company declare?

4. I own 500 shares ($100) in a stock company. If I have to pay $250 on an assessment, what is the rate?

CASE III.

252. Given the dividend or assessment and the rate, to find the stock.

FORMULA.—$\frac{\textit{Dividend or Assessment}}{\textit{Rate}} = \textit{Stock}.$

EXAMPLES FOR PRACTICE.

1. An insurance company earns $18000, and declares a 15% dividend: what is its stock account?

2 A man gets \$94.50 as a 7% dividend: how many shares of stock (\$50 each) has he?

3 Received 5 shares (\$50 each), and \$26 of another share, as an 8% dividend on stock: how many shares had I?

CASE IV.

253. Given the rate and the stock plus the dividend, or the stock minus the assessment, to find the stock.

$$\text{Formulas.— } Stock = \begin{cases} \dfrac{Stock + Dividend}{1 + Rate.} \\ \dfrac{Stock - Assessment}{1 - Rate.} \end{cases}$$

Examples for Practice.

1. Received 10% stock dividend, and then had 102 shares (\$50 each), and \$15 of another share: how many shares had I before the dividend?

2. Having received two dividends in stock, one of 5%, another of 8%, my stock has increased to 567 shares: how many had I at first?

III. PREMIUM AND DISCOUNT.

254. 1. **Premium, Discount,** and **Par** are mercantile terms applied to money, stocks, bonds, drafts, etc.

2. **Drafts, Bills of Exchange, or Checks** are written orders for the payment of money at some definite place and time.

3. The **Par Value** of money, stocks, drafts, etc., is the nominal value on their face.

4. The **Market Value** is the sum for which they sell.

5. **Discount** is the excess of the par value of money, stocks, drafts, etc., over their market value.

6. **Premium** is the excess of their market value over their par value.

7. **Rate of Premium** or **Rate of Discount** is the rate per cent the premium or discount is of the face.

255. Problems involving premium or discount give rise to **four cases,** corresponding to those of Percentage.

The **quantities involved** are the *Par Value,* the *Rate,* the *Premium* or *Discount,* and the *Market Value.*

The par value corresponds to the *Base;* the premium or discount, to the *Percentage*; and the market value, to the *Amount* or *Difference.*

CASE I.

256. Given the par value and the rate, to find the premium or discount.

FORMULAS.— $\begin{cases} \textit{Premium} = \textit{Par Value} \times \textit{Rate.} \\ \textit{Discount} = \textit{Par Value} \times \textit{Rate.} \end{cases}$

NOTE.—If the result is a *premium,* it must be *added* to the *par* to get the *market* value; if it is a *discount,* it must be *subtracted.*

EXAMPLES FOR PRACTICE.

1. Bought 54 shares of railroad stock ($100 each) at 4% discount: find the discount and cost.

2. Buy 18 shares of stock ($100 each) at 8% discount: find the discount and cost.

3. Sell the same at $4\frac{1}{2}$% premium: find the premium, the price, and the gain.

4. Bought 62 shares of railroad stock ($50 each) at 28% premium: what did they cost?

5. What is the cost of 47 shares of railroad stock ($50 each) at 30% discount?

6. Bought $150 in gold, at $\frac{3}{4}$% premium: what is the premium and cost?

7. Sold a draft on New York of $2568.45, at $\frac{1}{2}$% premium: what do I get for it?

8. Sold $425 uncurrent money, at 3% discount: what did I get, and lose?

9. What is a $5 note worth, at 6% discount?

10. Exchanged 32 shares of bank stock ($50 each), 5% premium, for 40 shares of railroad stock ($50 each), 10% discount, and paid the difference in cash: what was it?

11. Bought 98 shares of stock ($50 each), at 15% discount; gave in payment a bill of exchange on New Orleans for $4000, at $\frac{5}{8}$% premium, and the balance in cash: how much cash did I pay?

12. Bought 56 shares of turnpike stock ($50 each), at 69%; sold them at $76\frac{1}{2}$%: what did I gain?

13. Bought telegraph stock at 106%; sold it at 91%: what was my loss on 84 shares ($50 each)?

14. What is the difference between a draft on Philadelphia of $8651.40, at $1\frac{1}{4}$% premium, and one on New Orleans for the same amount, at $\frac{1}{2}$% discount?

CASE II.

257. Given the face and the discount or premium, to find the rate.

FORMULAS.—*Rate* = $\begin{cases} \dfrac{\textit{Discount or Premium}}{\textit{Par Value.}} \\ \dfrac{\textit{Difference between Market and Par Value}}{\textit{Par Value.}} \end{cases}$

NOTES.—1. If the *par* and the *market value* are known, take their difference for the discount or premium.

2. If the rate of *profit* or *loss* is required, the *market value* or *cost* is the standard of comparison, not the *face.*

EXAMPLES FOR PRACTICE.

1. Paid $2401.30 for a draft of $2360 on New York: what was the rate of premium?

2. Bought 112 shares of railroad stock ($50 each) for $3640: what was the rate of discount?

3. If the stock in the last example yields 8% dividend. what is my rate of profit?

4. I sell the same stock for $5936: what rate of premium is that? what rate of profit?

5. If I count my dividend as part of the profit, what is my rate of profit?

6. Exchanged 12 Ohio bonds ($1000 each), 7% premium, for 280 shares of railroad stock ($50 each): what rate of discount were the latter?

7. Gave 266.66\frac{2}{3}$ of notes, 4% discount, for $250 of gold: what rate of premium was the gold?

8. Bought 58 shares of mining stock ($50 each), at 40% premium, and gave in payment a draft on Boston for $4000: what rate of premium was the draft?

9. Received $4.60 for an uncurrent $5 note: what was the rate of discount?

10. Paid $2508.03 for 26 shares of stock ($100 each), and brokerage, $25.03: what is the rate of discount?

CASE III.

258. Given the discount or premium and the rate, to find the face.

FORMULA.— $\textit{Par Value} = \dfrac{\textit{Discount or Premium}}{\textit{Rate.}}$

NOTES.—1. After the *face* is obtained, add to it the premium, or subtract the discount, to get the market value or cost.

2. If the *profit* or *loss* is given, and the rate per cent of the *face* corresponding to it, work by Case III, Percentage.

EXAMPLES FOR PRACTICE.

1. Paid 36 ct. premium for gold $\frac{3}{4}\%$ above par: how much gold was there?

2. Took stock at par; sold it for $2\frac{1}{4}\%$ discount, and lost $117: how many shares ($50 each) had I?

3. The discount, at $7\frac{1}{2}\%$, on stocks, was $93.75: how many shares ($50 each) were sold?

4. Buy stock at $4\frac{1}{2}\%$ premium; sell at $8\frac{1}{4}\%$ premium; profit, $345: how many shares ($100 each)?

5. Buy stocks at 14% discount; sell at $3\frac{1}{2}\%$ premium; profit, $192.50: how many shares ($50 each)?

6. The premium on a draft, at $\frac{7}{8}\%$, was $10.36: what was the face?

7. Buy stocks at 6% discount; sell at 42% discount; loss, $666: how many shares ($50 each)?

8. Bought stock at 10% discount, which rose to 5% premium, and sold for cash; paying a debt of $33, I invested the balance in stock at 2% premium, which, at par, left me $11 less than at first: how much money had I at first?

CASE IV.

259. Given the market value and the rate, to find the par value.

FORMULAS.—
$$\textit{Par Value} = \begin{cases} \dfrac{\textit{Market Value}}{1 + \textit{Rate of Premium.}} \\ \dfrac{\textit{Market Value}}{1 - \textit{Rate of Discount.}} \end{cases}$$

NOTES.—1. After the *face* is known, take the difference between it and the market value, to find the discount or premium.

2. Bear in mind that the rate of premium or discount, and the rate of profit or loss, are entirely different things; the former is referred to the par value or *face*, as a standard of comparison, the latter to the market value or *cost*.

EXAMPLES FOR PRACTICE.

1. What is the face of a draft on Baltimore costing $2861.45, at $1\frac{1}{2}\%$ premium?

2. Invested $1591 in stocks, at 26% discount: how many shares ($50 each) did I buy?

3. Bought a draft on New Orleans, at $\frac{1}{2}\%$ discount, for $6398.30: what was its face?

4. Bought notes at 5% discount, for $2375: what is their face?

5. Exchanged 17 railroad bonds ($500 each) 25% below par, for bank stock at $6\frac{1}{4}\%$ premium: how many shares ($100 each) did I get?

6. How much gold, at $\frac{5}{8}\%$ premium, will pay a check for $7567?

7. How much silver, at $1\frac{1}{4}\%$ discount, can be bought for $3172.64 of currency?

8. How large a draft, at $\frac{1}{4}\%$ premium, is worth 54 city bonds ($100 each), at 12% discount?

9. Exchanged 72 Ohio State bonds ($1000 each), at $6\frac{1}{4}\%$ premium, for Indiana bonds ($500 each), at 2% premium: how many of the latter did I get?

IV. COMMISSION AND BROKERAGE.

DEFINITIONS.

260. 1. A **Commission-Merchant, Agent, or Factor** is a person who sells property, makes investments, collects debts, or transacts other business for another.

2. The **Principal** is the person for whom the commission-merchant transacts the business.

3. **Commission** is the percentage paid to the commission-merchant for doing the business.

4. A **Consignment** is a quantity of merchandise sent to a commission merchant to be sold.

5. The person sending the merchandise is the **Consignor** or **Shipper,** and the commission merchant is the **Consignee.** When living at a distance from his principal, the consignee is spoken of as the **Correspondent.**

6. The **Net Proceeds** is the sum left after all charges have been paid.

7. A **Guaranty** is a promise to answer for the payment of some debt, or the performance of some duty in the case of the failure of another person, who, in the first instance, is liable. Guaranties are of two kinds: of *payment*, and of *collection.*

8. In a **Guaranty of Payment,** the guarantor makes an absolute agreement that the instrument shall be paid at maturity.

9. The usual form of a guaranty, written on the back of a note or bill, is:

"*For value received, I hereby guaranty the payment of the within.* JOHN SAUNDERS."

10. A **Broker** is a person who deals in money, bills of credit, stocks, or real estate, etc.

11. The commission paid to a broker is called **Brokerage.**

261. Commission and Brokerage involve **four cases,** corresponding to those of Percentage.

The **quantities involved** are the *Amount Bought* or *Sold,* the *Rate of Commission* or *Brokerage,* the *Commission* or *Brokerage,* and the *Cost* or *Net Proceeds.*

The amount of sale or purchase corresponds to the *Base;* the commission or brokerage, to the *Percentage;* the cost, to the *Amount;* and the net proceeds, to the *Difference.*

CASE I.

262. Given the amount of sale, purchase, or collection and the rate, to find the commission.

FORMULA.— *Amount of Sale or Purchase* × *Rate* = *Commission.*

EXAMPLES FOR PRACTICE.

1. I collect for A $268.40, and have 5% commission: what does A get?

2. I sell for B 650 barrels of flour, at $7.50 a barrel, 28 barrels of whisky, 35 gal. each, at $1.25 a gal.: what is my commission, at $2\frac{1}{4}$%?

3. Received on commission 25 hhd. sugar (36547 lb.), of which I sold 10 hhd. (16875 lb.), at 6 ct. a lb., and 6 hhd. (8246 lb.) at 5 ct. a lb., and the rest at $5\frac{1}{2}$ ct. a lb.: what is my commission, at 3%?

4. A lawyer charged 8% for collecting a note of $648.75: what are his fee and the net proceeds?

5. A lawyer, having a debt of $1346.50 to collect, compromises by taking 80%, and charges 5% for his fee: what are his fee, and the net proceeds?

6. Bought for C, a carriage for $950, a pair of horses for $575, and harness for $120; paid charges for keeping, packing, shipping, etc., $18.25; freight, $36.50: what was my commission, at $3\frac{1}{3}$%, and what was the whole amount of my bill?

7. An architect charges $3\frac{1}{2}$% for designing and superintending a building, which cost $27814.60: to what does his fee amount?

8. A factor has $2\frac{3}{4}$% commission, and $3\frac{1}{2}$% for guarantying payment: if the sales are $6231.25, what does he get?

9. Sold 500000 lb. of pork at $5\frac{1}{2}$ ct. a lb.: what is my commission at $1\frac{1}{4}\%$?

10. An architect charges $1\frac{1}{4}\%$ for plans and specifications, and $2\frac{7}{8}\%$ for superintending: what does he make, if the building costs $14902.50?

11. A sells a house and lot for me at $3850, and charges $\frac{5}{8}\%$ brokerage: what is his fee?

12. I have a lot of tobacco on commission, and sell it through a broker for $4642.85: my commission is $2\frac{1}{2}\%$, the brokerage $1\frac{1}{8}\%$: what do I pay the broker, and what do I keep?

CASE II.

263. Given the commission and the amount of the sale, purchase, or collection, to find the rate.

FORMULA.— $\frac{\textit{Commission}}{\textit{Amount of Sale or Purchase}} = \textit{Rate.}$

EXAMPLES FOR PRACTICE.

1. An auctioneer's commission for selling a lot was $50, and the sum paid the owner was $1200: what was the rate of commission?

2. A commission-merchant sells 800 barrels of flour, at 6.43\frac{3}{4}$ a barrel, and remits the net proceeds, $5021.25: what is his rate of commission?

3. The cost of a building was $19017.92, including the architect's commission, which was $553.92: what rate did the architect charge?

4. Bought flour for A; my whole bill was $5802.57, including charges, $76.85, and commission, $148.72: find the rate of commission.

5. Charged $52.50 for collecting a debt of $1050: what was my rate of commission?

6. An agent gets $169.20 for selling property for $8460: what was his rate of brokerage?

7. My commission for selling books was $6.92, and the net proceeds, $62.28: what rate did I charge?

8. Paid $38.40 for selling goods worth $6400: what was the rate of brokerage?

9. Paid a broker $24.16, and retained as my part of the commission $42.28, for selling a consignment at $2416: what was the rate of brokerage, and my rate of commission?

CASE III.

264. Given the commission and rate, to find the sum on which commission is charged.

FORMULA.— $\frac{\textit{Commission}}{\textit{Rate}} = \textit{Amount of Sale or Purchase.}$

NOTE.—After finding the sum on which commission is charged, subtract the commission to find the *net proceeds*, or add it to find the *whole cost*, as the case may be.

EXAMPLES FOR PRACTICE.

1. My commissions in 1 year, at $2\frac{1}{2}\%$, are $3500: what were the sales, and the whole net proceeds?

2. An insurance agent's income is $1733.45, being 10% on the sums received for the company: what were the company's net receipts?

3. A packing-house charged $1\frac{1}{2}\%$ commission, and cleared $2376.15, after paying out $1206.75 for all expenses of packing: how many pounds of pork were packed, if it cost $4\frac{1}{2}$ ct. a pound?

4. Paid $64.05 for selling coffee, which was $\frac{7}{8}\%$ brokerage: what are the net proceeds?

5. An agent purchased, according to order, 10400 bushels of wheat; his commission, at $1\frac{1}{4}\%$, was \$156, and charges for storage, shipping, and freight, \$527.10: what did he pay a bushel, and what was the whole cost?

6. Received produce on commission, at $2\frac{1}{4}\%$; my surplus commission, after paying $\frac{1}{2}\%$ brokerage, is \$107.03: what was the amount of the sale, the brokerage, and net proceeds?

CASE IV.

265. Given the rate of commission and the net proceeds or the whole cost, to find the sum on which commission is charged.

FORMULAS.— $$\begin{cases} \dfrac{Cost}{(1 + Rate)} = \textit{Amount of Sale or Purchase.} \\ \dfrac{Net\ Proceeds}{(1 - Rate)} = \textit{Amount of Sale or Purchase.} \end{cases}$$

NOTE.—After the sum on which commission is charged is known, find the commission by subtraction.

EXAMPLES FOR PRACTICE.

1. A lawyer collects a debt for a client, takes 4% for his fee, and remits the balance, \$207.60: what was the debt and the fee?

2. Sent \$1000 to buy a carriage, commission $2\frac{1}{2}\%$: what must the carriage cost?

SUGGESTION.—$100\% + 2\frac{1}{2}\% = 102\frac{1}{2}\% = \$1000$; find 1%, then 100%.

3. A buys per order a lot of coffee; charges, \$56.85; commission, $1\frac{1}{4}\%$; the whole cost is \$539.61: what did the coffee cost?

4. Buy sugar at $2\frac{1}{4}\%$ commission, and $2\frac{1}{2}\%$ for guaranteeing payment: if the whole cost is $1500, what was the cost of the sugar?

5. Sold 2000 hams (20672 lb.); commission, $2\frac{1}{2}\%$, guaranty, $2\frac{3}{4}\%$, net proceeds due consignor, $2448.34: what did the. hams sell for a lb.?

6. Sold cotton on commission, at 5%; invested the net proceeds in sugar; commission, 2%; my whole commission was $210: what was the value of the cotton and sugar?

SUGGESTION.—Note carefully the different processes required here for commission in buying and commission in selling.

7. Sold flour at $3\frac{1}{2}\%$ commission; invested $\frac{2}{3}$ of its value in coffee, at $1\frac{1}{2}\%$ commission; remitted the balance, $432.50: what was the value of the flour, the coffee, and my commissions?

8. Sold a consignment of pork, and invested the proceeds in brandy, after deducting my commissions, 4% for selling, and $1\frac{1}{4}\%$ for buying. The brandy cost $2304.00: what did the pork sell for, and what were my commissions?

9. Sold 1400 barrels of flour, at $6.20 a barrel; invested the proceeds in sugar, as per order, reserving my commissions, 4% for selling and $1\frac{1}{2}\%$ for buying, and the expense of shipping, $34.16: how much did I invest in sugar?

10. An agent sold my corn; and, after reserving his commission, invested all the proceeds in corn at the same price, his commission, buying and selling, was 3%, and his whole charge $12: for what was the corn first sold?

11. My agent sold my flour at 4% commission; increasing the proceeds by $4.20, I bought wheat, paying 2% commission; wheat declining $3\frac{1}{3}\%$, my loss, including commissions, was $5: what was the flour worth?

V. STOCK INVESTMENTS.

DEFINITIONS.

266. 1. A **Stock Exchange** is an association of brokers and dealers in stocks, bonds, and other securities.

REMARKS.—1. The name "Stock Exchange" is also applied to the building in which the association meets to transact business.

2. New York city is the commercial center of the United States, and the transactions of the New York Stock Exchange, as telegraphed throughout the country, determine the market value of nearly all stocks sold.

2. **United States Government Bonds** are of two kinds,—coupon and registered.

REMARKS.—1. *Coupon bonds* may be transferred like bank-notes: the interest is represented by certificates, called coupons, printed at the bottom of the bond, which may be presented for payment when due.

2. *Registered bonds* are recorded in the name of the owner in the U. S. Treasurer's office, and the interest is sent directly to the owner. Registered bonds must be indorsed, and the record must be changed, to effect a transfer.

3. The United States also issues *legal tender* notes, known as "Greenbacks," which are payable in coin on demand, and bear no interest.

3. The various kinds of United States bonds are distinguished, 1st. By the rate of interest; 2d. By the date at which they are payable. Most of the bonds are payable in coin; a few are payable in currency.

EXAMPLE.—Thus, "U. S. 4½'s, 1891," means bonds bearing interest at 4½%, and payable in 1891. "U. S. cur. 6's, 1899," means bonds bearing interest at 6%, and payable in currency in 1899. Quotations of the principal bonds are given in the leading daily papers.

4. Bonds are also issued by the several states, by cities and towns, by counties, and by corporations.

REMARKS.—Legitimate stock transactions involve the following terms and abbreviations, which need explanation:

1. A person who anticipates a decline, and contracts to deliver stocks at a future day, at a fixed price which is lower than the present market price, expecting to buy in the interval at a still lower price, is said to *sell short.*

Short sales are also made for *cash,* deliverable on the same day, or in the *regular* way, where the certificates are delivered the day after the sale. In these cases the seller borrows the stock from a third party, advancing security equivalent to the market price, and waits a decline; he buys at what he considers the lowest point, returns his borrowed stock, and reclaims his security.

2. A person who buys stock in anticipation of a rise is said to *buy long.*

3. Those who sell short are interested, of course, in forcing the market price down, and are called *bears;* while those who buy long endeavor to force the market price up, and are known as *bulls.*

4. The following are the principal abbreviations met with in stock quotations: c., means *coupon;* r., *registered;* pref'd, or pf., *preferred,* applied to stock which has advantages over common stock of the same company in the way of dividends, etc.; xd., *without dividend,* meaning that the buyer is not entitled to the dividend about to be declared; c., *cash;* s3, s30, s60, *seller's option, three, thirty,* or *sixty days,* as the case may be, means that the seller has the privilege of closing the transaction at any time within the specified limit; b3, b30, b60, *buyer's option, three days,* etc., giving the buyer the privilege; bc., *between calls,* means that the price was fixed between the calls of the whole list of stocks, which takes place in the New York Exchange twice a day; opg., *for delivery at the opening of the books of transfer.*

5. The usual rate of brokerage is $\frac{1}{8}$% on the par value of the stock, either for a purchase or a sale.

267. The **quantities involved** in problems in stock investments are: the *Amount Invested,* the *Market Value,* the *Rate of Dividend* or *Interest* paid on the par value of the stock, the *Rate of Income* on the amount invested, and the *Income* itself.

These quantities give rise to **five cases,** all of which may be solved by the principles of Percentage.

NOTATION.

268. The following notation may be adopted to advantage in the formulas:

Amount Invested	= A. I.
Market Value	= M. V.
Rate of Dividend or Interest	= R. D.
Rate of Income	= R. I.
Income	= I.

CASE I.

269. Given the amount invested, the market value, and the rate of dividend or interest, to find the income.

FORMULA.—$\frac{\text{A. I.}}{\text{M. V.}} \times \text{R. D.} = \text{I.}$

PROBLEM.—A person invests \$5652.50 in Mutual Insurance Company stock at 95 cents: what will be his income if the stock pay 10% dividend annually?

OPERATION.

\$5652.50 ÷ .95 = \$5950 = par value of stock purchased.
\$5950 × .1 = \$595 = income.

Or, 95% = $\frac{19}{20}$ of the par value = \$5652.50
5% = $\frac{1}{20}$ " " " " = \$ 297.50
10% = $\frac{2}{20}$ " " " " = \$ 595, *Ans.*

SOLUTION.—As many dollars' worth of stock can be bought as \$.95 is contained times in \$5652.50, which is 5950 times. Therefore \$5950 is the par value of the stock purchased; and 10% dividend on \$5950 is \$595, the income on the investment.

EXAMPLES FOR PRACTICE.

1. A invests \$28000 in Lake Shore Railroad stock, at 70%. If the stock yields 8% annually, what is the amount of his income?

2. B invests \$100962 in U. S. cur. 6's, 1899, at $106\frac{1}{2}\%$. If gold is at $\frac{1}{8}$ premium, what does the government save by paying him his interest in greenbacks?

3. If I invest \$10200 in Tennessee 6's, new, at 30%, what is my annual income?

4. A broker invested \$36000 in quicksilver preferred stock, at 40%: if the stock pays 4%, what is the income derived?

5. Which is the better investment, stock paying $6\frac{1}{2}\%$ dividend, at a market value of $106\frac{1}{2}\%$, or stock paying $4\frac{1}{2}\%$ dividend, at $104\frac{1}{8}\%$?

6. Which is the more profitable, to invest \$10000 in 6% stock purchased at 75%, or in 5% stock purchased at 60%, allowing brokerage $\frac{1}{2}\%$?

CASE II

270. Given the amount invested, the market value, and the income, to find the rate of dividend or interest.

FORMULA.— $\text{I.} \div \frac{\text{A. I.}}{\text{M. V.}} = \text{R. D.}$

PROBLEM.—Invested \$10132.50 in railroad stock at 105%, which pays me annually \$965: what is the rate of dividend on the stock?

SOLUTION.—By Art. **259**, \$10132.50 ÷ 1.05 = \$9650 = par value of the stock; and by Art. **251**, \$965 ÷ \$9650 = .1, or 10%, *Ans.*

EXAMPLES FOR PRACTICE.

1. A has a farm, valued at \$46000, which pays him 5% on the investment. Through a broker, who charges \$56.50 for his services, he exchanges it for insurance stock at 9% premium, and this increases his annual income by \$1072: what dividend does the stock pay?

2. What dividend must stock pay, in order that my rate of income on an investment of \$64968.75 shall be $4\frac{4}{33}\%$, provided the stock can be bought at $103\frac{1}{8}\%$?

3. My investment was \$9850, my income is \$500, and the market value of the stock $108\frac{1}{10}\%$, brokerage $\frac{1}{4}\%$: what is the rate of dividend?

4. The sale of my farm cost me \$500, but I gave the proceeds to a broker, allowing him $\frac{1}{2}\%$, to purchase railroad stock then in market at 102%; the farm paid a 5% income, equal to \$2075, but the stock will pay \$2025 more: what is the rate of dividend?

5. Howard has at order \$122400, and can allow brokerage $\frac{1}{2}\%$, and buy insurance stock at $101\frac{1}{2}\%$, yielding $4\frac{1}{6}\%$; but if he send to the broker \$100 more for investment, and buy rolling-mill stock at $103\frac{1}{2}\%$, the income will only be half so large: what rate does the higher stock pay?

CASE III.

271. Given the income, rate of dividend, and market value, to find the amount invested.

FORMULA.—$\frac{\text{I.}}{\text{R. D.}} \times \text{M. V.} = \text{A. I.}$

PROBLEM.—If U. S. bonds, paying 5% interest, are selling at $108\frac{1}{2}\%$, how much must be invested to secure an annual income of \$2000?

OPERATION.

$\$2000 \div \$.05 = 40000;\ \$40000 \times 1.085 = \$43400$, *Ans.*

SOLUTIONS.—1. To produce an income of \$2000 it will require as many dollar's worth of stock at par as 5 ct. is contained times in \$2000, which is 40000; and, at $108\frac{1}{2}\%$, it will require $\$40000 \times 1.08\frac{1}{2} = \$43400$.

2. \$1 of stock will give 5 cents income, and \$2000 income will require \$40000 worth of stock at par. \$40000 of stock, at $108\frac{1}{2}\%$, will cost $\$40000 \times 1.085 = \$43400$.

EXAMPLES FOR PRACTICE.

1. What amount is invested by A, whose canal stock, yielding 4%, brings an income of $300, but sells in market for 92%?

2. If I invest all my money in 5% furnace stock, salable at 75%, my income will be $180: how much must I borrow to make an investment in 6% state stock, selling at 102%, to have that income?

3. If railroad stock be yielding 6%, and is 20% below par, how much would have to be invested to bring an income of $390?

4. A banker owns $2\frac{1}{2}$% stocks, at 10% below par, and 3% stocks, at 15% below par. The income from the former is $66\frac{2}{3}$% more than from the latter, and the investment in the latter is $11400 less than in the former: required the whole investment and income.

5. Howard M. Holden sold $21600 U. S. 4's, 1907, registered at $99\frac{3}{8}$%, and immediately invested a certain amount of the proceeds in Illinois Central Railroad stock, at 80%, which pays an annual dividend of 6%; he receives $840 from the railroad investment; with the remainder of his money he bought a farm at $30 an acre: required the amount invested in railroad stock, and the number of acres in the farm?

6. W. T. Baird, through his broker, invested a certain sum of money in Philadelphia 6's, at $115\frac{1}{2}$%, and three times as much in Union Pacific 7's, at $89\frac{1}{2}$%, brokerage $\frac{1}{2}$% in both cases: how much was invested in each kind of stock if his annual income is $9920?

7. Thomas Reed, bought 6% mining stock at $114\frac{1}{2}$%, and 4% furnace stock at 112%, brokerage $\frac{1}{2}$%; the latter cost him $430 more than the former, but yielded the same income: what did each cost him?

CASE IV.

272. Given the market value and the rate of dividend or interest, to find the rate of income.

FORMULA.—$\frac{\text{R. D.}}{\text{M. V.}} = \text{R. I.}$

PROBLEM.—What per cent of his money will a man realize in buying 6% stock at 80%?

OPERATION.

$$\$.06 \div \$.80 = \tfrac{6}{80} = \tfrac{3}{40} = 7\tfrac{1}{2}\%.$$

SOLUTION.—The expenditure of 80 ct. buys a dollar's worth of stock, giving an income of 6 ct. The rate per cent of income on investment is $\$.06 \div \$.80 = .07\frac{1}{2}$, or $7\frac{1}{2}\%$.

EXAMPLES FOR PRACTICE.

1. What is the rate of income on Pacific Mail 6's, bought at 30%?

2. What is the rate of income on Union Pacific 6's, bought at 110%?

3. Which is the better investment: U. S. new 4's, registered, at $99\frac{3}{8}\%$; or U. S. new $4\frac{1}{2}$'s, coupons, at 106%?

4. Thomas Sparkler has an opportunity of investing $30000 in North-western preferred stock, at 76%, which pays an annual premium of 5%; in Panama stock, at 125%, which pays a premium annually of $8\frac{1}{2}\%$; or he can lend his money, on safe security, at $6\frac{1}{2}\%$ per annum. Prove which is the best investment for Mr. Sparkler.

5. Thomas Jackson bought 500 shares of Adams Express stock, at $105\frac{1}{2}\%$, and paid $1\frac{1}{2}\%$ brokerage: what is the rate of income on his investment per annum if the annual dividend is 8%?

CASE V.

273. Given the rate of income and the rate of dividend or interest, to find the market value.

FORMULA.—$\frac{\text{R. D.}}{\text{R. I.}} = \text{M. V.}$

PROBLEM.—What must I pay for Lake Shore 6's ($100 a share), that the investment may yield 10% ?

OPERATION.

$\$.06 \div .1 = \frac{6}{10} = 60\% = \$60$ a share.

SOLUTION.—If bought for $100, or par, it will yield 6%; to yield 1% it must be bought for $\$100 \times 6 = \$600$; to yield 10% it must be bought for $\frac{1}{10}$ of $\$600 = \$60$.

EXAMPLES FOR PRACTICE.

1. What must I pay for Chicago, Burlington & Quincy Railroad stock that bears 6%, that my annual income on the investment may yield 5% ?

2. Which is the best permanent investment: 4's at 70%, 5's at 80%, 6's at 90%, or 10's at 120% ? Why?

3. The rate of income being 7% on the investment, and the dividend rate 4%, what is the market value of $3430 of the stock?

4. In a mutual insurance company one capitalist has an investment paying 8%: what is the premium on the stock, the dividend being 9% ?

5. Suppose 10% state stock 20% better in market than 4% railroad stock; if A's income be $500 from each, how much money has he paid for each, the whole investment bringing $6\frac{2}{333}$% ?

6. At what figure must be government 5 per cent's to make my purchase pay 9% ?

VI. INSURANCE.

DEFINITIONS.

274. 1. **Insurance** is indemnity against loss or damage.

2. There are two kinds of insurance, viz.: *Property Insurance* and *Personal Insurance.*

3. Under **Property Insurance** the two most important divisions are: *Fire Insurance* and *Marine Insurance.*

4. **Fire Insurance** is indemnity against loss by fire.

5. **Marine Insurance** is indemnity against the dangers of navigation.

NOTES.—1. *Transit Insurance* is applied to risks which are taken when property is transferred by railroad, or by railroad and water routes combined.

2. There are several minor forms of property insurance, also, such as *Live Stock Insurance, Steam Boiler Insurance, Plate Glass Insurance,* etc., the special purposes of which are indicated by their names.

6. **Personal Insurance** is of three kinds, viz.: *Life Insurance, Accident Insurance,* and *Health Insurance.* Personal insurance will be discussed in another chapter.

7. The **Insurer** or **Underwriter** is the party or company that undertakes to pay in case of loss.

8. The **Risk** is the particular danger against which the insurer undertakes.

9. The **Insured** is the party protected against loss.

10. A **Contract** is an agreement between two or more competent parties, based on a sufficient consideration, each promising to do or not to do some particular thing possible to be done, which thing is not enjoined nor prohibited by law.

11. The **Primary Elements** of a contract are: the *Parties*, the *Consideration*, the *Subject Matter*, the *Consent of the Parties*, and the *Time*.

12. The written contract between the two parties in insurance is called a **Policy.**

13. The **Premium** is the sum paid for insurance. It is a certain per cent of the amount insured.

14. The **Rate** varies with the nature of the risk.

15. The **Amount** or **Valuation** is the sum for which the premium is paid.

NOTES.—1. Whoever owns or has an interest in property, may insure it to the full amount of his interest or liability.

2. Only the *actual loss* can be recovered by the insured, whether there be one or several insurers.

3. Usually property is insured for about two thirds of its value.

16. Insurance business is usually transacted by incorporated companies.

17. These companies are either **joint-stock** companies, or **mutual** companies.

REMARKS.—1. In *joint-stock companies* the capital is owned by individuals who are the stockholders. They share the profits and losses.

2. In *mutual companies* the profits and losses are divided among the insured.

275. The operations in insurance are included under the principles of Percentage.

The **quantities involved** are, the *Amount* insured, the *Per Cent* of premium, and the *Premium*.

The amount corresponds to the *Base*, and the premium, to the *Percentage*.

CASE I.

276. Given the rate of insurance and the amount insured, to find the premium.

FORMULA.—*Amount Insured* $\times$ *Rate* = *Premium.*

EXAMPLES FOR PRACTICE.

1. Insured $\frac{5}{8}$ of a vessel worth \$24000, and $\frac{2}{3}$ of its cargo worth \$36000, the former at $2\frac{1}{4}\%$, the latter at $1\frac{1}{8}\%$: what is the premium?

2. Insured a house for \$2500, and furniture for \$600, at $\frac{6}{10}\%$: what is the premium?

3. What is the premium on a cargo of railroad iron worth \$28000, at $1\frac{3}{4}\%$?

4. Insured goods invoiced at \$32760, for three months, at $\frac{8}{10}\%$: what is the premium?

5. My house is permanently insured for \$1800, by a deposit of 10 annual premiums, the rate per year being $\frac{3}{4}\%$: how much did I deposit, and if, on terminating the insurance, I receive my deposit less 5%, how much do I get?

6. A shipment of pork, costing \$1275, is insured at $\frac{5}{9}\%$, the policy costing 75 cents: what does the insurance cost?

7. An insurance company having a risk of \$25000, at $\frac{9}{10}\%$, re-insured \$10000, at $\frac{4}{5}\%$, with another office, and \$5000, at 1%, with another: how much premium did it clear above what it paid?

CASE II.

277. Given the amount insured and premium, to find the rate of insurance.

FORMULA.—$\frac{\textit{Premium}}{\textit{Amount Insured}} = \textit{Rate.}$

EXAMPLES FOR PRACTICE.

1. Paid \$19.20 for insuring $\frac{2}{3}$ of a house, worth \$4800: what was the rate?

2. Paid $234, including cost of policy, $1.50, for insuring a cargo worth $18600: what was the rate?

3. Bought books in England for $2468; insured them for the voyage for $46.92, including the cost of the policy, $2.50: what was the rate?

4. A vessel is insured for $42000; $18000 at $2\frac{1}{2}\%$, $15000 at $3\frac{4}{5}\%$, and the rest at $4\frac{2}{3}\%$: what is the rate on the whole $42000?

5. I took a risk of $45000; re-insured at the same rate, $10000 each, in three offices, and $5000 in another; my share of the premium was $262.50: what was the rate?

6. I took a risk at $1\frac{1}{2}\%$; re-insured $\frac{2}{5}$ of it at 2%, and $\frac{1}{4}$ of it at $2\frac{1}{2}\%$: what rate of insurance do I get on what is left?

CASE III.

278. Given the premium and rate of insurance, to find the amount insured.

FORMULA.—$\frac{\textit{Premium}}{\textit{Rate.}} = \textit{Amount Insured.}$

EXAMPLES FOR PRACTICE.

1. Paid $118 for insuring, at $\frac{4}{5}\%$: what was the amount insured?

2. Paid 411.37\frac{1}{2}$ for insuring goods, at $1\frac{1}{2}\%$: what was their value?

3. Paid $42.30 for insuring $\frac{5}{8}$ of my house, at $\frac{9}{10}\%$: what is the house worth?

4. Took a risk at $2\frac{1}{4}\%$; re-insured $\frac{3}{5}$ of it at $2\frac{1}{2}\%$; my share of the premium was $197.13: how large was the risk?

5. Took a risk at $1\frac{3}{5}\%$; re-insured half of it at the same rate, and $\frac{1}{3}$ of it at $1\frac{1}{2}\%$; my share of the premium was $58.11: how large was the risk?

6. Took a risk at 2%; re-insured $10000 of it at $2\frac{1}{8}\%$, and $8000 at $1\frac{3}{4}\%$; my share of the premium was $207.50: what sum was insured?

7. The Mutual Fire Insurance Company insured a building and its stock for $\frac{2}{3}$ of its value, charging $1\frac{3}{4}\%$. The Union Insurance Company relieved them of $\frac{1}{4}$ of the risk, at $1\frac{1}{2}\%$. The building and stock being destroyed by fire, the Union lost forty-nine thousand dollars less than the Mutual: what amount of money did the owners of the building and stock lose?

VII. TAXES.

DEFINITIONS.

279. 1. A **Tax** is a sum of money levied on persons, or on persons and property, for public use.

2. **Taxes** in this country are: (1.) *State* and *Local Taxes*; (2.) *Taxes for the National Government.*

3. Taxes are further classified as *Direct* and *Indirect.*

4. A **Direct Tax** is one which is levied on the person or property of the individual.

5. An **Indirect Tax** is a tax levied on articles of consumption, for which each person pays in proportion to the quantity or number of such articles consumed.

6. A **Poll-Tax,** or **Capitation Tax,** is a direct tax levied on each male citizen liable to taxation.

7. A **Property Tax** is a direct tax levied on property.

NOTE.—In legal works property is treated of under two heads; viz., *Real Property*, or *Real Estate*, including houses and lands; and *Personal Property*, including money, bonds, cattle, horses, furniture,—in short, all kinds of movable property.

8. **State** and **Local Taxes** are generally *direct*, while the **United States Taxes** are *indirect*.

9. An **Assessor** is a public officer elected or appointed to prepare the *Assessment Roll*.

10. An **Assessment Roll** is a list of the names of the taxable inhabitants living in the district assessed, and the valuation of each one's property.

11. The **Collector** is the public officer who receives the taxes.

NOTE.—In some states all the taxes are collected by the counties; in others, the towns collect; while in others the collections are made by separate collectors. Generally a number of different taxes are aggregated,—such as state, county, road, school, etc.

280. The **quantities involved** in problems under taxation are, the *Assessed Value* of the property, the *Rate of Taxation*, the *Tax*, and the *Amount Left* after taxation.

They require the application of the four cases of Percentage, the assessed value corresponding to the *Base;* the tax, to the *Percentage;* and the amount left after taxation, to the *Difference.*

CASE I.

281. Given the taxable property and the rate, to find the property-tax.

FORMULA.—*Taxable Property* $\times$ *Rate* $=$ *Amount of Tax.*

NOTE.—If there be a poll-tax, the sum produced by it should be added to the property-tax, to give the whole tax.

EXAMPLES FOR PRACTICE.

1. The taxable property of a county is \$486250, and the rate of taxation is 78 ct. on \$100; that is, $\frac{78}{100}\%$: what is the tax to be raised?

REMARK.—The rate of taxation being usually small, is expressed most conveniently as so many cents on $100, or as so many mills on $1.

2. A's property is assessed at $3800; the rate of taxation is 96 ct. on $100 ($\frac{96}{100}$%): what is his whole tax, if he pays a poll-tax of $1?

REMARKS.—1. In making out bills for taxes, a table is used, containing the units, tens, hundreds, thousands, etc., of property, with the corresponding tax opposite each.

2. To find the tax on any sum by the table, take out the tax on each figure of the sum, and add the results. In this table, the rate is $1\frac{1}{4}$%, or 125 ct. on $100.

TAX TABLE.

Prop.	Tax.	Prop.	Tax.	Prop.	Tax.	Prop.	Tax.	Prop.	Tax.
$1	.0125	$10	.125	$100	$1.25	$1000	$12.50	$10000	$125
2	.025	20	.25	200	2.50	2000	25.	20000	250
3	.0375	30	.375	300	3.75	3000	37.50	30000	375
4	.05	40	.50	400	5.	4000	50.	40000	500
5	.0625	50	.625	500	6.25	5000	62.50	50000	625
6	.075	60	.75	600	7.50	6000	75.	60000	750
7	.0875	70	.875	700	8.75	7000	87.50	70000	875
8	.10	80	1.	800	10.	8000	100.	80000	1000
9	.1125	90	1.125	900	11.25	9000	112.50	90000	1125

3. What will be the tax by the table, on property assessed at $25349?

SOLUTION.—The tax for $20000 is $250; for $5000 is 62.50; for $300 is $3.75; for $40 is .50; for $9 is .1125; which, added, give $316.86, the tax on $25349.

4. Find the tax for $6815.30

5. What is the tax on property assessed at $10424.50, and two polls, at $1.50 each?

6. A's property is assessed at $251350, and B's at $25135. What is the difference in their taxes?

CASE II.

282. Given the taxable property and the tax, to find the rate.

FORMULA.—$\frac{Tax}{Taxable\ Property} = Rate.$

EXAMPLES FOR PRACTICE.

1. Property assessed at $2604, pays $19.53 tax: what is the rate of taxation?

2. The taxable property in a town of 1742 polls, is $6814320; a tax of $66913.54 is proposed: if a poll-tax of $1.25 is levied, what should be the rate of taxation?

3. An estate of $350000 pays a tax of $5670: what is the rate of taxation?

4. A's tax is $50.46; he pays a poll-tax of $1.50, and owns $8704 taxable property: what is the rate of taxation?

CASE III.

283. Given the tax and the rate, to find the assessed value of the property.

FORMULA.—$\frac{Tax}{Rate} = Taxable\ Property.$

NOTE.—If any part of the tax arises from polls, it should be first deducted from the given tax.

EXAMPLES FOR PRACTICE.

1. What is the assessed value of property taxed $66.96, at $1\frac{4}{5}\%$?

2. A corporation pays $564.42 tax, at the rate of $\frac{46}{100}$%, or 46 ct. on $100 : find its capital.

3. A is taxed $71.61 more than B; the rate is $1\frac{8}{25}$%, or $1.32 on $100: how much is A assessed more than B?

4. A tax of $4000 is raised in a town containing 1024 polls, by a poll-tax of $1, and a property-tax of $\frac{24}{100}$% (24 ct. on $100): what is the value of the taxable property in it?

5. A's income is 16% of his capital; he is taxed $2\frac{1}{2}$% of his income, and pays $26.04: what is his capital?

CASE IV.

284. Given the amount left after payment of tax and the rate, to find the assessed value of the property.

FORMULA.— $\frac{\textit{Amount Left after Payment}}{1 - \textit{Rate \%}} = \textit{Taxable Property.}$

EXAMPLES FOR PRACTICE.

1. A pays a tax of $1\frac{7}{20}$% ($1.35 on $100) on his capital, and has left $125127.66: what was his capital, and his tax?

2. Sold a lot for $7599, which covered its cost and 2% beside, paid for tax: what was the cost?

VIII. UNITED STATES REVENUE.

DEFINITIONS.

285. 1. The Revenue of the United States arises from the *Internal Revenue*, from the *Customs*, and from *Sales of Public Lands.*

2. The **Internal Revenue** is derived from taxes on spirits, tobacco, fermented liquors, banks, and from the sale of stamps, etc.

3. The **Customs** or **Duties** are taxes imposed by Government on imported goods. They are of two kinds,—*ad valorem* and *specific.*

4. **Ad Valorem Duties** are levied at a certain per cent on the cost of the goods as shown by the invoice.

5. **Specific Duties** are certain sums collected on each gallon, bushel, yard, ton, or pound, whatever may be the cost of the article.

NOTE.—Problems involving specific duty only are not solved, of course, by the principles of Percentage.

6. An **Invoice** is a detailed statement of the quantities and prices of goods purchased.

7. A **Tariff** is a schedule of the rates of duties, as fixed by law.

NOTES.—1. The collection of duties is made at the custom-houses established at ports of entry and ports of delivery. The principal custom-house officers are collectors, naval officers, surveyors, and appraisers.

2. *Tare* is an allowance for the weight of whatever contains the goods. Duty is collected only on the quantities passed through the custom-house. The ton, for custom-house purposes, consists of 20 cwt., of 112 lb. each.

3. On some classes of goods, both specific and ad valorem duties are collected. The cost price, if given in foreign money, must be changed to United States currency.

286. Problems in United States Customs, where the duty is wholly or in part ad valorem, are solved by the principles of Percentage.

The **quantities involved** are: the *Invoice Price* corresponding to the *Base;* the *Duty,* corresponding to the *Percentage;* and the *Total Cost* of the importation, corresponding to the *Amount.*

CASE I.

287. Given the invoice price and the rate, to find the duty.

FORMULA.—*Invoice Price* $\times$ *Rate* = *Duty.*

EXAMPLES FOR PRACTICE.

1. Import 24 trunks, at $5.65 each, and 3 doz. leather satchels, at $2.25 each; the rate is 35% ad valorem: what is the duty?

2. What is the duty of 45 casks of vinegar, of 36 gal. each, invoiced at $1.25 a gal., at 40 ct. a gal. specific duty?

3. There is a specific duty of $3 per gallon, and an ad valorem duty of 50% on cologne-water: what is the total amount of duty paid on 25 gallons, invoiced at $16.50 per gallon?

4. What is the total duty on 36 boxes of sugar, each weighing 6 cwt. 2 qr. 18 lb., invoiced at $2\frac{1}{2}$ ct. per lb., the specific duty being 2 ct. per lb., and the ad valorem duty 25%?

5. What is the duty on 575 yards of broadcloth, weighing 1154 lb., invoiced at $2.56 per yd., the specific duty being 50 ct. per lb., and the ad valorem duty 35%? If the freight, charges, and losses amount to $160.80, how much a yard must I charge to gain 15%?

6. A merchant imported a ton of manilla, invoiced at 5 ct. per lb.; he paid a specific duty per ton, which, on this shipment, was equivalent to an ad valorem duty of $22\frac{9}{28}$%: what is the specific duty?

7. Received a shipment of 3724 lb. of wool, invoiced at 23 ct. per lb.; the duty is 10 ct. per lb., and 11% ad valorem, less 10%: what is the total amount of duty?

8. A dry-goods merchant imports 1120 yards of dress goods, $1\frac{1}{4}$ yd. wide, invoiced at 23 ct. a sq. yd.; there is a specific duty of 8 ct. per sq. yd., and an ad valorem duty of 40%: what must he charge per yard, cloth measure, to clear 25% on the whole?

CASE II.

288. Given the invoice price and the duty, to find the rate.

FORMULA.— $\frac{\textit{Duty}}{\textit{Invoice Price}} = \textit{Rate}.$

EXAMPLES FOR PRACTICE.

1. If goods invoiced at \$3684.50 pay a duty of \$1473.80, what is the rate of duty?

2. If laces invoiced at \$7618.75, cost, when landed, \$10285.31$\frac{1}{4}$, what is the rate of duty?

3. Forty hhd. (63 gal. each) of molasses, invoiced at 52 ct. a gallon, pay \$453.60 duty. The specific duty is 5 ct. a gallon: what is the additional ad valorem duty?

CASE III.

289. Given the duty and the rate, to find the invoice price.

FORMULA.— $\frac{\textit{Duty}}{\textit{Rate}} = \textit{Invoice Price}.$

EXAMPLES FOR PRACTICE.

1. Paid \$575.80 duty on watches, at 25%: at what were they invoiced, and what did they cost me in store?

2. The duty on 1800 yards of silk was \$2970, at 60%

ad valorem: what was the invoice price per yard, and what must I charge per yard to clear 20%?

3. The duty on 15 gross, qt. bottles of vanilla, at a tax of 35 ct. a gallon, was $151.20: if this were equivalent to an ad valorem duty of $20\frac{5}{26}$% on the entire purchase, how many bottles were allowed for a gallon, and at how much per bottle must the whole be sold to clear 20%?

CASE IV.

290. Given the entire cost and the rate, to find the invoice price.

FORMULA.—$\frac{\textit{Whole Cost}}{1 + \textit{Rate}} = \textit{Invoice Price.}$

EXAMPLES FOR PRACTICE.

1. 1000 boxes (100 each) of cigars, weighing 1200 lb., net, cost in store $13675. There is a specific duty of $2.50 per lb., an ad valorem duty of 25%, and an internal revenue tax of 60 ct. a box: freight and charges amount to $75; find the invoice price per thousand cigars.

2. Supposing No. 1 pig-iron, American manufacture, to be of equal quality with Scotch pig-iron: at what price must the latter be invoiced, to compete in our markets, if American iron sells for $45 a ton; freight and charges amounting to $10 a ton, and the specific duty being equivalent, in this instance, to an ad valorem duty of 25%?

3. A marble-cutter imports a block of marble 6 ft. 6 in. long, 3 ft. wide, 2 ft. $9\frac{3}{5}$ in. thick; the whole cost to him being $130; he pays a specific duty of 50 ct. per cu. ft. and an ad valorem duty of 20%: freight and charges being $20.80, what was the invoice price per cu. ft.?

Topical Outline.

Applications of Percentage.

(*Without Time.*)

1. Profit and Loss	1. Definitions:—Cost, Selling Price, Profit, Loss. 2. Four Cases.
2. Stocks and Bonds	1. Definitions:—Company, Corporation, Charter, Stock, Shares, Scrip, Bond, Assessment, Dividend. 2. Four Cases.
3. Premium and Discount	1. Definitions:—Drafts, Par Value, Market Value, Discount, Premium, Rate. 2. Four Cases.
4. Commission and Brokerage.	1. Definitions:—Commission-Merchant, Principal, Commission, Consignment, Consignor, Consignee, Correspondent, Net Proceeds, Guaranty, Broker, Brokerage. 2. Four Cases.
5. Stock Investments	1. Definitions:—Stock Exchange, Government Bonds, State Bonds, etc. 2. Notation. 3. Five Cases.
6. Insurance	1. Definitions:—Fire Insurance, Marine Insurance, Personal Insurance, Insurer, Risk, Insured, Contract, Primary Elements, Policy, Premium, Rate, Amount. Joint-Stock and Mutual Companies. 2. Three Cases.
7. Taxes	1. Definitions:—Tax, State and Government, Direct and Indirect, Poll-tax, Property-tax, Assessor, Assessment Roll, Collector. 2. Four Cases.
8. United States Revenue	1. Definitions:—Internal Revenue, Customs, Ad Valorem Duties, Specific Duties, Invoice, Tariff, Tare. 2. Four Cases.

XVI. PERCENTAGE.—APPLICATIONS.

I. INTEREST.

DEFINITIONS.

291. 1. **Interest** is money charged for the use of money.

NOTE.—The profits accruing at regular periods on permanent investments, such as dividends or rents, are called interest, since they are the increase of capital, unaided by labor.

2. The **Principal** is the sum of money on which interest is charged.

NOTE.—The principal is either a sum loaned; money invested to secure an income; or a debt, which not being paid when due, is allowed by agreement or by law to draw interest.

3. The **Rate of Interest** is the number of per cent the yearly interest is of the principal.

4. The **Amount** is the sum of the principal and interest.

5. Interest is **Payable** at regular intervals, *yearly*, *half-yearly*, or *quarterly*, as may be agreed: if there is no agreement, it is understood to be *yearly*.

NOTE.—If interest is payable half-yearly, or quarterly, the rate is still the rate *per annum*, or rate per year. In short loans, the rate per month is generally given; but the rate per year, being 12 times the rate per month, is easily found; thus, 2% a month = 24% a year.

6. The **Legal Rate** is the highest rate allowed by the law.

7. If interest be charged at a rate higher than the law allows, it is called **Usury**; and, in some states, the person offending is subject to a penalty.

Remark.—The legal interest (1893) in the different states is given below. It is subject to change by State Legislatures.

TABLE.

State or Territory.	Rate.		State or Territory.	Rate.	
Alabama	8%	8%	Montana	10%	Any
Alaska	8%	10%	Nebraska	7%	10%
Arkansas	6%	10%	Nevada	7%	Any
Arizona	7%	Any	New Hampshire	6%	6%
California	7%	Any	New Jersey	6%	6%
Colorado	8%	Any	New Mexico	6%	12%
Connecticut	6%	6%	New York	6%	6%
Delaware	6%	6%	North Carolina	6%	8%
Dist. of Columbia	6%	....	North Dakota	7%	12%
Florida	8%	10%	Ohio	6%	8%
Georgia	7%	8%	Oklahoma	7%	12%
Idaho	10%	18%	Oregon	8%	10%
Illinois	5%	7%	Pennsylvania	6%	6%
Indian Ter.	10%	15%	Rhode Island	6%	Any
Indiana	6%	8%	South Carolina	7%	8%
Iowa	6%	8%	South Dakota	7%	12%
Kansas	6%	10%	Tennessee	6%	6%
Kentucky	6%	6%	Texas	6%	10%
Louisiana	5%	8%	Utah	8%	Any
Maine	6%	Any	Vermont	6%	6%
Maryland	6%	6%	Virginia	6%	6%
Massachusetts	6%	Any	Washington	8%	Any
Michigan	6%	8%	West Virginia	6%	6%
Minnesota	7%	10%	Wisconsin	6%	10%
Mississippi	6%	10%	Wyoming	12%	Any
Missouri	6%	8%			

Note.—When the per cent of interest is not mentioned in the note or contract, the first column gives the per cent that may be collected by law. If stipulated in the note, a per cent of interest as high as that in the second column may be collected.

8. Interest is either *Simple* or *Compound.*

9. **Simple Interest** is interest which, even if not paid

when due, is not convertible into principal, and therefore can not accumulate in the hands of the debtor by drawing interest itself, however long it may be retained.

NOTE.—*Compound Interest* is interest which, not being paid when due, is convertible into principal, and from that time draws interest itself and accumulates in the hands of the debtor, according to the time it is retained. It will be treated of in a separate chapter.

292. Simple Interest differs from the applications of Percentage in Chapter XV, by taking *time* as an element in the calculation, which they do not.

293. The **five quantities** embraced in questions of interest are: the *Principal*, the *Interest*, the *Rate*, the *Time*, and the *Amount*,—or the sum of the principal and interest. Any three of these being given, the others may be found. They give rise to **five cases.**

The principal corresponds to the *Base*, and the interest to the *Percentage.*

NOTATION.

294. The following notation can be adopted to advantage in the formulas:

Principal = P.
Rate = R.
Interest = I.
Amount = A.
Time = T.

CASE I.

295. Given the principal, the rate, and the time, to find the interest and the amount.

PRINCIPLE.—*The interest is equal to the continued product of the principal, rate, and time.*

Formulas.— $\begin{cases} P \times R \times T = I. \\ P + I = A. \end{cases}$

Note.—The time is expressed in years or parts of years. or both.

COMMON METHOD.

Problem.—What is the interest of \$320 for 3 yr. 5 mo. 18 da., at 4% ?

OPERATION.

$$\$320 \times .04 \times 3\tfrac{7}{15} = \$44.37\tfrac{1}{3}, \text{ Int.}$$

Solution.—The interest of \$320 for 1 year, at 4%, is \$12.80; and for $3\frac{7}{15}$ years, it is $3\frac{7}{15}$ times as much as it is for 1 year, or $\$12.80 \times 3\frac{7}{15} = \$44.37\frac{1}{3}$.

Remark.—Unless otherwise specified, 30 days are considered to make a month; hence, 5 mo. 18 da. $= \frac{7}{15}$ of a year.

General Rule.—1. *Multiply the principal by the rate, and that product by the time expressed in years; the product is the interest.*

2. *Add the principal and interest, to find the amount.*

Examples for Practice.

Find the simple interest of:

1. \$178.63 for 2 yr. 5 mo. 26 da., at 7%.
2. \$6084.25 for 1 yr. 3 mo., at $4\frac{1}{2}$%.
3. \$64.30 for 1 yr. 10 mo. 14 da., at 9%.
4. \$1052.80 for 28 da., at 10%.
5. \$419.10 for 8 mo. 16 da., at 6%.
6. \$1461.85 for 6 yr. 7 mo. 4 da., at 10%.
7. \$2601.50 for 72 da., at $7\frac{1}{2}$%.
8. \$8722.43 for $5\frac{1}{2}$ yr., at 6%.
9. \$326.50 for 1 mo. 8 da., at 8%.
10. \$1106.70 for 4 yr. 1 mo. 1 da., at 6%.
11. \$10000 for 1 da., at 6%.

METHOD BY ALIQUOT PARTS.

296. Many persons prefer computing interest by the method of *Aliquot Parts*. The following illustrates it:

PROBLEM.—What is the simple interest of $354.80 for 3 yr. 7 mo. 19 da., at 6%?

SOLUTION.—The yearly interest, being 6% of the principal, is found, by Case I of Percentage; the interest for 3 yr. 7 mo. 19 da. is then obtained by aliquot parts; each item of interest is carried no lower than mills, the next figure being neglected if less than 5; but if 5 or over, it is counted 1 mill.

OPERATION.

	$354.80
	.06
1 yr. =	21.2880
3 yr. = 3	63.864
6 mo. = $\frac{1}{2}$	10.644
1 mo. = $\frac{1}{6}$	1.774
18 da. = $\frac{1}{10}$	1.064
1 da. = $\frac{1}{30}$	.059
	$77.41, *Ans.*

SIX PER CENT METHODS.

FIRST METHOD.

297. The *six per cent method* possesses many advantages, and is readily and easily applied, as the following will show:

At 6% per annum, the interest on $1

for 1 year	is 6 ct., or	.06	of the principal.
" 2 mo.	" 1 " "	.01	" " "
" 1 "	" $\frac{1}{2}$ " "	.005	" " "
" 6 da. = $\frac{1}{5}$ mo.	" $\frac{1}{10}$ " "	.001	" " "
" 1 " = $\frac{1}{30}$ "	" $\frac{1}{60}$ " "	.000$\frac{1}{6}$	" " "

PROBLEM.—What is the interest of $560 for 3 yr. 8 mo. 12 da., at 6%?

SOLUTION.—The interest on $1 for 1 yr., at 6%, is 6 ct., and for 3 yr. is 18 ct.; the interest on $1 for 8 mo. is 4 ct., and the interest for 12 da. is 2 mills; hence, the interest for the entire time is $.222: therefore, the interest on $560 for 3 yr. 8 mo. 12 da. is equal to $.222 × 560 = $124.32

OPERATION.

$560 × .222 = $124.32, *Ans.*

Six Per Cent Rule.—1. *Take six cents for every year, $\frac{1}{2}$ a cent for every month, and $\frac{1}{6}$ of a mill for every day; their sum is the interest on \$1, at 6%, for the given time.*

2. *Multiply the interest on \$1 for the given time by the principal; the product is the interest required.*

REMARKS.—1. To find the interest at any other rate than 6%, increase or decrease the interest at 6% by such a part of it as the given rate is greater or less than 6%.

2. After finding the interest at 6%, observe that the interest at 5% = interest at 6% — $\frac{1}{6}$ of itself. And the interest

at $4\frac{1}{2}$% = int. at 6% — $\frac{1}{4}$ of it.	at 1 % = $\frac{1}{6}$ int. at 6%.
at 4 % = int. at 6% — $\frac{1}{3}$ of it.	at 7 % = int. at 6% + $\frac{1}{6}$ of it.
at 3 % = $\frac{1}{2}$ int. at 6%.	at $7\frac{1}{2}$% = " 6% + $\frac{1}{4}$ of it.
at 2 % = $\frac{1}{3}$ int. at 6%.	at 8 % = " 6% + $\frac{1}{3}$ of it.
at $1\frac{1}{2}$% = $\frac{1}{4}$ int. at 6%.	at 9 % = " 6% + $\frac{1}{2}$ of it.

at 12%, 18%, 24% = 2, 3, 4 times interest at 6%.

at 5%, 10%, 15%, 20% = $\frac{1}{12}$, $\frac{1}{6}$, $\frac{1}{4}$, $\frac{1}{3}$ interest at 6%, after moving the point one figure to the right.

3. Or, having the interest at 6%, multiply by the given rate and divide by 6.

SECOND METHOD.

PROBLEM.—What is the simple interest of \$354.80 for 3 yr. 7 mo. 19 da., at 6%?

OPERATION.

$(\$354.80 \div 2) \times .436\frac{1}{3} =$ \$77.40553, *Ans.*

SOLUTION.—The interest of any sum (\$354.80), at 6%, equals the interest of half that sum (\$177.40), at 12%. But 12% a year equals 1% a month, and for 3 yr. 7 mo., or 43 mo., the rate is 43%, and for 19 da., which is $\frac{19}{30}$ of a month, the rate is $\frac{19}{30}$%; hence, the rate for the whole time is $43\frac{19}{30}$%, and $43\frac{19}{30}$% of the principal will be the interest. To get $43\frac{19}{30}$%, multiply by $43\frac{19}{30}$ hundredths = $.43\frac{19}{30}$ = $.436\frac{1}{3}$

REMARK.—In the multiplier $.436\frac{1}{3}$, the hundredths (43) are the number of months, and the thousandths ($6\frac{1}{3}$) are $\frac{1}{3}$ of the days (19 da.), in the given time, 3 yr. 7 mo. 19 da.

Rule.—*Reduce the years, if any, to months, and write the whole number of months as decimal hundredths; after which, place $\frac{1}{3}$ of the days, if any, as thousandths; multiply half the principal by this number, the product is the interest.*

NOTE.—In applying the rule, when the number of days is 1 or 2, place a cipher to the right of the months, and write the $\frac{1}{3}$ or $\frac{2}{3}$; otherwise, they will not stand in the thousandths' place: thus, if the time is 1 yr. 4 mo. 1 da., the multiplier is $.160\frac{1}{3}$

REMARK.—*Exact or Accurate Interest* requires that the common year should be 365, and leap year 366 days; hence, *exact* interest is $\frac{1}{73}$ less for common years, and $\frac{1}{61}$ less for leap years than the ordinary interest for 360 days.

EXAMPLES FOR PRACTICE.

Find the simple interest of:

1. \$1532.45 for 9 yr. 2 mo. 7 da., at 12%.
2. \$78084.50 for 2 yr. 4 mo. 29 da., at 18%.
3. \$512.60 for 8 mo. 18 da., at 7%.
4. \$1363.20 for 39 da., at $1\frac{1}{4}$% a month.
5. \$402.50 for 100 da., at 2% a month.
6. \$6919.32 for 7 yr. 6 mo., at 6%.
7. \$990.73 for 9 mo. 19 da., at 7%.
8. \$4642.68 for 5 mo. 17 da., at 15%.
9. \$13024 for 9 mo. 13 da., at 10%.
10. \$615.38 for 4 yr. 11 mo. 6 da., at 20%.
11. \$2066.19 for 3 yr. 6 mo. 2 da., at 30%.
12. \$92.55 for 3 mo. 22 da., at 5%.

Find the amount of:

13. \$757.35 for 117 da., at $1\frac{1}{2}$% a month.
14. \$1883 for 1 yr. 4 mo. 21 da., at 6%.
15. \$262.70 for 53 da., at 1% a month
16. \$584.48 for 133 da., at $7\frac{1}{2}$%.
17. \$392.28 for 71 da., at $2\frac{1}{2}$% a month.

18. Find the interest of \$7302.85 for 365 da., at 6%, counting 360 da. to a year.

19. If I borrow \$1000000 at 7%, accurate interest, and lend the same amount at 7% interest, to be computed by the ordinary method, what do I gain in 180 da.?

20. Find the accurate interest of \$5064.30 for 7 mo. 12 da., at 7%?

21. If I borrow \$12500 at 6%, and lend it at 10%, what do I gain in 3 yr. 4 mo. 4 da.?

22. If \$4603.15 is loaned July 17, 1881, at 7%, what is due March 8, 1883?

NOTE.—When the time is to be found by subtraction, and the *days* in the subtrahend exceed those in the minuend, disregard days in subtracting, and take the months in the minuend one less than the actual count. Having found the years and months by subtraction, find the days by actual count, beginning in the month preceding the later of the two dates. Thus, in the above example, we find 1 yr. 7 *mo.* (not 8), and count from *February* 17th to March 8th 19 da.

23. Find the interest, at 8%, of \$13682.45, borrowed from a minor 13 yr. 2 mo. 10 da. old, and retained till he is of age (21 years).

24. In one year a broker loans \$876459.50 for 63 da., at $1\frac{1}{2}$% a mo., and pays 6% on \$106525.20 deposits: what is his gain?

25. What is a broker's gain in 1 yr., on \$100, deposited at 6%, and loaned 11 times for 33 da., at the rate of 2% a month?

26. Find the simple interest of £493 16s. 8d. for 1 yr. 8 mo., at 6%.

NOTE.—In England the actual number of days is counted.

27. Find the simple interest of £24 18s. 9d. for 10 mo., at 6%.

28. Of £25 for 1 yr. 9 mo., at 5%.

29. Of £648 15s. 6d. from June 2 to November 25, at 5%.

CASE II.

298. Given the principal, the rate, and the interest, to find the time.

FORMULA.— $\frac{I}{P \times R} = T.$

PROBLEM.—John Thomas loaned \$480, at 5%, till the interest was \$150; required the time.

OPERATION.

$\$150 \div (\$480 \times .05) = 6\frac{1}{4}$ years.

SOLUTION.—The interest on \$480 for 1 yr., at 5%, is \$24. If the principal produce \$24 interest in 1 year, it will require as many years to produce \$150 interest as \$24 is contained times in \$150, which is $6\frac{1}{4}$ years, or 6 yr. 3 mo. From the preceding, the following rule is derived:

Rule.—*Divide the given interest by the interest of the principal for one year; the quotient is the time.*

REMARK.—If the principal and amount are given, take their difference for the interest.

EXAMPLES FOR PRACTICE.

In what time will:

1. \$1200 amount to \$1800, at 10%?
2. \$415.50 to \$470.90, at 10%?
3. \$3703.92 to \$4122.15, at 8%?

NOTE.—A part of a day, not being recognized in interest, is omitted in the *answer*, but must *not* be omitted in the *proof*.

4. In what time will any sum, as \$100, double itself by simple interest, at $4\frac{1}{2}$, 6, $7\frac{1}{2}$, 9, 10, 12, 20, 25, 30%?

5. In what time will any sum treble itself by simple interest, at 4, 10, 12%?

6. How long must I keep on deposit $1374.50, at 10%, to pay a debt of $1480.78?

7. How long will it take $3642.08 to amount to $4007.54, at 12%?

8. How long would it take $175.12 to produce $6.43 interest, at 6%?

9. How long would it take $415.38 to produce $10.69 accurate interest, at 7%?

CASE III.

299. Given the principal, interest, and time, to find the rate.

FORMULA.— $\frac{I}{P \times 1\% \times T} = R.$

PROBLEM.—At what rate per cent will $4800 gain $840 interest in $2\frac{1}{2}$ years?

OPERATION.

Int. of $4800 at 1% for $2\frac{1}{2}$ yr. = $120;
∴ $840 ÷ $120 = 7; or,
1% × 7 = 7%, *Ans.*

SOLUTION.—The interest of $4800, at 1%, for $2\frac{1}{2}$ yr. is $120; hence, the rate is as many times 1% as $120 is contained times in $840, which is 7 times; or, 7%. Hence the rule.

Rule.—*Divide the given interest by the interest of the principal for the given time, at 1%.*

EXAMPLES FOR PRACTICE.

1. At what rate per annum will any sum treble itself, at simple interest, in 5, 10, 15, 20, 25, 30 years, respectively?

2. At what rate of interest per annum will any sum quadruple itself, at simple interest, in 6, 12, 18, 24, and 30 years, respectively?

3. What is the rate of interest when \$35000 yields an income of \$175 a month?

4. What is the rate of interest when \$29200 produces \$6.40 a day?

5. What is the rate of interest when \$12624.80 draws \$315.62 interest quarterly.

6. Find the rate when stock, bought at 40% discount, yields a semi-annual dividend of 5%? $16\frac{2}{3}$% per annum.

7. A house that cost \$8250, rents for \$750 a year; the insurance is $\frac{6}{10}$%, and the repairs $\frac{1}{2}$%, every year: what rate of interest does it pay?

CASE IV.

300. Given the interest, rate, and time, to find the principal.

FORMULA.—$\frac{I}{R \times T} = P.$

PROBLEM.—A man receives \$490.84 interest annually on a mortgage, at 7%: what is the amount of the mortgage?

SOLUTION.—Since \$.07 is the interest of \$1 for one year, \$490.84 is the interest of as many dollars as \$.07 is contained times in \$490.84, which is 7012; therefore, \$7012 is the amount of the mortgage.

OPERATION.

\$490.84 ÷ \$.07 = 7012
\$1 × 7012 = \$7012, *Ans.*
Or, 7% = \$490.84
1% = \$70.12
100% = \$7012.

Or, thus: the time being 1 year, \$490.84 is 7% of the principal; 1% of it is $\frac{1}{7}$ of \$490.84, which is \$70.12, and hence 100% of it, or the whole principal, is \$7012.

Rule.—*Multiply the rate by the time, and divide the interest by the product; the quotient will be the principal.*

EXAMPLES FOR PRACTICE.

What principal will produce:

1. \$1500 a year, at 6%?
2. \$1830 in 2 yr. 6 mo., at 5%?
3. \$45 a mo., at 9%?
4. \$17 in 68 da., at 1% a month?
5. \$656.25 in 9 mo., at $3\frac{1}{2}$%?
6. \$86.15 in 9 mo. 11 da., at 10%?
7. \$313.24 in 112 da., at 7%?
8. \$146.05 in 7 mo. 14 da., at 6%?
9. \$58.78 in 1 yr. 3 mo. 20 da., at 4%?
10. \$79.12 in 5 mo. 25 da., at 7%.?

CASE V.

301. Given the amount, rate, and time, to find the principal.

FORMULA.— $A \div (1 + \overline{R \times T}) = P$.

PROBLEM.—What is the par value of a bond which, in 8 yr. 8 mo., at 6%, will amount to \$15200?

OPERATION.

$\$15200 \div \$1.52 = 10000$;
$\$1 \times 10000 = \$10000$, *Ans.*
Or, 100% = bond;
100% + 52% = \$15200;
1% = \$100
100% = \$10000, *Ans.*

SOLUTION.—There are as many dollars in the principal as \$1.52 is contained times in \$15200, which is 10000; therefore, \$10000 is the par value of the bond. Or, the amount is $\frac{152}{100}$ of the principal; hence, $\frac{152}{100}$ of the principal is \$15200; and $\frac{1}{100}$ of the principal is \$100, and $\frac{100}{100}$ of the principal is \$10000.

Rule.—*Divide the amount by the amount of* \$1 *for the given time and rate; the quotient will be the principal.*

EXAMPLES FOR PRACTICE.

1. What principal in 2 yr. 3 mo. 12 da., at 6%, will amount to $1367.84?

2. What principal in 10 mo. 26 da., will amount to $2718.96, at 10% interest?

3. What principal, at $4\frac{1}{2}$%, will amount to $4613.36 in 3 yr. 1 mo. 7 da.?

4. What principal, at 7%, will amount to $562.07 in 79 da. (365 da. to a year)?

PROMISSORY NOTES.

DEFINITIONS.

302. 1. A **Promissory Note** is a written promise by one party to pay a specified sum to another.

2. The **Face** of a note is the sum promised to be paid.

3. The **Maker** is the party who binds himself to pay the note by signing his name to it.

4. The **Payee** is the party to whom the payment is promised.

NOTE.—The maker and the payee are called the *original parties* to a note.

5. The **Holder** of a note is the person who owns it: he may or may not be the payee.

6. The **Indorser** of a note is the person who writes his name across the back of it.

7. A **Time Note** is one in which a specified time is set for payment.

REMARK.—There are various forms used in time notes, the principal ones being as follows.

1. Ordinary Form.

$450.50 Boston, Mass., June 30, 1880.

Three months after date, I promise to pay Thomas Ford, or order, Four Hundred and Fifty, and $\frac{50}{100}$, Dollars, for value received, with interest.

Edward E. Morris.

2. Joint Note.

$350. Denver, Col., Jan. 2, 1881.

Twelve months after date, we, or either of us, promise to pay Frank R. Harris, or bearer, Three Hundred and Fifty Dollars, for value received, with interest from date.

James West.
Daniel Tate.

Remarks.—1. This note is called a "joint note," because James West and Daniel Tate are *jointly liable* for its payment.

2. If the note read, "We jointly and severally promise to pay, etc., it would then be called a *joint and several note*, and the makers would be both jointly and singly liable for its payment.

3. Principal and Surety Note.

$125.00 St. Louis, Mo., Nov. 20, 1880.

Ninety days after date, I promise to pay James Miller, or order, One Hundred and Twenty-five Dollars, for value received, negotiable and payable without defalcation or discount.

Surety, James Miller. H. A. White.

Remark.—The maker of this note, H. A. White, is the *principal*, and the note is made payable to the order of the *surety*, James Miller, who becomes security for its payment, indorsing it on the back to the order of Mr. White's creditor.

8. A **Demand Note** is one in which no time is specified, and the payment must be made whenever demanded by the holder. The following is an example:

DEMAND NOTE.

$1100.00 CINCINNATI, O., Mar. 18, 1881.

For value received, I promise to pay David Swinton, or order, on demand, Eleven Hundred Dollars, with interest.

HENRY RUDOLPH.

9. A promissory note is **negotiable**—that is, transferable to another party by indorsement,—when the words "or order," or "or bearer," follow the name of the payee, as in the above examples; otherwise, the note is not negotiable.

10. If the note is payable to "bearer," no indorsement is required on transferring it to another party, and the maker only is responsible for its payment.

11. If the note is payable to "order," the payee and each holder in his turn must, on transferring it, indorse the note by writing his name on its back, thus becoming liable for its payment, in case the maker fails to meet it when due.

REMARKS.—1. An indorser may free himself from responsibility for payment if he accompany his signature with the words, "without recourse;" in which case his indorsement simply signifies the transfer of ownership.

2. If the indorser simply writes his name on the back of the note, it is called a "blank" indorsement; but if he precedes his signature by the words, "Pay to the order of John Smith," for example, it is called a "special" indorsement, and John Smith must indorse the note before it can legally pass to a third holder.

12. It is essential to a **valid** promissory note, that it contain the words "value received," and that the sum of money to be paid should be written in words.

13. If a note contains the words "with interest," it draws interest from date, and, if no rate is mentioned, the legal rate prevails.

REMARKS.—1. The face of such a note is the sum mentioned with its interest from date to the day of payment.

2. If a note does not contain the words "with interest," and is not paid when due, it draws interest from the day of maturity, at the legal rate, till paid.

14. A note **matures** on the day that it is legally due. However, in many of the states, three days, called *Days of Grace*, are allowed for the payment of a note after it is nominally due.

REMARKS.—1. The day before "grace" begins, the note is *nominally* due; it is *legally* due on the last day of grace.

2. The days of grace are rejected by writing "without grace" in the note.

3. Notes falling due on Sunday or on a legal holiday, are due the day before or the day after, according to the special statute of each state or territory.

4. If a note is payable "on demand," it is legally due when presented.

15. When a note in bank is not paid at maturity, it **goes to protest**—that is, a written notice of this fact, made out in legal form by a notary public, is served on the indorsers, or security.

REMARK.—Some of the states require that certain words shall be inserted in every negotiable note in addition to the usual forms. For instance, in Pennsylvania, the words "without defalcation" are required; in New Jersey, "without defalcation or discount;" and in Missouri, the statute requires the insertion of "negotiable and payable without defalcation or discount." In Indiana, in order to *evade* certain provisions of the law, the following words are usually inserted, "without any relief from valuation or appraisement laws." This constitutes what is known as the "iron-clad" note.

303. If a note be payable a certain time "after date," proceed thus to find the day it is legally due:

Rule.—*Add to the date of the note, the number of years and months to elapse before payment; if this gives the day of a month higher than that month contains, take the last day in that month; then, count the number of days mentioned in the note and 3 more: this will give the day the note is legally due; but if it is a Sunday or a national holiday, it must be paid the day previous.*

REMARKS.—1. When counting the days, do not reckon the one *from* which the counting begins. (For exceptions, see Art. **313 16, Rem.2.**)

2. The months mentioned in a note are calendar months. Hence, a 3 mo. note would run longer at one time than at another; one dated Jan. 1st, will run 93 days; one dated Oct. 1st, will run 95 days: to avoid this irregularity, the time of short notes is generally given in days instead of months; as, 30, 60, 90 days, etc.

3. When the time is given in days, it is convenient to use the following table in determining the *day of maturity* of a note:

TABLE

Showing the Number of Days from any Day of:

Jan.	Feb.	Mar.	Apr.	May.	June.	July.	Aug.	Sept.	Oct.	Nov.	Dec.	To the same day of next
365	334	306	275	245	214	184	153	122	92	61	31	Jan.
31	365	337	306	276	245	215	184	153	123	92	62	Feb.
59	28	365	334	304	273	243	212	181	151	120	90	Mar.
90	59	31	365	335	304	274	243	212	182	151	121	April.
120	89	61	30	365	334	304	273	242	212	181	151	May.
151	120	92	61	31	365	335	304	273	243	212	182	June.
181	150	122	91	61	30	365	334	303	273	242	212	July.
212	181	153	122	92	61	31	365	334	304	273	243	Aug.
243	212	184	153	123	92	62	31	365	335	304	274	Sept.
273	242	214	183	153	122	92	61	30	365	334	304	Oct.
304	273	245	214	184	153	123	92	61	31	365	335	Nov.
334	303	275	244	214	183	153	122	91	61	30	365	Dec.

REMARK.—In leap years, if the last day of February be included in the time, 1 day must be added to the number obtained from the table

EXAMPLES FOR PRACTICE.

Find the maturity and amount of each of the following notes:

$560.60 TACOMA, WASH., June 3, 1872.

1. For value received, sixty days after date, I promise to pay Madison Wilcox five hundred and sixty $\frac{60}{100}$ dollars, with interest at 7%. JAMES DAILY.

$430.00 PRESCOTT, ARIZONA, Jan. 30, 1879.

2. Six months after date, I promise to pay Christine Ladd four hundred and thirty dollars, for value received, with interest at 12% per annum. AMOS LYLE.

$4650.80 RENO, NEV., August 10, 1875.

3. For value received, three months after date, I promise to pay Oliver Davis, or order, four thousand six hundred and fifty $\frac{80}{100}$ dollars, with interest at the rate of 10% per annum, negotiable and payable without defalcation or discount. MILTON MOORE.

Surety, OLIVER DAVIS.

ANNUAL INTEREST.

304. Annual Interest is interest on the **principal, and** on each annual interest after it is due.

Annual interest is legal in some of the states.

EXPLANATION.—If Charles Heydon gives a note and agrees to pay the interest annually, but fails to do so, and lets the note run for three years before settlement, the first year's interest would draw interest for two years, and the second year's interest would draw interest for one year. The last year's interest would be paid at the time of settlement, or as it fell due.

PROBLEM.—Find the amount of $10500 for 4 yr. 6 mo., interest 6%, payable annually.

OPERATION.

Int. of $10500 for $4\frac{1}{2}$ yr., at 6%	=$2835;
" " $630 " 8 " " "	=$ 302.40
Total annual interest,	=$3137.40

Amount=$10500+$3137.40=$13637.40

SOLUTION.—The interest of $10500 for 1 yr., at 6%, is $630, and for $4\frac{1}{2}$ yr. is $2835. The first *annual* interest draws interest $3\frac{1}{2}$ yr., the second $2\frac{1}{2}$, the third $1\frac{1}{2}$, and the fourth $\frac{1}{2}$ yr. This is the same as one *annual* interest for ($3\frac{1}{2} + 2\frac{1}{2} + 1\frac{1}{2} + \frac{1}{2} = 8$) eight years. But the interest of $630, for 8 yr., at 6%, is $302.40; therefore, the amount = $10500 + $2835 + $302.40 = $13637.40

Rule.—1. *Find the interest on the principal for the entire time, and on each year's interest till the time of settlement.*

2. *The sum of these interests is the interest required.*

NOTE.—In Ohio and several other states, interest on *unpaid annual interest* is calculated at the *legal* rate only.

EXAMPLES FOR PRACTICE.

1. Find the amount of a note for $1500, interest 6%, payable annually, given Sept. 3, 1870, and not paid till March 1, 1874.

2. A gentleman holds six $1000 railroad bonds, due in 3 years, interest 6%, payable semi-annually: no interest having been paid, what amount is owing him when the bonds mature?

3. $2500.00 ST. PAUL, MINN., Jan. 11, 1869.

For value received, I promise to pay Morgan Stuart, or order, twenty-five hundred dollars, with interest at 7%, payable annually. LEONARD DOUGLAS.

What was due on this note March 17, 1873, if the interest was paid the first two years?

II. PARTIAL PAYMENTS.

305. A **Partial Payment** is a *payment in part* of a note or other obligation.

Whenever a partial payment is made, the holder should write the date and amount of the payment on the back of the note. This is called an **Indorsement.**

306. The following decision by Chancellor Kent, "Johnson's Chancery Rep., Vol. I, p. 17," has been adopted by the Supreme Court of the United States, and, with few exceptions, by the several states of the Union, and is known as the "United States Rule:"

U. S. Rule.—1. "*The rule for casting interest when partial payments have been made, is to apply the payment, in the first place, to the discharge of the interest due.*

2. "*If the payment exceeds the interest, the surplus goes towards discharging the principal, and the subsequent interest is to be computed on the balance of principal remaining due.*

3. "*If the payment be less than the interest, the surplus of interest must not be taken to augment the principal; but interest continues on the former principal until the period when the payments, taken together, exceed the interest due, and then the surplus is to be applied towards discharging the principal, and interest is to be computed on the balance as aforesaid.*

307. This rule is based upon the following principles:

PRINCIPLES.—1. *Payments must be applied first to the discharge of interest due, and the balance, if any, toward paying the principal.*

2. *Interest must, in no case, become part of the principal.*

3. *Interest or payment must not draw interest.*

REMARKS.—1. It is worthy of remark, that the whole *aim* and *tenor* of legislative enactments and judicial decisions on questions of interest, have been to favor the debtor, by disallowing compound interest, and yet *this very rule* fails to secure the end in view, and really maintains and enforces the principle of compound interest in a most objectionable shape; *for it makes interest due* (not every year as compound interest ordinarily does), *but as often as a payment is made;* by which it happens that *the closer the payments are together, the greater the loss of the debtor,* who thus suffers a penalty for his very promptness.

To illustrate, suppose the note to be for $2000, drawing interest at 6%, and the debtor pays every month $10, which just meets the interest then due; at the end of the year he would still owe $2000. But if he had invested the $10 each month, at 6%, he would have had, at the end of the year, $123.30 available for payment, while the debt would have increased only $120, being a difference of $3.30 in his favor, and leaving his debt $1996.70, instead of $2000.

2. To find the difference of time between two dates on the note, reckon by years and months as far as possible, and then count the days.

Problem.

$850. CINCINNATI, April 29, 1880.

Ninety days after date, I promise to pay Stephen Ludlow, or order, eight hundred and fifty dollars, with interest; value received. CHARLES K. TAYLOR.

Indorsements.—Oct. 13, 1880, $40; June 9, 1881, $32; Aug. 21, 1881, $125; Dec. 1, 1881, $10; March 16, 1882, $80.

What was due Nov. 11, 1882?

SOLUTION.—Interest on principal ($850) from April 29 to Oct. 13, 1880, 5 mo. 14 da., at 6% per annum,	$23.233
	850.
Whole sum due Oct. 13, 1880, ...	873.233
Payment to be deducted, ...	40.
Balance due Oct. 13, 1880, ...	833.233
Interest on balance ($833.233) from Oct. 13, 1880, to June 9, 1881, being 7 mo. 27 da., ...	32.913
Payment not enough to meet the interest, ...	32.
Surplus interest not paid June 9, 1881, ...	.913
Interest on former principal ($833.233) from June 9, 1881, to Aug. 21, 1881, being 2 mo. 12 da., . . .	9.999
Whole interest due Aug. 21, 1881, ...	**10.912**

(Brought forward) Int. due Aug. 21, 1881, . .	10.912
	833.233
Whole sum due Aug. 21, 1881, ...	844.145
Payment to be deducted, ...	125.
Balance due Aug. 21, 1881, ...	719.145
Interest on the above balance ($719.145) from Aug. 21, 1881, to Dec. 1, 1881, being 3 mo. 10 da., . . .	11.986
Payment not enough to meet the interest, ...	10.
Surplus interest not paid Dec. 1, 1881, ...	1.986
Interest on former principal ($719.145) from Dec. 1, 1881, to March 16, 1882, being 3 mo. 15 da., . .	12.585
Whole interest due March 16, 1882, ...	14.571
	719.145
Whole sum due March 16, 1882, ...	733.716
Payment to be deducted, ...	80.
Balance due March 16, 1882, ...	653.716
Interest on balance ($653.716) from March 16, 1882, to Nov. 11, 1882, being 7 mo. 26 da., ...	25.713
Balance due on settlement, Nov. 11, 1882, . . .	$679.43

Examples for Practice.

1. 304\frac{75}{100}$. Little Rock, Ark., March 10, 1882.

For value received, six months after date, I promise to pay G. Riley, or order, three hundred and four $\frac{75}{100}$ dollars.

No payments. H. McMakin.

What was due Nov. 3. 1883?

2. 429\frac{30}{100}$. Indianapolis, April 13, 1873.

On demand, I promise to pay W. Morgan, or order, four hundred and twenty-nine $\frac{30}{100}$ dollars, value received, with interest. R. Wilson.

Indorsed: Oct. 2, 1873, $10; Dec. 8, 1873, $60; July 17, 1874, $200.

What was due Jan. 1, 1875?

3. \$1750. CINCINNATI, Nov. 22, 1872.

For value received, two years after date, I promise to pay to the order of Spencer & Ward, seventeen hundred and fifty dollars, with interest at 7 per cent.

JACOB WINSTON.

Indorsed: Nov. 25, 1874, \$500; July 18, 1875, \$50; Sept. 1, 1875, \$600; Dec. 28, 1875, \$75.

What was due Feb. 10, 1876?

4. A note of \$312 given April 1, 1872, 8% from date, was settled July 1, 1874, the exact sum due being \$304.98.

Indorsed. April 1, 1873, \$30.96; Oct. 1, 1873, \$——; April 1, 1874, \$20.40

Restore the lost figures of the second payment.

5. There have been two equal annual payments on a 6% note for \$175, given two years ago this day. The balance is \$154.40: what was each payment?

6. A merchant gives his note, 10% from date, for \$2442.04: what sum paid annually will have discharged the whole at the end of 5 years?

308. In Connecticut and Vermont the laws provide the following special rules relative to partial payments:

Connecticut Rule.—1. *Compute the interest to the time of the first payment, if that be one year or more from the time that interest commenced; add it to the principal, and deduct the payment from the sum total.*

2. *If there be after payments made, compute the interest on the balance due to the next payment, and then deduct the payment as above; and in like manner from one payment to another, till all the payments are absorbed:* PROVIDED, *the time between one payment and another be one year or more.*

3. BUT *if any payment be made before one year's interest hath accrued, then compute the interest on the sum then due on the obligation, for one year, add it to the principal, and compute the interest on the sum paid, from the time it was*

paid, up to the end of the year; add it to the sum paid, and deduct that sum from the principal and interest added as above. (See Note.)

4. *If any payment be made of a less sum than the interest arisen at the time of such payment, no interest is to be computed, but only on the principal sum, for any period.*

NOTE.—*If a year does not extend beyond the time of payment; but if it does, then find the amount of the principal remaining unpaid up to the time of settlement, likewise the amount of the payment or payments from the time they were paid to the time of settlement, and deduct the sum of these several amounts from the amount of the principal.*

What is due on the 2d and 3d of the preceding notes, by the Connecticut rule?

Vermont Rule.—1. *On all notes, etc., payable* WITH INTEREST, *partial payments are applied, first, to liquidate the interest that has accrued at the time of such payments; and, secondly, to the extinguishment of the principal.*

2. *When notes, etc., are drawn with* INTEREST PAYABLE ANNUALLY, *the annual interests that remain unpaid are subject to simple interest from the time they become due to the time of final settlement.*

3. PARTIAL PAYMENTS, *made after annual interest begins to accrue,* ALSO DRAW SIMPLE INTEREST *from the time such payments are made to the end of the year; and their amounts are then applied, first, to liquidate the simple interest that has accrued from the unpaid annual interests; secondly, to liquidate the annual interests that have become due; and, thirdly, to the extinguishment of the principal.*

$1480. WOODSTOCK, VT., April 12, 1879.

For value received, I promise to pay John Jay, or order, fourteen hundred and eighty dollars, with interest annually.

JAMES BROWN.

Indorsed: July 25, 1879, $40; May 20, 1880, $50; June 3, 1881, $350.

What was due April 12, 1882?

309. Business men generally settle notes and accounts payable within a year, by the Mercantile Rule.

Mercantile Rule.—1. *Find the amount of the principal from the date of the note to the date of settlement.*

2. *Find the amount of each payment from its date to the date of settlement.*

3. *From the amount of the principal subtract the sum of the amounts of the payments.*

NOTE.—In applying this rule, the time should be found for the exact number of days.

EXAMPLES FOR PRACTICE.

1. $950.00 NEW ORLEANS, LA., Jan. 25, 1876.

For value received, nine months after date, I promise to pay Peter Finley nine hundred and fifty dollars, with 7% accurate interest. THOMAS HUNTER.

The following payments were made on this note: March 2, 1876, $225; May 5, 1876, $174.19; June 29, 1876, $187.50; Aug. 1, 1876, $79.15

Required—the balance at settlement.

2. A note for $600 was given June 12, 1865, 6% interest, and the following indorsements were made: Aug. 12, 1865, $100; Nov. 12, 1865, $250; Jan. 12, 1866, $120: what was due Feb. 12, 1866, counting by months instead of days?

III. TRUE DISCOUNT.

DEFINITIONS.

310. 1. **Discount** on a debt payable by agreement at some future time, is a deduction made for "cash," or present payment.

It arises from the consideration of the *present worth* of the debt.

2. **Present Worth** is that sum of money which, at a given rate of interest, will amount to the same as the debt at its maturity.

3. **True Discount,** then, is the difference between the present worth and the whole debt.

REMARKS.—1. That is, it is the simple interest on the *present worth*, from the day of discount until the day of maturity.

2. Discount on a debt must be carefully distinguished from *Commercial* Discount, which is simply a deduction from the regular price or value of an article; the latter is usually expressed as such a "per cent off."

311. The different cases of true discount may be solved like those of simple interest: the present worth corresponding to the *Principal;* the discount, to the *Interest;* and the face of the debt, to the *Amount.* The following case is the only one much used:

312. Given the face, time, and rate, to find the present worth and true discount.

PROBLEM.—Find the present worth and discount of \$5101.75, due in 1 yr. 9 mo. 19 da., at 6%.

OPERATION.

Amount of \$1 for 1 yr. 9 mo. 19 da., at 6% = \$1.108$\frac{1}{6}$,
and \$5101.75 ÷ \$1.108$\frac{1}{6}$ = 4603.775;
\$1 × 4603.775 = \$4603.775, present worth;
\$5101.75 − \$4603.775 = \$497.975, true discount.

SOLUTION.—The amount of \$1 for 1 yr. 9 mo. 19 da., at 6%, is \$1.108$\frac{1}{6}$, and \$5101.75 is the amount of as many dollars as \$1.108$\frac{1}{6}$ is contained times in \$5101.75, which is 4603.775 times; therefore, the present worth is \$4603.775, and the true discount is \$497.975

Rule.—1. *Divide the debt by the amount of* \$1 *for the given time and rate; the quotient is the present worth.*

2. *The difference between the debt and the present worth is the true discount.*

Examples for Practice.

1. Find the true discount on a debt for $5034.15 due in 3 yr. 5 mo. 20 da., without grace, at 7%.

2. What is the present worth of a note for $2500, bearing 6% interest, and payable in 2 yr. 6 mo. 15 da., discounted at 8%.

IV. BANK DISCOUNT.

DEFINITIONS.

313. 1. A **Bank** is an institution anthorized by law to deal in money.

2. The **three chief provinces** of banks are: the receiving of deposits, the loaning of money, and the issuing of bank-bills. Any or all of these provinces may be exercised by the same bank.

REMARKS.—1. A *Bank of Deposit* is one which takes charge of money, stocks, etc., belonging to its customers.

Money so intrusted is called a *deposit,* and the customers are known as *depositors.*

2. A *Bank of Discount* is one which loans money by discounting notes, drafts, etc.

3. A *Bank of Issue* is one which issues notes, called "bank-bills," that circulate as money.

3. The banks of the United States may be divided into two general classes,—**National Banks** and **Private Banks.**

REMARKS.—1. National Banks are organized under special legislation of Congress. They must conform to certain restrictions as to the number of stockholders, amount of capital stock, reserve, circulation, etc. In return, they have privileges which give them certain advantages over Private Banks: they are banks of issue, of discount, and of deposit.

2. Private Banks are unable to issue their own bank-notes to advantage, owing to the heavy tax on their circulation; they are, therefore, confined to the provinces of deposit and discount.

4. A **Check** is a written order on a bank by a depositor for money. The following is a usual form:

> No. 1032. CINCINNATI, *Nov.* 27, 1881.
>
> BANK OF CINCINNATI,
>
> PAY TO ______ *Rufus B. Shaffer,* __________ OR ORDER
>
> *Four Hundred and Twenty-five* ________ $\frac{\times}{100}$ DOLLARS.
>
> \$425$\frac{\times}{}$ *George Potter Hollister.*

REMARKS.—1. Before this check is paid, it must be indorsed by Rufus B. Shaffer. He may either draw the money from the bank itself, or he may transfer the check to another party, who must also indorse it, as in the case of a promissory note. (See Art. **302**, 11.)

2. If the words *or Bearer* are used in place of *or Order,* no indorsement is necessary, and any one holding the check may draw the money for it.

5. A **Draft** is a written order of one person or company upon another for the payment of money. The following is a customary form:

> \$1453$\frac{\times}{}$. NEW ORLEANS, LA., *July* 25, 1880.
>
> ________________ *At sight* __________ PAY TO THE
>
> ORDER OF ________ *First National Bank* ______________
>
> *Fourteen Hundred Fifty-three and* ________ $\frac{\times}{100}$ DOLLARS
>
> VALUE RECEIVED AND CHARGE TO ACCOUNT OF
>
> TO *John R. Williams & Co.,*
No. 2136. *Memphis, Tenn.* } *Robert James.*

6. Drafts may be divided into two classes, *Sight Drafts* and *Time Drafts.*

7. A **Sight Draft** is one payable "at sight." (See the form above.)

8. A **Time Draft** is payable a specified time "after sight," or "after date."

9. The signer of the draft is the **maker** or **drawer.**

10. The one to whom the draft is addressed, and who is requested to pay it, is the **drawee.**

11. The one to whom the money is ordered to be paid, is the **payee.**

12. The one who has possession of the draft, is called the **owner** or **holder**; when he sells it, and becomes an **indorser,** he is liable for its payment.

13. The **Indorsement** of a draft is the writing upon the back of it, by which the payee transfers the payment to another.

REMARKS.—1. A *special indorsement* is an order to pay the draft to a particular person named, who is called the *indorsee,* as "Pay to F. H. Lee.—W. Harris," and no one but the *indorsee* can collect the bill. Drafts are usually collected through banks.

2. When the *indorsement* is *in blank,* the payee merely writes his name on the back, and any one who has lawful possession of the draft can collect it.

3. If the drawee promises to pay a draft at maturity, he writes across the face the word "Accepted," with the date, and signs his name, thus: "Accepted, July 11, 1881.—H. Morton." The *acceptor* is first responsible for payment, and the draft is called an *acceptance.*

4. A draft, like a promissory note, may be payable "to order," or "bearer," and is subject to protest in case the payment or acceptance is refused.

5. A *time* draft is universally entitled to the three days grace: but, in about half of the states, no grace is allowed on *sight* drafts.

14. When a bank loans money, it discounts the time notes,

drafts, etc., offered by the borrower at a *rate of discount* agreed upon, but not in accordance with the principles of true discount.

15. **Bank Discount** is simple interest on the face of a note, calculated from the day of discount to the day of maturity, and paid in advance.

16. The **Proceeds** of a note is the amount which remains after deducting the discount from the face.

REMARKS.—1. Since the face of every note is a debt due at a future time, its cost ought to be the present worth of that debt, and the bank discount should be the same as the true discount. As it is, the former is greater than the latter; for, *bank discount is interest on the face of the note, while true discount is the interest on the present worth, which is always less than the face.* Hence, their difference is the interest of the difference between the present worth and face; that is, *the interest of the true discount.* (See Art. **310**, 3.)

2. In discounting notes, the three days of grace are usually taken into account; and in Delaware, the District of Columbia, Maryland, Missouri, and Pennsylvania, the *day of discount* and the *day of maturity* are both included in the time. But Alas., Cal., Colo., Conn., D. C., Fla., Ida., Ill., Me., Md., Mass., Mont., N. H., N. J., N. Y., N. D., O., Ore., Pa., Utah, Vt. and Wis. have abolished days of grace.

3. If the borrower is not the maker of the notes or drafts, he must be the last indorser, or holder of them.

314. Problems in bank discount are solved like those of simple interest. The face of the note corresponds to the *Principal;* the bank discount, to the *Interest;* the proceeds, to the *Principal less the Interest;* and the term of discount, to the *Time.*

CASE I.

315. Given the face of the note, the rate, and time, to find the discount and proceeds.

PROBLEM.—What is the bank discount of $770 for 90 days, at 6%?

OPERATION.

Int. of \$1 for 93 da., at 6% = \$.0155 = Rate of Discount.
\$770 × .0155 = \$11.935, Discount.
\$770 − \$11.935 = \$758.065, Proceeds.

SOLUTION.—Find the interest of \$770 for 93 da., at 6%; this is \$11.935, which is the bank discount. The proceeds is the difference between \$770 and \$11.935, or \$758.065

PRINCIPLES.—1. *The interest on the sum discounted for the given time (plus the three days of grace) at the given rate per cent, is the bank discount.*

2. *The proceeds is equal to the sum discounted minus the discount.*

Rule.—1. *Find the interest on the sum discounted for three days more than the given time, at the given rate; it is the discount.*

2. *Subtract the discount from the sum discounted, and the remainder is the proceeds.*

REMARKS.—1. As with promissory notes, the three days of grace are not counted on a draft bearing the words "without grace."

2. The following problems present, each, two dates,—one showing when the note is nominally due, and the other when it is legally due. Thus, in an example which reads, "Due May 3/6," the first is the nominal, and the second the legal date.

EXAMPLES FOR PRACTICE.

Find the day of maturity, the time to run, and the proceeds of the following notes:

1. \$792.50 TOPEKA, KAN., Jan. 3, 1870.

Six months after date, I promise to pay to the order of Jones Brothers seven hundred and ninety-two $\frac{50}{100}$ dollars, at the First National Bank, value received.

Discounted Feb. 18, at 6%. ALBERT L. TODD.

2. $\$1962\frac{45}{100}$ RENO, NEV., July 26, 1879.

Value received, four months after date, I promise to pay B. Thoms, or order, one thousand nine hundred sixty-two $\frac{45}{100}$ dollars, at the Chemical National Bank.

E. WILLIAMS.

Discounted Aug. 26, at 7%.

3. $\$2672\frac{18}{100}$. WILMINGTON, DEL., March 10, 1872.

Nine months after date, for value received, I promise to pay Edward H. King, or order, two thousand six hundred seventy-two $\frac{18}{100}$ dollars, without defalcation.

JEREMIAH BARTON.

Discounted July 19, at 6%.

4. $3886. CHEYENNE, WYO., Jan. 31, 1879.

One month after date, we jointly and severally promise to pay C. McKnight, or order, three thousand eight hundred eighty-six dollars, value received, negotiable and payable, and without defalcation or discount.

T. MONROE,
I. FOSTER.

Discounted Jan. 31, at 1½% a month.

REMARK.—A note, drawn by two or more persons, "jointly and severally," may be collected of either of the makers, but if the words "jointly and severally" are not used, it can only be collected of the makers as a firm or company.

5. $2850. AUSTIN, TEX., April 11, 1879.

For value received, eight months after date, we promise to pay Henry Hopper, or order, twenty-eight hundred and fifty dollars, with interest from date, at six per cent per annum.

HANNA & TUTTLE.

Discounted June 15, at 6%.

6. $\$737\frac{40}{100}$. SEATTLE, WASH., Feb. 14, 1880.

Value received, two months after date, I promise to pay to J. K. Eaton, or order, seven hundred thirty-seven $\frac{40}{100}$ dollars, at the Suffolk National Bank.

WILLIAM ALLEN.

Discounted Feb. 23, at 10%.

7. $\$4085\frac{20}{100}$. NEW ORLEANS, Nov. 20, 1875.

Value received, six months after date, I promise to pay John A. Westcott, or order, four thousand eighty-five $\frac{20}{100}$ dollars, at the Planters' National Bank.

E. WATERMAN.

Discounted Dec. 31, 1875, at 5%.

CASE II.

316. Given the proceeds, time, and the rate of discount, to find the face of the note.

PROBLEM.—For what sum must a 60 da. note be drawn, to yield $1000, when discounted, at 6% per annum?

OPERATION.

The bank discount of $1 for 63 da., at 6% = $.0105

Proceeds of $1 = $.9895; $1000 ÷ .9895 = $1010.61, the face of the note.

SOLUTION.—For every $1 in the face of the note, the proceeds, by Case I, is $.9895; hence, there must be as many dollars in the face as this sum, $.9895, is contained times in the given proceeds, $1000; this gives $1010.61 for the face of the note.

Rule.—*Divide the given proceeds by the proceeds of* $1 *for the given time and rate; or, divide the given discount by the discount of* $1 *for the given time and rate; the quotient is equal to the face of the note.*

EXAMPLES FOR PRACTICE.

1. Find the face of a 30 da. note, which yields \$1650, when discounted at $1\frac{1}{2}\%$ a mo.

2. The face of a 60 da. note, which, discounted at 6% per annum, will yield \$800.

3. The face of a 90 da. note, bought for \$22.75 less than its face, discounted at 7%.

4. The face of a 4 mo. note, which, discounted at 1% a month, yields \$3375.

5. The face of a 6 mo. note, which, discounted at 10% per annum, yields \$4850.

6. The face of a 60 da. note, discounted at 2% a month, to pay \$768.25

7. The face of a 40 da. note, which, discounted at 8%, yields \$2072.60

8. The face of a 30 da. and 90 da. note, to net \$1000 when discounted at 6%.

CASE III.

317. Given the rate of bank discount, to find the corresponding rate of interest.

PROBLEM.—What is the rate of interest when a 60 da. note is discounted at 2% a month?

OPERATION.

Assume \$100 as the face of the note.
Then, 63 days = time.
\$4.20 = discount.
\$95.80 = proceeds.
\$.00175 = interest of \$1, at 1%, for the given time.
And $\$95.80 \times .00175 = \$.16765$, interest of proceeds, at 1%, for the given time; then,

$$\$4.20 \div \$.16765 = 25\tfrac{25}{479}\%, \text{ rate.}$$

SOLUTION.—The discount of $100 for 63 days, at 2% a month, is $4.20, and the proceeds is $95.80 The interest of $1 for the given time, at 1%, is $.00175, and of $95.80 is $.16765 To find the rate, we proceed thus: $4.20 ÷ $.16765 = $25\frac{25}{479}$.

Rule.—1. *Find the discount and proceeds of* $100 *or* $1 *for the time the note runs.*

2. *Divide the discount by the interest of the proceeds, at* 1% *for the same time; the quotient is the rate.*

EXAMPLES FOR PRACTICE.

What is the rate of interest:

1. When a 30 da. note is discounted at 1, $1\frac{1}{4}$, $1\frac{1}{2}$, 2% a month?

2. When a 60 da. note is discounted at 6, 8, 10% per annum?

3. When a 90 da. note is discounted at 2, $2\frac{1}{2}$, 3% a month?

4. When a note running 1 yr. without grace, is discounted at 5, 6, 7, 8, 9, 10, 12%?

5. My note, which will be legally due in 1 yr. 4 mo. 20 da., is discounted by a banker, at 8%: what rate of interest does he receive?

REMARK.—It may seem unnecessary to regard the time the note has to run, in determining the rate of interest; but, a comparison of examples 1 and 3, shows that a 90 da. note, discounted at 2% a month, yields a higher rate of *interest* than a 30 da. note of the same face, discounted at 2% a month. The discount, at the same rate, on all notes of the same face, *varies* as the time to run, and if in each case, it was referred to the *same principal*, the rate of interest would be the same; but when the discount becomes *larger*, the proceeds or principal to which it is referred, becomes *smaller*, and therefore *the rate of interest corresponding to any rate of discount increases with the time the note has to run.* Hence, the profit of the discounter is greater proportionally on *long* notes than on *short* ones, at the same rate.

CASE IV.

318. Given the rate of interest, to find the corresponding rate of discount.

PROBLEM.—What is the rate of discount when a 60 day note yields 2% interest a month?

OPERATION.

Assume $100 as the proceeds.
Then, 63 days = the time;
$4.20 = the interest;
and $104.20 = the amount or face of note;
also, $.18235 = interest of amount, (discount) for the given time at 1% per annum.
$4.20 ÷ $.18235 = $23\frac{17}{521}$%, rate (per annum).

REMARK.—Note the difference between this solution and that under the preceding case. In that, the interest was found on the proceeds; in this, the discount is computed on the amount.

Rule.—1. *Find the interest and the amount of* $100 *or* $1 *for the given time.*

2. *Divide the interest by the interest of the amount at* 1% *for the given time; the quotient is the corresponding rate of discount.*

EXAMPLES FOR PRACTICE.

What rates of discount:

1. On 30 da. notes, yield 10, 15, 20% interest?

2. On 60 da. notes, yield 6, 8, 10% interest?

3. On 90 da. notes, yield 1, 2, 4% a month interest?

4. On notes running 1 yr. without grace, yield 5, 6, 7, 8, 9, 10% interest?

V. EXCHANGE.

DEFINITIONS.

319. 1. **Exchange** is the method of paying a debt in a distant place by the transfer of a credit.

REMARK.—This method of transacting business is adopted as a convenience: it avoids the danger and expense of sending the money itself.

2. The payment is effected by means of a *Bill of Exchange.*

3. A **Bill of Exchange** is a written order on a person or company in a distant place for the payment of money.

REMARK.—The term includes both drafts and checks.

4. Exchange is of two kinds: *Domestic*, or *Inland*, and *Foreign.*

5. **Domestic Exchange** treats of drafts payable in the country where they are made. (See page 269 for form.)

6. **Foreign Exchange** treats of drafts made in one country and payable in another.

7. A foreign bill of exchange is usually drawn in triplicate, called a **Set of Exchange**; the different copies, termed respectively the *first*, *second*, and *third* of exchange, are then sent by different mails, that miscarriage or delay may be avoided. When one is paid, the others are void. The following is a common form:

FOREIGN BILL OF EXCHANGE.

1. Exchange for NEW YORK, July 2, 1878.

£1000. *Thirty days after sight of this First of Exchange (Second and Third of same tenor and date unpaid), pay to the order of James S. Rollins One Thousand Pounds Sterling, value received, and charge to account of*

To JOHN BROWN & CO., J. S. CHICK.

No. 1250. LIVERPOOL, ENGLAND.

8. The **Rate of Exchange** is a rate per cent of the face of the draft.

9. The **Course of Exchange** is the current price paid in one place for bills of exchange on another.

10. The **Par of Exchange** is the estimated value of the monetary unit of one country compared with that of another country, and is either *Intrinsic* or *Commercial.*

11. **Intrinsic Par of Exchange** is based on the comparative weight and purity of coins.

12. **Commercial Par of Exchange** is based on commercial usage, or the price of coins in different countries.

DOMESTIC EXCHANGE.

320. Where time is involved, problems in Domestic Exchange are solved in accordance with the principles of Bank Discount.

Examples for Practice.

1. What is paid for $3805.40 sight exchange on Boston, at $\frac{1}{2}\%$ premium?
2. What is the cost of a check on St. Louis for $1505.40, at $\frac{1}{4}\%$ discount?
3. What must be the face of a sight draft to cost $2000, at $\frac{5}{8}\%$ premium?
4. What must be the face of a sight draft to cost $4681.25, at $1\frac{1}{4}\%$ discount?
5. What will a 30 day draft on New Orleans for $7216.85 cost, at $\frac{3}{8}\%$ discount, interest 6%?

OPERATION.

$\$1 - \$.00\frac{3}{8} = \$.99625$, rate of exchange.
$\$.0055 \quad =$ bank discount of $1 for 33 da.
$\$.99075 =$ cost of exchange for $1.
$\$7216.85 \times .99075 = \$7150.094+$, *Ans.*

SOLUTION.—From the rate of exchange subtract the bank discount of $1 for 33 days, at 6%; the remainder is the cost of exchange for $1. Then multiply the face of the draft by the cost of exchange for $1, which gives $7150.094, the cost of the draft.

6. What will a 60 day draft on Mobile for $12692.50 cost, at $\frac{3}{4}$% premium, interest 6%?

7. What must be the face of an 18 day draft, costing $5264.15, at $\frac{1}{2}$% premium, interest 6%?

8. What must be the face of a 21 day draft, costing $6836.75, at $\frac{7}{8}$% discount, interest 6%?

9. A commission merchant in St. Louis sold 5560 lb. of bacon, at 11$\frac{1}{2}$ ct. a lb.; his commission is 2$\frac{1}{2}$%, and the course of exchange 98$\frac{1}{2}$%: what is the amount of the draft that he remits to his consignor?

10. A grain merchant in Detroit sold 11875 bu. of corn, at 40 cents a bushel; deducting 3% as commission, he purchased a 60 day draft with the proceeds, at 2% premium; required the face of the draft?

11. Sold lumber to the amount of $20312.50, charging 1$\frac{1}{2}$% commission, and remitted the proceeds to my consignor by draft; required the face of the draft, exchange $\frac{1}{2}$% discount?

12. If a 45 day draft for $5500 costs $5538.50, find the rate of exchange.

OPERATION.

The bank discount of $5500 for 48 days, at 6% = $44.
Then $5538.50 + $44 — $5500 = $82.50 premium;

And $\frac{\$82.50}{\$5500} = .015 = 1\frac{1}{2}\%$, rate of exchange.

13. A father sent to his son, at school, a draft for $250, at 3 mo., interest 6%, paying $244.25 for it; find the rate of exchange.

14. An agent owing his principal $1011.84, bought a draft with this sum and remitted it; the principal received $992: find the rate of exchange.

FOREIGN EXCHANGE.

321. 1. In Foreign Exchange it is necessary to find the value of money in one country in terms of the monetary unit of another country. We reduce foreign coins by comparison with U. S. money, to find their value.

REMARK.—By an Act of Congress, March 3, 1873, the Director of the Mint is authorized to estimate *annually* the values of the standard coins, in circulation, of the various nations of the world. In compliance with this law, the Secretary of the Treasury issued the following estimate of values of foreign coins, January 1, 1884:

COUNTRY.	MONETARY UNIT.	VALUE IN U. S. MONEY.
Austria	Florin	$ .383
Bolivia	Boliviano	.795
Brazil	Milreis of 1000 reis	.546
Chili	Peso	.912
Cuba	Peso	.932
Denmark, Norway, Sweden	Crown	.268
Egypt	Piaster	.049
France, Belgium, Switz	Franc	.193
Great Britain	Pound sterling	4.866½
German Empire	Mark	.238
India	Rupee of 16 annas	.378
Japan	Yen (silver)	.858
Mexico	Dollar	.864
Netherlands	Florin	.402
Portugal	Milreis of 1000 reis	1.08
Russia	Rouble of 100 copecks.	.636
Tripoli	Mahbub of 20 piasters.	.717
Turkey	Piaster	.044

NOTE.—The *Drachma* of Greece, the *Lira* of Italy, the *Peseta* (100 centimes) of Spain, and the *Bolivar* of Venezuela, are of the same value as the Franc. The *Dollar*, of the same value as our own, is the standard in the British Possessions, N. A., Liberia, and the Sandwich Islands. The *Peso* of Ecuador and the United States of Colombia. and the *Sol* of Peru, are the same in value as the Boliviano.

2. Bills of Exchange on England, Ireland, and Scotland are bought and sold without reference to the *par of exchange.*

3. It is customary in foreign exchange to deal in drafts on the various commercial centers. London exchange is the most common, and is received in almost all parts of the civilized world.

REMARK.—London exchange is quoted in the newspapers in several grades: as, "Prime bankers' sterling bills," or bills upon banks of the highest standing; "good bankers," next in rank; "prime" and "good commercial," or bills on merchants, etc. The prices quoted depend upon the standing of the drawee and upon the demand for the several classes of bills. Usually a double quotation, one for 60 day bills, and the other for 3 day bills, is given for each class.

EXAMPLES FOR PRACTICE.

1. Find the cost of a bill of exchange on London, at 3 days sight, for £530 12s., exchange being quoted at $4.88

OPERATION.

£530 12s. = £530.6

$\$4.88 \times 530.6 = \$2589.328$

2. Find the cost of a bill of exchange on London, at sight, for £625 10s. 10d., when exchange is $4.87 pound sterling.

3. What is the cost of a bill on Paris for 1485 francs, $\$1 = 5.15$ francs.

4. What is the cost of a bill on Amsterdam for 4800 guilders, quoted at $41\frac{1}{2}$ cents, brokerage $\frac{1}{2}\%$?

OPERATION.

$\$.415 \times 4800 = \$1992.$

$\$1992 \times .00\frac{1}{2} = \$9.96$, brokerage.

$\$1992 + \$9.96 = \$2001.96$, cost of bill.

5. What will a draft in St. Petersburg on New York for $5000 cost, if a rouble be worth $.74?

6. What will a draft on Charleston for $4500 cost at Rio Janeiro (milreis = $.54), discount 2%?

7. A gentleman sold a 60 day draft, which was drawn on Amsterdam for 1000 guilders; he discounted it at 6%, and brokerage was $\frac{1}{8}$%: what did he get for it, a guilder being valued at $40\frac{1}{4}$ cents?

ARBITRATION OF EXCHANGE.

DEFINITIONS.

322. 1. Owing to the constant variation of exchange, it is sometimes advantageous to draw *through* an intermediate point, or points, in place of drawing directly. This is called **Circular Exchange.**

2. The process of *finding* the proportional exchange between two places by means of Circular Exchange, is called the **Arbitration of Exchange.**

3. **Simple Arbitration** is finding the proportional exchange when there is but one intermediate point.

4. **Compound Arbitration** is finding the proportional exchange through two or more intermediate points.

PROBLEM.—A merchant wishes to remit $2240 to Lisbon: is it more profitable to buy a bill directly on Lisbon, at 1 milreis = $1.10; or to remit through London and Paris, at £1 = $4.88 = 25.20 francs, to Lisbon, 1 milreis = 5.5 francs?

OPERATION.

$\$2240 \div 1.10 = 2036.364$ milreis, direct exchange.

( ) milreis = $2240.
$4.88 = 25.20 francs.
5.5 francs = 1 milreis.

$$\frac{2240 \times 25.20 \times 1}{4.88 \times 5.5} = 2103.129 \text{ milreis, circular exchange.}$$

Hence, the gain is 66.765 milreis in buying circular exchange.

SOLUTION.—By direct exchange $2240 will buy 2036.364 milreis. By circular exchange we have a set of equations, all of whose members are known except one. Multiplying the right-hand members together for a dividend, and dividing the product by the product of the left-hand members, the quotient is the required member which, in this case, is 2103.129 milreis. The difference in favor of the circular method is 66.765 milreis.

PROBLEM.—A merchant in Memphis wishes to remit $8400 to New York; exchange on Chicago is $1\frac{1}{2}\%$ premium, between Chicago and Detroit 1% discount, and between Detroit and New York $\frac{1}{2}\%$ discount: what is the value of the draft in New York, if sent through Chicago and Detroit?

OPERATION.

($) N. Y. = $8400, Memphis.
$$1.01\frac{1}{2}$ Mem. = $1, Chicago.
$.99 Chi. = $1, Detroit.
$$.99\frac{1}{2}$ Det. = $1, N. Y.

$$\frac{8400 \times 1 \times 1 \times 1}{1.015 \times .99 \times .995} = \$8401.46, \textit{ Ans.}$$

SOLUTION.—By the problem, $1.015 in Memphis = $1 in Chicago; $.99 in Chicago = $1 in Detroit; and $.995 in Detroit = $1 in New York. Hence, multiplying and dividing, as in the preceding problem, we obtain the answer, $8401.46

Rule 1.—1. *Form a series of equations, expressing the conditions of the question; the first containing the quantity given equal to an unknown number of the quantity required, and all arranged in such a way that the right-hand quantity of each equation and the left-hand quantity in the equation next following, shall be of the same denomination, and also the right-hand quantity of the last and the left-hand quantity of the first.*

2. *Cancel equal factors on opposite sides, and divide the product of the quantities in the column which is complete by the product of those in the other column. The quotient will be the quantity required.*

Rule 2.—*Reduce the given quantity to the denomination with which it is compared; reduce this result to the denomination with which it is compared; and so on, until the required denomination is reached, indicating the operations by writing as multipliers the proper unit values. The compound fraction thus obtained, reduced to its simplest form, will be the amount of the required denomination.*

NOTES.—1. The first is sometimes called the *chain* rule, because each equation and the one following, as well as the last and first, are connected as in a chain, having the right-hand quantity in one of the same denomination as the left-hand quantity of the next.

2. In applying this rule to exchange, if any currency is at a premium or a discount, its unit value in the currency with which it is compared, must be multiplied by such a mixed number, or decimal, as will increase or diminish it accordingly.

3. If commission or brokerage is charged at any point of the circuit, for effecting the exchange on the next point, the sum to be transmitted must be diminished accordingly, by multiplying it by the proper number of decimal hundredths.

EXAMPLES FOR PRACTICE.

1. A, of Galveston, has \$6000 to pay in New York. The direct exchange is $\frac{1}{2}\%$ premium; but exchange on New Orleans is $\frac{1}{4}\%$ premium, and from New Orleans to New York is $\frac{1}{4}\%$ discount. By circular exchange, how much will pay his debt, and what is his gain?

2. A merchant of St. Louis wishes to remit \$7165.80 to Baltimore. Exchange on Baltimore is $\frac{1}{4}\%$ premium; but, on New Orleans it is $\frac{1}{8}\%$ premium; from New Orleans to Havana, $\frac{1}{8}\%$ discount; from Havana to Baltimore, $\frac{1}{4}\%$ discount. What will be the value in Baltimore by each method, and how much better is the circular?

3. A Louisville merchant has \$10000 due him in Charleston. Exchange on Charleston is $\frac{1}{8}\%$ premium. Instead of drawing directly, he advises his debtor to remit to his agent in New York at $\frac{1}{5}\%$ premium, on whom he immediately draws at 12 da., and sells the bill at $\frac{3}{4}\%$ premium, interest off at 6%. What does he realize, and what gain over the direct exchange?

SUGGESTION.—Interest at 6% per annum is $\frac{1}{2}\%$ a month, or $\frac{1}{4}\%$ for 15 days, which diminishes the rate of premium to $\frac{1}{2}\%$.

4. A Cincinnati manufacturer receives, April 18th, an account of sales from New Orleans; net proceeds \$5284.67, due June $^4/_7$. He advises his agent to discount the debt at 6%, and invest the proceeds in a 7 day bill on Nashville, interest off at 6%, at $\frac{1}{2}\%$ discount, and remit it to Cincinnati. The agent does this, April 26. The bill reaches Cincinnati May 3, and is sold at $\frac{1}{4}\%$ premium. What is the proceeds, and how much greater than if a bill had been drawn May 3, on New Orleans, due June 7, sold at $\frac{1}{8}\%$ premium, and interest off at 6%?

5. A merchant in Louisville wishes to pay \$10000, which he owes in Berlin. He can buy a bill of exchange in Louisville on Berlin at the rate of \$.96 for 4 reichmarks; or he is offered a circular bill through London and Paris, brokerage at each place $\frac{1}{8}\%$, at the following rates: £1 = \$4.90 = 25.38 francs, and 5 francs = 4 reichmarks. What is the difference in the cost?

VI. EQUATION OF PAYMENTS.

DEFINITIONS.

323. 1. **Equation of Payments** is the process of finding *the time* when two or more debts, due at different times, may be paid without loss to either debtor or creditor.

2. The time of payment is called the **Equated Time.**

3. The **Term of Credit** is the time to elapse before a debt is due.

4. The **Average Term of Credit** is the time to elapse before several debts, due at different dates, may be paid at once.

5. To **Average an Account** is to find the equitable time of payment of the balance.

6. **Settling** or **Closing an Account** is finding how much is due between the parties at a particular time. It is sometimes called **Striking a Balance.**

7. A **Focal Date** is a date from which we begin to reckon in averaging an account.

324. Equation of Payments is based upon the following principles:

PRINCIPLES.—1. *That any sum of money paid before it is due, is balanced by keeping an equal sum of money for an equal time after it is due.*

2. *That the interest is the measure for the use of any sum of money for any time.*

REMARK.—The first statement is not strictly correct, although it has the sanction of able writers, and is very generally accepted. For short periods, it will make no appreciable difference.

PROBLEM.—A buys an invoice of $250 on 3 mo. credit; $800 on 2 mo.; and $1000 on 4 mo.; and gives his note for the whole amount: how long should the note run?

OPERATION.

Debts.		Terms.		Equivalents.
250	×	3	=	750
800	×	2	=	1600
1000	×	4	=	4000
2050				6350

$$\therefore \text{ time to run} = \frac{6350}{2050} \text{ mo.} = 3 \text{ mo. } 3 \text{ da., } \textit{Ans.}$$

SOLUTION.—The use of $1 for 6350 months is balanced by the use of $2050 for 3 months and 3 days. Or, the interest of $1 for 6350 months equals the interest of $2050, 3 months 3 days, at the same rate.

PROBLEM.—I buy of a wholesale dealer, at 3 mo. credit, as follows: Jan. 7, an invoice of $600; Jan. 11, $240; Jan. 13, $400; Jan. 20, $320; Jan. 28, $1200; I give a note at 3 mo. for the whole amount: when is it dated?

OPERATION.

$600 × 0	= $1	for	0 days
$240 × 4	= $1	"	960 "
$400 × 6	= $1	"	2400 "
$320 × 13	= $1	"	4160 "
$1200 × 21	= $1	"	25200 "
$2760	$1	"	32720 "

Whence $32720 \div 2760 = 11.8+$ days; hence the time is **12** days, nearly, after Jan. 7; that is, Jan. 19.

SOLUTION.—Start at the date of the first purchase, and proceed as in the preceding solution.

325. When the terms of credit begin at the same date, we have the following rule:

Rule.—*Multiply each debt by its term of credit, and divide the sum of the products by the sum of the debts; the quotient will be the equated time.*

326. If the account has credits as well as debits, it is called a *compound equation*, and may be averaged on the same principle as the simple equation.

The following problem will illustrate the difference between simple and compound equations, as it involves both debits and credits:

PROBLEM.—What is the balance in the following account, and when is it due?

A in acc't with B.

Dr. **Cr.**

1875.			*Da.*	*Dol.*	*Ct.*	*Prod.*	1875.			*Da.*	*Dol.*	*Ct.*	*Prod.*
Jan.	14	To invoice, 2 mo.	13	400		5200	Mar.	1	By cash,	0	500		0
Feb.	1	" " "	31	200		6200	Apr.	15	" remittance May 1,	61	250		15250
Mar.	10	" " "	70	500		35000	May	20	" cash,	80	375		30000
"	25	" " "	85	800		68000	June	4	" remittance June 25,	116	450		52200
May	16	" " "	137	300		41100							
				2200		155500					1575		97450
				1575		97450							
				625)		58050							
							93 days after March 1 = June 2.						

SOLUTION.—Start with March 1, the earliest day upon which a payment is made or falls due. Find the products, both on the Dr. and Cr. side of the account, using the dates when the items are due, as cash. The balance due, $625, is found by subtracting the amount of the credits from the amount of the debits. The balance of products, 58050, divided by the balance of items, 625, determines the mean time to be 93 days after March 1 = June 2.

Rule.—1. *Find when each item is due, and take the earliest or latest date as the focal date.*

2. *Find the difference between the focal date and the remaining dates, and multiply each item by its corresponding difference.*

3. *Find the difference between the Dr. and Cr. items, and also between the Dr. and Cr. products, and divide the difference of the products by the difference of the items. Add the quotient to the focal date if it be the first date, or subtract if it be the last date; the result in either case will be the equated time.*

4. *If the two balances be on opposite sides of the account, the quotient must be subtracted from the focal date if the first, or added if it be the last date.*

REMARK.—The solutions thus far have been by the *product method.* The same result will be obtained by dividing the interest of the product balance for 1 day by the interest of the item balance for 1 day. This depends upon the principle that if both dividend and divisor are multiplied or divided by the same number the quotient is unchanged.

EXAMPLES FOR PRACTICE.

What is the mean time of the following invoices:

1. *A to B.* DR.

1877.			Dol.	Ct.	When	due	Days after	Products
May	15	To invoice at 4 months.	800					
June	1	" " " 4 "	700					
"	10	" " " 4 "	900					
July	20	" " " 4 "	600					
Aug.	1	" " " 4 "	500					
"	15	" " " 4 "	1000					
		Sum total,	4500					

Oct. 30, 1877.

REMARK.—Start with Sept. 15, the day the first debt is due.

2. DR. *E in acc't current with F.* CR.

Feb.	4	To invoice 3 mo.	$550	—	*Prod.*	May	8	By cash,	$150	—	*Prod.*
Mar.	20	" " 3 "	260	—		"	26	" remit., June 5,	420	—	
Apr.	1	" " 3 "	150	—		June	3	" note, 1 month,	340	—	
"	5	" " 3 "	325	—		July	1	" " 2 "	170	—	

3. *H. Wright to Mason & Giles.* DR.

1876.			Dol.	Ct.	When	due	Days after April 8.	Products.
Feb.	1	To invoice 3 months.	900					
"	20	" " 3 "	700					
Mar.	10	" " 3 "	600					
Apr.	8	" " cash,	500					
May	10	" " 3 "	900					
June	15	" " 3 "	400					
		Sum total,	4000					

REMARK.—Start with April 8, the time the first payment is due.

4. DR. *A in acc't current with B.* CR.

Mar.	19	To invoice cash,	$900	—	*Prod.*	Feb.	20	By cash,	$400	—	*Prod.*
Apr.	20	" " "	800	—		Mar.	5	" remit., Mar. 15,	300	—	
June	15	" " "	700	—		June	20	" cash,	200	—	
May	10	" " "	600	—		July	10	" "	500	—	

REMARK.—Start with Feb. 20, and date the remittance, March 15, the day it is received.

5. DR. *C in acc't current with D.* CR.

					Prod.						*Prod.*
Jan.	4	To invoice 2 mo.	$250	—		Mar.	10	By cash,	$350	—	
Feb.	3	" " 1 "	140	—		"	21	" "	200	—	
"	15	" " 2 "	450	—		Apr.	4	" note, 2 mo.	240	—	
Apr.	2	" " cash,	100	—		May	20	" remit., May 25,	120	—	
						June	16	" accep. 16 da. sight	500		

6. I owe $912, due Oct. 16, and $500, due Dec. 20. If I pay the first, Oct. 1, 15 days before due, when should I pay the last?

7. Oct. 3, I had two accounts, amounting to $375, one due Nov. 6, and the other Dec. 6, but equated for Nov. 16: what was each in amount?

8. A owes $840, due Oct. 3; he pays $400, July 1; $200, Aug. 1: when will the balance be due?

9. I owe $3200, Oct. 25; I pay $400, Sept. 15; $800, Sept. 30: when should the balance be paid?

10. An account of $2500 is due Sept. 16; $500 are paid Aug. 1; $500, Aug. 11; $500, Aug. 21: when will the balance be due?

11. Exchange the five following notes for six others, each for the same amount, and payable at equal intervals: one of $1200, due in 41 days; one of $1500, due in 72 days; one of $2050, due in 80 days; one of $1320, due in 110 days; one of $1730, due in 125 days; total, $7800.

12. Burt owed in two accounts $487; neither was to draw interest till after due,—one standing a year, and the other two years. He paid both in 1 yr. 5 mo., finding the true discount of the second, at 6%, exactly equal to the interest of the first: what difference of time would the common rule have made?

VII. SETTLEMENT OF ACCOUNTS.

DEFINITIONS.

327. 1. An **Account** is a written statement of debit and credit transactions between two parties, giving the date, quantity, and price of each item.

2. An **Account Current** is a written statement of the business transactions between two parties, for a definite time.

328. In settling an account, one may wish to find:

(1). *When the balance is equitably due;*

(2). *What sum, at a given time, should be paid to balance the account.*

NOTE.—The first process is *Averaging the Account* (Art. **324**); the second is *Finding the Cash Balance.* When both equated time and cash balance are required, the methods of Art 324-326 should be applied.

Henry Armor in acc't with City Bank.

DR. CR.

1875.			*Dol.*	*Ct.*	1874.			*Dol.*	*Ct.*
Jan.	5	To check,	300		Dec.	31	By balance old account,	500	
"	20	" "	500		1875.				
"	21	" "	100		Jan.	7	" cash,	50	
"	27	" "	850		"	15	" "	400	
"	31	" "	75		"	24	" "	1000	

EXPLANATION.—The two parties to this account are Henry Armor and City Bank. The left-hand, or Dr. side, shows, with their dates, the sums paid by the bank on the checks of Henry Armor, for which he is their debtor. The right-hand, or Cr. side, shows, with their dates, the sums deposited in the bank by Henry Armor, for which he is their creditor.

329. Generally, in an account current, each item draws interest from its date to the day of settlement.

Problem.—Find the interest due, and balance this account:

F. H. Willis in acc't with E. S. Kennedy.

Dr. Cr

1879.			Dol.	Ct.	1878.			Dol.	Ct.
Jan.	6	To check,	170		Dec.	31	By balance forward,	325	
"	13	" "	480		1879.				
"	16	" "	96		Jan.	7	" cash deposit,	800	
"	25	" "	500		"	20	" " "	175	
"	28	" "	50		"	30	" " "	240	

Interest to Jan. 31, at 6 per cent.

OPERATION.

Dr.		Cr.	
\$170, 6 %, 25 da.	= \$.708	\$325, 6 %, 31 da.	= \$1.679
480, " 18 "	= 1.44	800, " 24 "	= 3.20
96, " 15 "	= .24	175, " 11 "	= .32
500, " 6 "	= .50	240, " 1 "	= .04
50, " 3 "	= .025	\$1540	\$5.239
\$1296	\$2.913	\$1296	\$2.913
		\$244	\$2.33

∴ Interest due = \$2.33; and cash balance due Willis is \$244 + \$2.33 = \$246.33, *Ans.*

Solution.—The interest is found on each item from its date to the day of settlement; then the sum of the items and the sum of the interests are found on each side. The difference between the sums of the interests is equal to the interest due, which, added to the difference between the sums of the items, gives the balance due.

Rule.—1. *Find the number of days to elapse between the date when each item is due and the date of settlement.*

2. *Find the sum of the items on the Dr. side, and then add to each item its interest, if the item is due before the date of settlement, or subtract it if due after the date of settlement; do the same with Cr. side.*

3. *The difference between the amounts on the two sides of the account is the cash balance.*

1. Find the equated time and cash balance of the following account, July 7, 1876; also April 30, 1876, interest 6% per annum:

Henry Hammond.

Dr.							**Cr.**
1876.							
Jan.	4	To Mdse, at 3 mo.	$900.10				
Feb.	1	" " " 4 "	400.00				
"	18	" " " 3 "	700.50				
Mar.	7	" " " 4 "	600.40				
April	8	" Cash,	500.20				
May	10	" Mdse, at 30 da.	400.00				
June	1	" " " 1 mo.	100.60				

REMARKS.—1. When, in forming the several products, the cents are 50 or less, reject them; more than 50, increase the dollars by 1.

2. The cash balance is the sum that Henry Hammond will be required to pay in settling his account *in full* at any given date.

2. Find the equated time and cash balance of the following account, Oct. 4, 1876, money being worth 10% per annum:

William Smith.

Dr.			*Dol.*	*Ct.*			**Cr.**	*Dol.*	*Ct.*
1876.			*Dol.*	*Ct.*	1876.			*Dol.*	*Ct.*
Jan.	1	To Mdse,	800	00	Jan.	10	By Cash,	400	00
"	16	" " at 30 da.	180	30	"	28	" "	200	00
Feb.	14	" " " 60 "	400	60	Feb.	15	" Bills Rec. at 60 da.	180	30
Mar.	25	" "	500	00	"	28	" Cash,	100	00
April	1	" Cash,	800	00	Mar.	30	" Bills Rec. at 90 da.	450	00
May	7	" Mdse,	600	00	April	14	" Cash,	400	60
"	21	" " at 60 da.	700	00	May	1	" "	500	00
June	10	" "	200	00	"	15	" Bills Rec. at 1 mo.	680	00
"	15	" " " 90 da.	2000	00	June	16	" Cash,	300	00
July	12	" "	500	00	July	19	" "	700	00
Aug.	4	" " " 2 mo.	1000	10	Aug.	10	" Bills Rec. at 20 da.	200	00
					Sept.	1	" Cash,	150	00
					Oct.	3	" "	1100	00

REMARK.—All written obligations, of whatever form, for which a certain amount is to be received, are called *Bills Receivable.*

3. Find the equated time and the cash balance of account, March 31, 1876, interest 10% per annum.

George Cummings.

DR. CR.

1876.			Dol.	Ct.	1875.			Dol.	Ct.
Jan.	2	To Cash,	300	00	Dec.	1	By Mdse, at 1 mo.	583	00
					1876.				
"	21	" "	194	00	Feb.	3	" " " "	40	00
Mar.	4	" "	150	00	Mar.	30	" " " "	130	00

Find the interest due, and balance the following accounts:

A. L. Morris in acc't with T. J. Fisher & Co.

4. DR. CR.

1871.			Dol.	Ct.	1870.			Dol.	Ct.
Jan.	13	To check,	350		Dec.	31	By bal. from old acc't,	813	64
"	22	" "	275		1871.				
Feb.	25	" "	100		Feb.	4	" cash,	120	
May	1	" "	400		Mar.	17	" "	500	
June	23	" "	108	25	May	31	" "	84	50

Interest to June 30, at 6 per cent.

Wm. White in acc't with Beach & Berry.

5. DR. CR.

1875.			Dol.	Ct.	1875.			Dol.	Ct.
July	2	To check,	212	50	June	30	By bal. from old acc't,	1102	50
"	20	" "	66		July	6	" cash deposit,	50	
Aug.	7	" "	235		"	15	" " "	95	
"	25	" "	300		Aug.	9	" " "	168	75
Sept.	5	" "	110		Sept.	18	" " "	32	
"	11	" "	46	40	Oct.	3	" " "	79	90
"	27	" "	454	25					

Interest to Oct. 12, at 10 per cent.

ACCOUNT SALES.

330. 1. An **Account Sales** is a written statement made by an agent or consignee to his employer or consignor, of the quantity and price of goods sold, the charges, and the net proceeds.

2. **Guaranty** is a charge made to secure the owner against loss when the goods are sold on credit.

3. **Storage** is a charge made for keeping goods, and is usually reckoned by the week or month on each piece or article.

4. In Account Sales the charges for freight, commission, etc., are the *Debits*, and the proceeds of sales are the *Credits;* the *Net Proceeds* is the difference between the sums of the credits and debits.

331. Account Sales are averaged by the following rule:

Rule.—1. *Average the sales alone; this result will be the date to be given to the commission and guaranty.*

2. *Make the sales the Cr. side, and the charges the Dr. side, and find the equated time for paying the net proceeds.*

1. Find the equated time of paying the proceeds on the following account of Charles Maynard:

Charles Maynard's Consignment.

1876.				*Dol.*	*Ct.*
Aug.	12	By J. Barnes, at 10 da.		50	60
"	14	" " " Cash		800	00
"	24	" " " Bills Receivable, at 30 da.		850	00
"	29	" " " Cash		210	00
"	30	" " " George Hand		4900	00
"	31	" " " Bills Receivable, at 20 da.		1400	00
				8210	60
1876.		CHARGES.			
Aug.	10	To Cash paid, Freight	$ 75.00		
"	31	" " " Storage	10.00		
"	31	" " " Insurance ⅛ per cent.	10.26		
"	31	" " " Commission 2½ per cent.	205.25	300	51
		Net proceeds due Maynard		7910	09

2. Make an account sales, and find the net proceeds and the time the balance is due in the following:

William Thomas sold on account of B. F. Jonas 2000 bu. wheat July 8, 1876, for $2112.50; July 11, 300 bu. wheat, and took a 20 day note for $362.50; paid freight July 6, 1876, $150.00; July 11, storage, $6; drayage, $5; insurance, $4; commission, at $2\frac{1}{2}\%$, $61.87; loss and gain for his net gain, $11.57

3. *Mdse. Co. "B."*

1876.				*Dol.*	*Ct.*
July	18	By Note, at 20 da.		120	00
"	24	" " " 25 "		60	00
"	30	" Cash		55	00
1876.		CHARGES.			
July	15	To Cash paid, Freight	$ 33.00		
"	15	" " " Drayage	4.00		
"	30	" " " Insurance	1.50		
"	30	" " " Storage	3.00		
"	30	" " " Commission, at 2¼ per cent	5.29		
"	30	C. V. Carnes's net proceeds	62.73		
			$109.52		
		Less our ⅓ net loss	3.93		
			$105.59		

Find C. V. Carnes's net proceeds if paid Jan. 1, 1877, money being worth 10% per annum; also the equated time.

STORAGE ACCOUNTS.

332. Storage Accounts are similar to bank accounts, one side showing how many barrels, packages, etc., are received, and at what times; the other side showing how many have been delivered, and at what times.

Storage is generally charged at so much per month of 30 days on each barrel, package, etc.

1. Storage to Jan. 31, at 5 ct. a bbl. per month.

RECEIVED. DELIVERED.

1876.		*Bbl.*	*Balance on hand.*	*Days.*	*Products.*	1876.		*Bbl.*
January	2	200				January	10	110
	5	150					13	90
	7	30					17	20
	10	120					20	115
	14	80					25	140
	17	150					27	72
	20	75					30	100
	24	60					31	
	28	200						

2. Storage to Feb. 20, at 5 ct a bbl. per month.

RECEIVED. DELIVERED.

		Bbl.	*Balance on hand.*	*Days.*	*Products.*			*Bbl.*
January	31	418				February	5	100
February	4	250					10	80
	9	120					12	220
	12	100					14	140
	16	30					18	90
							20	288

VIII. COMPOUND INTEREST.

DEFINITIONS.

333. 1. **Compound Interest** is interest computed both upon the principal and upon each accrued interest as additional principal.

2. Annual Interest is the gain of a principal whose yearly interests have become debts at simple interest; but, distinct from this, *Compound Interest is the whole gain of a principal, increased at the end of each interval by all the interest drawn during that interval.*

3. The final amount in Compound Interest is called the **compound amount.**

EXAMPLE.—Let the principal be $1000, the rate per cent 6. The first year's interest is $60. If this be added as a new debt, the principal will become $1060, and the second year's interest $63.60; in like manner, the third principal is $1123.60, and a third year's interest $67.416; then, the whole amount is $1191.016, and the whole gain, $191.016

REMARK.—If the above debt be, at the same rate, on *annual interest* (Art. **304**), the whole amount will be $1190.80, and the whole gain $190.80; the difference is the interest of $3.60 (an interest upon an interest debt) for one year, $.216

Compound Interest has **four cases.**

CASE I.

334. Given the principal, rate, and time, to find the compound interest and amount.

PROBLEM.—Find the compound amount of $1000, in 4 years, at 2% per annum

SOLUTION.—Multiplying the principal, $1000, by 1.02, the number expressing the amount of $1 for a year, we have the first year's amount, $1020. Continuing the use of the factor 1.02 until the fourth product is obtained, we have for the required amount, $1082.43216 The same numerical result would have been obtained by taking 1.02 four times as a factor, and multiplying the product by 1000.

REMARK.—Compound Interest may be payable semi-annually or quarterly, and in such cases the computation is made by a multiplication similar to the last.

EXAMPLE.—Let the debt be $1000, and let the interest be compounded at 2% quarterly. In one year there are four intervals, and, as seen in the last process, the year's amount is $1082.43216 The real gain on each dollar is $.0824+, or, a fraction over $8\frac{6}{25}$%. According to the usual form of statement, this debt is compounding "*at 8% per annum, payable quarterly.*" But this must not be understood as an exact statement of the real gain; for, when the quarterly rate is 2%, the annual rate exceeds 8%, and when the annual rate is exactly 8%, the quarterly rate is 1.943%, nearly.

REMARK.—To ascertain the true rate for a smaller interval when the yearly rate is given, requires to separate the yearly multiplier into equal factors. Thus the true half-year rate, when the annual rate is 21%, is found by separating 1.21 into two equal factors, 1.10 × 1.10; and the quarterly rate, when the annual is 8%, is found by separating 1.08 into four factors, each nearly 1.01943 It is true that this process is rarely demanded, and that it is very tedious when the intervals are small; but, in a proper place, it will be a useful exercise. (See Art. **389**).

Rule.—*Find the amount of the principal for the first interval, at the rate for that interval, and in like manner treat the whole debt, at the end of each interval, as a principal at simple interest through the following interval or part of an interval; the result will be the compound amount. To find the compound interest, deduct the original debt from the compound amount.*

EXAMPLES FOR PRACTICE.

1. Find the compound amount and interest of $3850, for 4 yr. 7 mo. 16 da., at 5%, payable annually.

2. The compound interest of $13062.50, for 1 yr. 10 mo. 12 da., at 8%, payable quarterly.

3. The compound amount of $1000, for 3 yr., at 10%, payable semi-annually.

4. What sum, at simple interest, 6%, for 2 yr. 5 mo. 27 da., amounts to the same as $2000, at compound interest, for the same time and rate, payable semi-annually?

5. What is the difference between the annual and the compound interest of $5000 in 6 years, 6% per annum?

6. Required the amount of $1000, at compound interest, 21%, for 2 yr. 6 mo.

REMARKS.—1. In obtaining the answer to the last problem, the compound amount at the end of the second full interval is treated as a sum at simple interest for 6 months. But, calculated at a true half-year rate, the amount is only $1610.51, and *this* sum continued at the same true rate for the remaining half year will amount to the same sum which the debt would have reached in a full interval; for,

$1464.10 × 1.21 = $1771.561; and
$1464.10, for 2 intervals, at 10% comp. int. =
$1464.10 × 1.10 × 1.10 = $1771.561

2. Let the student carefully note that the interest drawn during a year *may* be considered as the *sum* of interests compounded through smaller intervals, at smaller rates. Thus, 6% a year may be regarded as the sum of interest compounded through quarterly intervals at 1.467%, through monthly intervals at .487%, or daily at .016%, approximately. Suppose 7 months have passed since interest began. The year may be taken as a period of 12 intervals, and the interest as having been compounded through 7 of them, at .487%. If the amount then reached be continued at compound interest through the other 5 intervals, the amount will be the same as that of the debt continued to the end of the year at the full rate.

3. In this view, the statement may be made general, *that the worth of the debt at any point in a year, is the principal, which, compounding at a true partial rate for the remaining fraction of a year, will amount to the same sum as the debt continued through that year at the annual rate.*

This is in strict accordance with the results obtained by Algebra, but in the common calculations of Arithmetic, the interest is added to the debt, and the rate divided, only according to the statements "payable annually," "payable quarterly," etc.

CALCULATION BY TABLES.

335. When the intervals are many, the actual multiplications become laborious; and, therefore, tables of compound interest have been prepared to shorten the work.

Amount of $1 at Compound Interest in any number of years. not exceeding fifty-five.

Yrs.	2 per cent.	2½ per cent.	3 per cent.	3½ per cent.	4 per cent.	4½ per cent.
1	1.0200 0000	1.0250 0000	1.0300 0000	1.0350 0000	1.0400 0000	1.0450 0000
2	1.0404 0000	1.0506 2500	1.0609 0000	1.0712 2500	1.0816 0000	1.0920 2500
3	1.0612 0800	1.0768 9062	1.0927 2700	1.1087 1787	1.1248 6400	1.1411 6612
4	1.0824 3216	1.1038 1289	1.1255 0881	1.1475 2300	1.1698 5856	1.1925 1860
5	1.1040 8080	1.1314 0821	1.1592 7407	1.1876 8631	1.2166 5290	1.2461 8194
6	1.1261 6242	1.1596 9342	1.1940 5230	1.2293 5533	1.2653 1902	1.3022 6012
7	1.1486 8567	1.1886 8575	1.2298 7387	1.2722 7926	1.3159 3178	1.3608 6183
8	1.1716 5938	1.2184 0290	1.2667 7008	1.3168 0904	1.3685 6905	1.4221 0061
9	1.1950 9257	1.2488 6297	1.3047 7318	1.3628 9735	1.4233 1181	1.4860 9514
10	1.2189 9442	1.2800 8454	1.3439 1638	1.4105 9876	1.4802 4428	1.5529 6942
11	1.2433 7431	1.3120 8666	1.3842 3387	1.4599 6972	1.5394 5406	1.6228 5305
12	1.2682 4179	1.3448 8882	1.4257 6089	1.5110 6866	1.6010 3222	1.6958 8143
13	1.2936 0663	1.3785 1104	1.4685 3371	1.5639 5606	1.6650 7351	1.7721 9610
14	1.3194 7876	1.4129 7382	1.5125 8972	1.6186 9452	1.7316 7645	1.8519 4492
15	1.3458 6834	1.4482 9817	1.5579 6742	1.6753 4883	1.8009 4351	1.9352 8244
16	1.3727 8570	1.4845 0562	1.6047 0644	1.7339 8604	1.8729 8125	2.0223 7015
17	1.4002 4142	1.5216 1826	1.6528 4763	1.7946 7555	1.9479 0050	2.1133 7681
18	1.4282 4625	1.5596 5872	1.7024 3306	1.8574 8920	2.0258 1652	2.2084 7877
19	1.4568 1117	1.5986 5019	1.7535 0605	1.9225 0132	2.1068 4918	2.3078 6031
20	1.4859 4740	1.6386 1644	1.8061 1123	1.9897 8886	2.1911 2314	2.4117 1402
21	1.5156 6634	1.6795 8185	1.8602 9457	2.0594 3147	2.2787 6807	2.5202 4116
22	1.5459 7967	1.7215 7140	1.9161 0341	2.1315 1158	2.3699 1879	2.6336 5201
23	1.5768 9926	1.7646 1068	1.9735 8651	2.2061 1448	2.4647 1555	2.7521 6635
24	1.6084 3725	1.8087 2595	2.0327 9411	2.2833 2849	2.5633 0417	2.8760 1383
25	1.6406 0599	1.8539 4410	2.0937 7793	2.3632 4498	2.6658 3633	3.0054 3446
26	1.6734 1811	1.9002 9270	2.1565 9127	2.4459 5856	2.7724 6979	3.1406 7901
27	1.7068 8648	1.9478 0002	2.2212 8901	2.5315 6711	2.8833 6858	3.2820 0956
28	1.7410 2421	1.9964 9502	2.2879 2768	2.6201 7196	2.9987 0332	3.4296 9999
29	1.7758 4469	2.0464 0739	2.3565 6551	2.7118 7798	3.1186 5145	3.5840 3649
30	1.8113 6158	2.0975 6758	2.4272 6247	2.8067 9370	3.2433 9751	3.7453 1813
31	1.8475 8882	2.1500 0677	2.5000 8035	2.9050 3148	3.3731 3341	3.9138 5745
32	1.8845 4059	2.2037 5694	2.5750 8276	3.0067 0759	3.5080 5875	4.0899 8104
33	1.9222 3140	2.2588 5086	2.6523 3524	3.1119 4235	3.6483 8110	4.2740 3018
34	1.9606 7603	2.3153 2213	2.7319 0530	3.2208 6033	3.7943 1634	4.4663 6154
35	1.9998 8955	2.3732 0519	2.8138 6245	3.3335 9045	3.9460 8899	4.6673 4781
36	2.0398 8734	2.4325 3532	2.8982 7833	3.4502 6611	4.1039 3255	4.8773 7846
37	2.0806 8509	2.4933 4870	2.9852 2668	3.5710 2543	4.2680 8986	5.0968 6049
38	2.1222 9879	2.5556 8242	3.0747 8348	3.6960 1132	4.4388 1345	5.3262 1921
39	2.1647 4477	2.6195 7448	3.1670 2698	3.8253 7171	4.6163 6599	5.5658 9908
40	2.2080 3966	2.6850 6384	3.2620 3779	3.9592 5972	4.8010 2063	5.8163 6454
41	2.2522 0046	2.7521 9043	3.3598 9893	4.0978 3381	4.9930 6145	6.0781 0094
42	2.2972 4447	2.8209 9520	3.4606 9589	4.2412 5799	5.1927 8391	6.3516 1548
43	2.3431 8936	2.8915 2008	3.5645 1677	4.3897 0202	5.4004 9527	6.6374 3818
44	2.3900 5314	2.9638 0808	3.6714 5227	4.5433 4160	5.6165 1508	6.9361 2290
45	2.4378 5421	3.0379 0328	3.7815 9584	4.7023 5855	5.8411 7568	7.2482 4843
46	2.4866 1129	3.1138 5086	3.8950 4372	4.8669 4110	6.0748 2271	7.5744 1961
47	2.5363 4351	3.1916 9713	4.0118 9503	5.0372 8404	6.3178 1562	7.9152 6849
48	2.5870 7039	3.2714 8956	4.1322 5188	5.2135 8898	6.5705 2824	8.2714 5557
49	2.6388 1179	3.3532 7680	4.2562 1944	5.3960 6459	6.8333 4937	8.6436 7107
50	2.6915 8803	3.4371 0872	4.3839 0602	5.5849 2686	7.1066 8335	9.0326 3627
51	2.7454 1979	3.5230 3644	4.5154 2320	5.7803 9930	7.3909 5068	9.4391 0490
52	2.8003 2819	3.6111 1235	4.6508 8590	5.9827 1327	7.6865 8871	9.8638 6463
53	2.8563 3475	3.7013 9016	4.7904 1247	6.1921 0824	7.9940 5226	10.3077 3853
54	2.9134 6144	3.7939 2491	4.9341 2485	6.4088 3202	8.3138 1435	10.7715 8677
55	2.9717 3067	3.8887 7303	5.0821 4859	6.6331 4114	8.6463 6692	11.2563 0817

Subtract $1 from the Amount in this Table to find the Interest.

Amount of $1 at Compound Interest in any number of years, not exceeding fifty-five.

Yrs.	5 per cent.	6 per cent.	7 per cent.	8 per cent.	9 per cent.	10 per cent.
1	1.0500 000	1.0600 000	1.0700 000	1.0800 000	1.0900 000	1.1000 000
2	1.1025 000	1.1236 000	1.1449 000	1.1664 000	1.1881 000	1.2100 000
3	1.1576 250	1.1910 160	1.2250 430	1.2597 120	1.2950 290	1.3310 000
4	1.2155 063	1.2624 770	1.3107 960	1.3604 890	1.4115 816	1.4641 000
5	1.2762 816	1.3382 256	1.4025 517	1.4693 281	1.5386 240	1.6105 100
6	1.3400 956	1.4185 191	1.5007 304	1.5868 743	1.6771 001	1.7715 610
7	1.4071 004	1.5036 303	1.6057 815	1.7138 243	1.8280 391	1.9487 171
8	1.4774 554	1.5938 481	1.7181 862	1.8509 302	1.9925 626	2.1435 888
9	1.5513 282	1.6894 790	1.8384 592	1.9990 046	2.1718 933	2.3579 477
10	1.6288 946	1.7908 477	1.9671 514	2.1589 250	2.3673 637	2.5937 425
11	1.7103 394	1.8982 986	2.1048 520	2.3316 390	2.5804 264	2.8531 167
12	1.7958 563	2.0121 965	2.2521 916	2.5181 701	2.8126 648	3.1384 284
13	1.8856 491	2.1329 283	2.4098 450	2.7196 237	3.0658 046	3.4522 712
14	1.9799 316	2.2609 040	2.5785 342	2.9371 936	3.3417 270	3.7974 983
15	2.0789 282	2.3965 582	2.7590 315	3.1721 691	3.6424 825	4.1772 482
16	2.1828 746	2.5403 517	2.9521 638	3.4259 426	3.9703 059	4.5949 730
17	2.2920 183	2.6927 728	3.1588 152	3.7000 181	4.3276 334	5.0544 703
18	2.4066 192	2.8543 392	3.3799 323	3.9960 195	4.7171 204	5.5599 173
19	2.5269 502	3.0255 995	3.6165 275	4.3157 011	5.1416 613	6.1159 090
20	2.6532 977	3.2071 355	3.8696 845	4.6609 571	5.6044 108	6.7275 000
21	2.7859 626	3.3995 636	4.1405 624	5.0338 337	6.1088 077	7.4002 499
22	2.9252 607	3.6035 374	4.4304 017	5.4365 404	6.6586 004	8.1402 749
23	3.0715 238	3.8197 497	4.7405 299	5.8714 637	7.2578 745	8.9543 024
24	3.2250 999	4.0489 346	5.0723 670	6.3411 807	7.9110 832	9.8497 327
25	3.3863 549	4.2918 707	5.4274 326	6.8484 752	8.6230 807	10.8347 059
26	3.5556 727	4.5493 830	5.8073 529	7.3963 532	9.3991 579	11.9181 765
27	3.7334 563	4.8223 459	6.2138 676	7.9880 615	10.2450 821	13.1099 942
28	3.9201 291	5.1116 867	6.6488 384	8.6271 064	11.1671 395	14.4209 936
29	4.1161 356	5.4183 879	7.1142 571	9.3172 749	12.1721 821	15.8630 930
30	4.3219 424	5.7434 912	7.6122 550	10.0626 569	13.2676 785	17.4494 023
31	4.5380 395	6.0881 006	8.1451 129	10.8676 694	14.4617 695	19.1943 425
32	4.7649 415	6.4533 867	8.7152 708	11.7370 830	15.7633 288	21.1137 768
33	5.0031 885	6.8405 899	9.3253 398	12.6760 496	17.1820 284	23.2251 544
34	5.2533 480	7.2510 253	9.9781 135	13.6901 336	18.7284 109	25.5476 699
35	5.5160 154	7.6860 868	10.6765 815	14.7853 443	20.4139 679	28.1024 369
36	5.7918 161	8.1472 520	11.4239 422	15.9681 718	22.2512 250	30.9126 805
37	6.0814 069	8.6360 871	12.2236 181	17.2456 256	24.2538 353	34.0039 486
38	6.3854 773	9.1542 524	13.0792 714	18.6252 756	26.4366 805	37.4043 434
39	6 7047 512	9.7035 075	13.9948 204	20.1152 977	28.8159 817	41.1447 778
40	7.0399 887	10.2857 179	14.9744 578	21.7245 215	31.4094 200	45.2592 556
41	7.3919 882	10.9028 610	16.0226 699	23.4624 832	34.2362 679	49.7851 811
42	7.7615 876	11.5570 327	17.1442 568	25.3394 819	37.3175 320	54.7636 992
43	8.1496 669	12.2504 546	18.3443 548	27.3666 404	40.6761 098	60.2400 692
44	8.5571 503	12.9854 819	19.6284 596	29.5559 717	44.3369 597	66.2640 761
45	8.9850 078	13.7646 108	21.0024 518	31.9204 494	48.3272 861	72.8904 837
46	9.4342 582	14.5904 875	22.4726 234	34.4740 853	52.6767 419	80.1795 321
47	9.9059 711	15.4659 167	24.0457 070	37.2320 122	57.4176 486	88.1974 853
48	10.4012 697	16.3938 717	25.7289 065	40.2105 731	62.5852 370	97.0172 338
49	10.9213 331	17.3775 040	27.5299 300	43.4274 190	68.2179 083	106.7189 572
50	11.4673 998	18.4201 543	29.4570 251	46.9016 125	74.3575 201	117.3908 529
51	12.0407 698	19.5253 635	31.5190 168	50.6537 415	81.0496 969	129.1299 382
52	12.6428 083	20.6968 853	33.7253 480	54.7060 408	88.3441 696	142.0429 320
53	13.2749 487	21.9386 985	36.0861 224	59.0825 241	96.2951 449	156.2472 252
54	13.9386 961	23.2550 204	38.6121 509	63.8091 260	104.9617 079	171.8719 477
55	14.6356 309	24.6503 216	41.3150 015	68.9138 561	114.4082 616	189.0591 425

Subtract $1 from the Amount in this Table to find the Interest.

How to use the table in finding the Compound Amount:

1. *Observe at what intervals interest is payable, and also the rate per interval.*

2. *If the number of full intervals can be found in the year column, note the sum corresponding to it in the column under the proper rate; multiply this sum, or its amount for any remaining fraction of an interval, by the principal.*

3. *If the number of intervals be not found in the table, separate the whole time into periods which are each within the limits of the table; find the amount of the principal for one of them, make that a principal for the next, and so on, till the whole time has been taken into the calculation.*

Examples for Practice.

1. Find the compound amount of $750 for 17 yr., at 6%, payable annually.
2. Of $5428 for 33 yr., 5% annually.
3. The compound interest of $1800 for 14 yr., at 8%, payable semi-annually.
4. If $1000 is deposited for a child, at birth, and draws 7% compound interest, payable semi-annually, till it is of age (21 yr.), what will be the amount?
5. Find the amount of $9401.50, at compound interest for 19 yr. 4 mo., 9%, payable semi-annually.
6. Find the compound amount of $1000 for 100 yr., at 10%, payable annually.
7. The compound interest of $3600 for 15 yr., at 8%, payable quarterly.
8. The compound interest of $4000 for 40 yr., at 5%, payable semi-annually.
9. The compound interest of $1200 for 27 yr. 11 mo. 4 da., at 12%, payable quarterly.

CASE II.

336. Given the principal, rate, and compound interest or amount, to find the time.

PROBLEM.—Find the time in which $750 will amount to $2000, the interest being 8 %, payable semi-annually.

SOLUTION.—Since a compound amount is found by multiplying the principal by the amount of $1, we here reverse that process, and say: $2000 ÷ 750 = $2.66666666, the amount of $1, at 4%. The number next lower, in the 4% column, is $2.66583633, the amount for 25 intervals, and is less than $2.66666666 by $.00083033

Since the amount for 25 intervals will, according to the table, gain $.10663343 in 1 interval, it will gain $.00083303 in such a fraction of an interval as the latter sum is of the former; .00083033 ÷ .10663346 = $\frac{1}{128}$, nearly; hence, the required period is $25\frac{1}{128}$ intervals of 6 mo., or, 12 yr. 6 mo. 1 da., *Ans.*

Rule.—1. *Divide the amount by the principal.*

2. *If the quotient be found in the table under the given rate, the years opposite will be the required number of intervals; but if not found exactly, in the table, take the number next less, noting its deficiency, its number of years, and its gain during a full interval.*

3. *Divide the deficiency by the interval gain, and annex the quotient to the number of full intervals; the result will be the required time.*

EXAMPLES FOR PRACTICE.

In what time, at compound interest, will:

1. $8000 amount to $12000, at 6%?

2. $5200 amount to $6508, 6%, payable semi-annually?

3. $12500 gain $5500, 10%, payable quarterly?

4. $1 gain $1, at 6, 8, 10%?

5. $9862.50 amount to $22576.15, 12%, payable semi-annually?

CASE III.

337. Given the principal, the compound interest or amount, and the time, to find the rate.

PROBLEM.—At what rate will $1000 amount to $2411.714 in 20 years?

SOLUTION.—Dividing $2411.714 by 1000, we have $2.411714, which, in the table, corresponds to the amount of $1 for the time, at 4½%.

Rule.—*Divide the amount by the principal; search in the table, opposite the given number of full intervals, for the exact quotient or the number nearest in value; if the time contain also a part of an interval, find the amount of the tabular sum for that time, before comparing with the quotient; the rate per cent at the head of the column will be the exact, or the approximate rate.*

EXAMPLES FOR PRACTICE.

At what rate, by compound interest,

1. Will $1000 amount to $1593.85 in 8 yr.?
2. $3600 amount to $9932.51 in 15 yr.?
3. $13200 amount to 48049.58, in 26 yr. 5 mo. 21 da.?

4. $2813.50 amount to $13276.03, in 17 yr. 7 mo. 14 da., interest payable semi-annually?

5. $7652.18 gain $17198.67, interest payable quarterly, in 11 yr. 11 mo. 3 da.?

6. Any sum double itself in 10, 15, 20 yr.?

CASE IV.

338. Given the compound interest or amount, the time, and the rate, to find the principal.

PROBLEM.—What principal will yield $31086.78 compound interest in 40 yr., at 8%?

SOLUTION.—In 40 yr. $1 will gain $20.7245215, at the given rate; the required principal must contain as many dollars as this interest is contained times in the given interest; $31086.78 ÷ 20.7245215 = $1500, *Ans.*

Rule.—*Divide the given interest or amount by the interest or amount of $1 for the given time and at the given rate; the quotient will be the required principal.*

REMARK.—If the amount be due at some future time, the principal is the present worth at compound interest, and the difference between the amount and present worth is the compound discount.

EXAMPLES FOR PRACTICE.

What principal, at compound interest,

1. Will yield $52669.93 in 25 yr., 6%?

2. Will gain $1625.75 in 6 yr. 2 mo., 7%, payable semi-annually?

3. Will yield $3598.61 in 3 yr. 6 mo. 9 da., 10%, payable quarterly?

4. Will yield $31005.76 in 9½ yr., at 8%, payable semi-annually?

5. Will amount to $27062.85 in 7 yr., at 4%?

6. What is the present worth of $14625.70, due in 5 yr. 9 mo., at 6% compound interest, payable semi-annually?

7. What is the compound discount on $8767.78, due in 12 yr. 8 mo. 25 da., 5%?

IX. ANNUITIES.*

DEFINITIONS.

339. 1. An **Annuity** is a sum of money payable at yearly or other regular intervals.

Annuities are—
1. Perpetual or Limited.
2. Certain or Contingent.
3. Immediate or Deferred.

2. A **Perpetual Annuity,** or a **Perpetuity,** is one that continues forever.

3. A **Limited Annuity** ceases at a certain time.

4. A **Certain Annuity** begins and ends at fixed times.

5. A **Contingent Annuity** begins or ends with the happening of a contingent event—as the birth or the death of a person.

6. An **Immediate Annuity** is one that begins at once.

7. A **Deferred Annuity,** or an **Annuity in Reversion,** is one that does not begin immediately; the term of the reversion may be definite or contingent.

---

*Since the problems in annuities may be classed under the Applications of Percentage, the subject is presented here, instead of being placed after Progression; however, those who prefer may omit this chapter until after Progression has been studied.

8. An annuity is **Forborne** or in **Arrears** if not paid when due.

9. The **Forborne** or **Final Value** of an annuity is the amount of the whole accumulated debt and interest, at the time the annuity ceases.

10. The **Present Value** of an annuity is that sum, which, put at interest for the given time and at the given rate, will amount to the final value.

11. The value of a deferred annuity at the time it commences, may be called its **Initial Value**; its **Present Value** is the present worth of its initial value, at an assumed rate of interest.

12. The rules for annuities are of great importance; their practical applications include leases, life-estates, rents, dowers, pensions, reversions, salaries, life insurance, etc.

REMARK.—An annuity begins, not at the time the first payment is made, but one interval before; if an annuity begin now, its first payment will be a year, half-year, or quarter of a year hence, accord ing to the interval named.

CASE I.

340. Given the payment, the interval, and the rate, to find the initial value of a perpetuity.

PROBLEM.—What is the initial value of a perpetual lease of $250 a year, allowing 6% interest?

SOLUTION.—The initial value must be the principal, which, at 6%, yields $250 interest every year; it is found, by Art. 300.

OPERATION.

$$\begin{array}{r} \$1 \\ \underline{.06} \\ .06)\overline{250.0000} \\ \$4166.66\tfrac{2}{3},\ \textit{Ans.} \end{array}$$

Rule.—*Divide the given payment by the interest of* **$1** *for one interval at the proposed rate.*

EXAMPLES FOR PRACTICE.

1. What is the initial value of a perpetual leasehold of $300 a year, allowing 6% interest?

2. What must I pay for a perpetual lease of $756.40 a year, to secure 8% interest?

3. Ground rents on perpetual lease, yield an income of $15642.90 a year: what is the present value of the estate, allowing 7% interest?

4. What is the initial value of a perpetual leasehold of $1600 a year, payable semi-annually, allowing 5% interest, payable annually?

SUGGESTION.—Here the yearly payment is $1620, by allowing 5% interest on the half-yearly payment first made.

5. What is the initial value of a perpetual leasehold of $2500 a year, payable quarterly, interest 6% payable semi-annually; 6% payable annually; 6% payable quarterly?

CASE II.

341. To find the present value of a deferred perpetuity, when the payment, the interval, the rate, and the time the perpetuity is deferred are known.

PROBLEM.—Find the present value of a perpetuity of $250 a year, deferred 8 yr., allowing 6% interest.

SOLUTION.—Initial value of perpetuity of $250 a year, by last rule = $4166.66⅔ The present value of $4166.66⅔, due 8 yr. hence, at 6% compound interest, = $4166.66⅔ ÷ 1.5938481 (Art. **335**). Use the contracted method, reserving 3 decimal places; the quotient, $2614.22, is the present value of the perpetuity.

Rule.—*Find the initial value of the perpetuity by the last rule; then find the present worth of this sum for the time the* *perpetuity is deferred, by Case IV of Compound Interest; this will be the present value required.*

EXAMPLES FOR PRACTICE.

1. Find the present value of a perpetuity of $780 a year, to commence in 12 yr., int. 5%.

2. Of a perpetual lease of $160 a year, to commence in 3 yr. 4 mo., int. 7%.

3. Of the reversion of a perpetuity of $540 a year, deferred 10 yr., int. 6%.

4. Of an estate which, in 5 yr., is to pay $325 a year forever: int. 8%, payable semi-annually.

5. Of a perpetuity of $1000 a year, payable quarterly, to commence in 9 yr. 10 mo. 18 da., int. 10%, payable semi-annually.

CASE III.

342. Given the rate, the payment, the interval, and the time to run, to find the present value of an annuity certain.

PROBLEM.—1. Find the present value of an immediate annuity of $250 continuing 8 years, 6% interest.

SOLUTION.

Present value of immediate perpetuity of $250, . . .	= $4166.67
Present value of perpetuity of $250, deferred 8 yr., . .	= 2614.22
Pres. val. of immediate annuity of $250, running 8 yr.,	= $1552.45

PROBLEM.—2. The present value of an annuity of $680, to commence in 7 yr. and continue 10 yr., 5% int.

SOLUTION.

Pres. val. of perpetuity of $680, deferred 7 yr., at 5%,	= $9665.27
Pres. val. of perpetuity of $680, deferred 17 yr., at 5%,	= 5933.64
Pres. val. of annuity of $680, deferred 7 yr., to run 10 yr.	= $3731.63

Rule.—*Find the present value of two perpetuities having the given rate, payment, and interval, one of them commencing when the annuity commences, and the other when the annuity ends. The difference between these values will be the present value of the annuity.*

NOTES.—1. This rule applies whether the annuity is immediate or deferred; in the latter, the time the annuity is deferred must be known, and used in getting the values of the perpetuities.

2. By using the *initial* instead of the *present* values of the perpetuities, the rule gives the *initial* value of the deferred annuity, which may be used in finding its final or forborne value. (Rem. 1, Case IV.)

EXAMPLES FOR PRACTICE.

1. Find the present value of an annuity of $125, to commence in 12 yr. and run 12 yr., int. 7%.

2. The present value of an immediate annuity of $400, running 15 yr. 6 mo., int. 8%.

3. The present value of an annuity of $826.50, to commence in 3 yr. and run 13 yr. 9 mo., int. 6%, payable semi-annually.

4. The present value of an annuity of $60, deferred 12 yr. and to run 9 yr., int. $4\frac{1}{2}$%.

5. Sold a lease of $480 a year, payable quarterly, having 8 yr. 9 mo. to run, for $2500: do I gain or lose, int. 8%, payable semi-annually?

CASE IV.

343. Given the payment, the interval, the rate, and time to run, to find the final or forborne value of an annuity.

PROBLEM.—Find the final or forborne value of an annuity of $250, continuing 8 yr., int. 6%.

SOLUTION.—The initial value of a perpetuity of $250, at 6%, = $4166.66⅔; its compound interest, at 6% for 8 yr., = $4166.66⅔ × .5938481 = $2474.37, the final or forborne value of the annuity.

Rule.—*Consider the annuity a perpetuity, and find its initial value by Case I. The compound interest of this sum, at the given rate for the time the annuity runs, will be the final or forborne value.*

NOTES.—1. The final or forborne value of an annuity may be obtained by finding first the initial value, as in Case III, and then the compound amount for the time the annuity runs.

2. The present value of an annuity can be obtained by finding first the forborne value, as in this case, and then the present worth for the time the annuity runs.

EXAMPLES FOR PRACTICE.

1. Find the forborne value of an immediate annuity of $300, running 18 yr., int. 5%.

2. A pays $25 a year for tobacco: how much better off would he have been in 40 yr. if he had invested it at 10% per annum?

3. Find the forborne value of an annuity of $75, to commence in 14 yr., and run 9 yr., int. 6%.

SUGGESTION.—The 14 yr. is not used.

4. A pays $5 a year for a newspaper: if invested at 9%, what will his subscription have produced in 50 yr.?

5. An annuity, at simple interest 6%, in 14 yr., amounted to $116.76: what would have been the difference had it been at compound interest 6%?

6. A boy just 9 yr. old, deposits $35 in a bank: if he deposit the same each year hereafter, and receive 10%, compound interest, what will be the entire amount when he is of age?

The present value of $1 per annum in any number of years, not exceeding fifty-five.

Yrs.	4 per cent.	5 per cent.	6 per cent.	7 per cent.	8 per cent.	10 per cent.
1	.961538	.952381	.943396	.934579	.925926	.909091
2	1.886095	1.859410	1.833393	1.808018	1.783265	1.735537
3	2.775091	2.723248	2.673012	2.624316	2.577097	2.486852
4	3.629895	3.545951	3.465106	3.387211	3.312127	3.169865
5	4.451822	4.329477	4.212364	4.100197	3.992710	3.790787
6	5.242137	5.075692	4.917324	4.766540	4.622880	4.355261
7	6.002055	5.786373	5.582381	5.389289	5.206370	4.868419
8	6.732745	6.463213	6.209794	5.971299	5.746639	5.334926
9	7.435332	7.107822	6.801692	6.515232	6.246888	5.759024
10	8.110896	7.721735	7.360087	7.023582	6.710081	6.144567
11	8.760477	8.306414	7.886875	7.498674	7.138964	6.495061
12	9.385074	8.863252	8.383844	7.942686	7.536078	6.813692
13	9.985648	9.393573	8.852683	8.357651	7.903776	7.103356
14	10.563123	9.898641	9.294984	8.745468	8.244237	7.366687
15	11.118387	10.379658	9.712249	9.107914	8.559479	7.606080
16	11.652296	10.837770	10.105895	9.446649	8.851369	7.823709
17	12.165669	11.274066	10.477260	9.763223	9.121638	8.021553
18	12.659297	11.689587	10.827603	10.059087	9.371887	8.201412
19	13.133939	12.085321	11.158116	10.335595	9.603599	8.364920
20	13.590326	12.462210	11.469921	10.594014	9.818147	8.513564
21	14.029160	12.821153	11.764077	10.835527	10.016803	8.648694
22	14.451115	13.163003	12.041582	11.061241	10.200744	8.771540
23	14.856842	13.488574	12.303379	11.272187	10.371059	8.883218
24	15.246963	13.798642	12.550358	11.469334	10.528758	8.984744
25	15.622080	14.093945	12.783356	11.653583	10.674776	9.077040
26	15.982769	14.375185	13.003166	11.825779	10.809978	9.160945
27	16.329586	14.643034	13.210534	11.986709	10.935165	9.237223
28	16.663063	14.898127	13.406164	12.137111	11.051078	9.306567
29	16.983715	15.141074	13.590721	12.277674	11.158406	9.369606
30	17.292033	15.372451	13.764831	12.409041	11.257783	9.426914
31	17.588494	15.592811	13.929086	12.531814	11.349799	9.479013
32	17.873552	15.802677	14.084043	12.646555	11.434999	9.526376
33	18.147646	16.002549	14.230230	12.753790	11.513888	9.569432
34	18.411198	16.192904	14.368141	12.854009	11.586934	9.608575
35	18.664613	16.374194	14.498246	12.947672	11.654568	9.644159
36	18.908282	16.546852	14.620987	13.035208	11.717193	9.676508
37	19.142579	16.711287	14.736780	13.117017	11.775179	9.705917
38	19.367864	16.867893	14.846019	13.193473	11.828869	9.732651
39	19.584485	17.017041	14.949075	13.264928	11.878582	9.756956
40	19.792774	17.159086	15.046297	13.331709	11.924613	9.779051
41	19.993052	17.294368	15.138016	13.394120	11.967235	9.799137
42	20.185627	17.423208	15.224543	13.452449	12.006699	9.817397
43	20.370795	17.545912	15.306173	13.506962	12.043240	9.833998
44	20.548841	17.662773	15.383182	13.557908	12.077074	9.849089
45	20.720040	17.774070	15.455832	13.605522	12.108402	9.862808
46	20.884654	17.880067	15.524370	13.650020	12.137409	9.875280
47	21.042936	17.981016	15.589028	13.691608	12.164267	9.886618
48	21.195131	18.077158	15.650027	13.730474	12.189136	9.896926
49	21.341472	18.168722	15.707572	13.766799	12.212163	9.906296
50	21.482185	18.255925	15.761861	13.800746	12.233485	9.914814
51	21.617485	18.338977	15.813076	13.832473	12.253227	9.922559
52	21.747582	18.418073	15.861393	13.862124	12.271506	9.929599
53	21.872675	18.493403	15.906974	13.889836	12.288432	9.935999
54	21.992957	18.565146	15.949976	13.915735	12.304103	9.941817
55	22.108612	18.633472	15.990543	13.939939	12.318614	9.947107

CALCULATIONS BY TABLE.

344. By the table on page 314, some interesting and important cases in annuities can be solved, among which are the following three:

CASE V.

345. Given the rate, time to run, and the present or final value of an annuity, to find the payment.

PROBLEM.—An immediate annuity running 11 yr., can be purchased for $6000: what is the payment, int. 6%?

SOLUTION.—The present value of an immediate annuity of $1 for 11 yr., at 6%, is $7.886875; $6000 divided by this, gives $760.76, the payment required.

Rule.—*Assume* $1 *for the payment; determine the present or final value on this supposition, and divide the given present or final value by it.*

EXAMPLES FOR PRACTICE.

1. How much a year should I pay, to secure $15000 at the end of 17 yr., int. 7%?

2. What is the payment of an annuity, deferred 4 yr., running 16 yr., and worth $4800, int. 6%?

CASE VI.

346. Given the payment, the rate, and present value of an annuity, to find the time it runs.

PROBLEM.—In what time will a debt of $10000, drawing interest at 6%, be paid by installments of $1000 a year?

SOLUTION.—The $10000 may be considered the present value of an annuity of $1000 a year at 6%; but $10000 ÷ 1000 — $10, the present value of an annuity of $1 for the same time and rate; by reference to the table, the time corresponding to this present value, under the head of 6%, is 15 yr.; the balance then due may be thus found:

Comp. amt. of $10000 for 15 yr., at 6% (Art. **335**), .	= $23965.58
Final val. of annuity $1000 for 15 yr., at 6% (Art. **343**), =	23275.97
Balance due at end of 15 yr.	$689.61

Rule.—*Divide the present value by the payment, and look in the table, under the given rate, for the quotient; the number of years corresponding to the quotient or to the tabular number next less, will be the number of full intervals required.*

NOTE.—The difference between the compound amount of the debt, and the forborne value of the annuity, for that number of intervals, will be the unpaid balance.

EXAMPLES FOR PRACTICE.

1. In how many years can a debt of $1000000, drawing interest at 6%, be discharged by a sinking fund of $80000 a year?

2. In how many years can a debt of $30000000, drawing interest at 5%, be paid by a sinking fund of $2000000?

3. In how many years can a debt of $22000, drawing 7% interest, be discharged by a sinking fund of $2500 a year?

4. Let the conditions be the same as those of the illustrative example, and let each $1000 payment be itself a year's accumulation of simple interest: what would be the whole time required to discharge the debt?

5. Suppose the national debt $2000000000, and funded at 4%: how many years would be required to pay it off, by a sinking fund of $100000000 a year?

CASE VII.

347. Given the payment, time to run, and present value of an annuity, to find the rate of interest.

Rule.—*Divide the present value by the payment; the quotient will be the present value of $1 for the given time and rate; look in the table and opposite the given number of years for the quotient or the tabular number of nearest value, and at the head of the column will be found the rate, or a number as near the true rate as the table can exhibit.*

Examples for Practice.

1. If $9000 is paid for an immediate annuity of $750, to run 20 yr., what is the rate?
2. If an immediate annuity of $80, running 14 yr., sells for $650, what is the rate?

CONTINGENT ANNUITIES.

DEFINITIONS.

348. 1. **Contingent Annuities** comprise *Life Annuities, Dowers, Pensions,* etc.

2. The value of such annuities depends upon the *expectation of life.*

3. **Expectation of Life** is the average number of years that a person of any age may be expected to live.

4. Tables, called "Mortality Tables," have been prepared in England and in this country for the purpose of ascertaining how many persons of a given number and of a certain

age would die during any one year, and in how many years the whole number would die.

REMARK.—These tables, though not absolutely accurate, are based upon so large a number of observations that their approximation is very close. Legal, medical, and scientific authorities use them in discussing vital statistics, and insurance companies make them a basis for the transaction of business.

349. The following table differs but slightly from those prepared in this country:

Carlisle Table

Of Mortality, based upon Observations at Carlisle (Eng.), showing the Rate of Extinction of 10,000 lives.

Age.	Number of Survivors.	Number of Deaths.	Age.	Number of Survivors.	Number of Deaths.	Age.	Number of Survivors.	Number of Deaths.
0	10000	1539	35	5362	55	70	2401	124
1	8461	682	36	5307	56	71	2277	134
2	7779	505	37	5251	57	72	2143	146
3	7274	276	38	5194	58	73	1997	156
4	6998	201	39	5136	62	74	1841	166
5	6797	121	40	5075	66	75	1675	160
6	6676	82	41	5009	69	76	1515	156
7	6594	58	42	4940	71	77	1359	146
8	6536	43	43	4869	71	78	1213	132
9	6493	33	44	4798	71	79	1081	128
10	6460	29	45	4727	70	80	953	116
11	6431	31	46	4657	69	81	837	112
12	6400	32	47	4588	67	82	725	102
13	6368	33	48	4521	63	83	623	94
14	6335	35	49	4458	61	84	529	84
15	6300	39	50	4397	59	85	445	78
16	6261	42	51	4338	62	86	367	71
17	6219	43	52	4276	65	87	296	64
18	6176	43	53	4211	68	88	232	51
19	6133	43	54	4143	70	89	181	39
20	6090	43	55	4073	73	90	142	37
21	6047	42	56	4000	76	91	105	30
22	6005	42	57	3924	82	92	75	21
23	5963	42	58	3842	93	93	54	14
24	5921	42	59	3749	106	94	40	10
25	5879	43	60	3643	122	95	30	7
26	5836	43	61	3521	126	96	23	5
27	5793	45	62	3395	127	97	18	4
28	5748	50	63	3268	125	98	14	3
29	5698	56	64	3143	125	99	11	2
30	5642	57	65	3018	124	100	9	2
31	5585	57	66	2894	123	101	7	2
32	5528	56	67	2771	123	102	5	2
33	5472	55	68	2648	123	103	3	2
34	5417	55	69	2525	124	104	1	1

Table

Showing the values of Annuities on Single Lives, according to the Carlisle Table of Mortality.

Age.	4 per ct.	5 per ct.	6 per ct.	7 per ct.	Age.	4 per ct.	5 per ct.	6 per ct.	7 per ct.
0	14.28164	12.083	10.439	9.177	52	12.25793	11.154	10.208	9.392
1	16.55455	13.995	12.078	10.605	53	11.94503	10.892	9.988	9.205
2	17.72616	14.983	12.925	11.342	54	11.62673	10.624	9.761	9.011
3	18.71508	15.824	13.652	11.978	55	11.29961	10.347	9.524	8.807
4	19.23133	16.271	14.042	12.322	56	10.96607	10.063	9.280	8.595
5	19.59203	16.590	14.325	12.574	57	10.62559	9.771	9.027	8.375
6	19.74502	16.735	14.460	12.698	58	10.28647	9.478	8.772	8.153
7	19.79019	16.790	14.518	12.756	59	9.96331	9.199	8.529	7.940
8	19.76443	16.786	14.526	12.770	60	9.66333	8.940	8.304	7.743
9	19.69114	16.742	14.500	12.754	61	9.39809	8.712	8.108	7.572
10	19.58339	16.669	14.448	12.717	62	9.13676	8.487	7.913	7.403
11	19.45857	16.581	14.384	12.669	63	8.87150	8.258	7.714	7.229
12	19.33493	16.494	14.321	12.621	64	8.59330	8.016	7.502	7.042
13	19.20937	16.406	14.257	12.572	65	8.30719	7.765	7.281	6.847
14	19.08182	16.316	14.191	12.522	66	8.00966	7.503	7.049	6.641
15	18.95534	16.227	14.126	12.473	67	7.69980	7.227	6.803	6.421
16	18.83636	16.144	14.067	12.429	68	7.37976	6.941	6.546	6.189
17	18.72111	16.066	14.012	12.389	69	7.04881	6.643	6.277	5.945
18	18.60656	15.987	13.956	12.348	70	6.70936	6.336	5.998	5.690
19	18.48649	15.904	13.897	12.305	71	6.35773	6.015	5.704	5.420
20	18.36170	15.817	13.835	12.259	72	6.02548	5.711	5.424	5.162
21	18.23196	15.726	13.769	12.210	73	5.72465	5.435	5.170	4.927
22	18.09386	15.628	13.697	12.156	74	5.45812	5.190	4.944	4.719
23	17.95016	15.525	13.621	12.098	75	5.23901	4.989	4.760	4.549
24	17.80058	15.417	13.541	12.037	76	5.02399	4.792	4.579	4.382
25	17.64486	15.303	13.456	11.972	77	4.82473	4.609	4.410	4.227
26	17.48586	15.187	13.368	11.904	78	4.62166	4.422	4.238	4.067
27	17.32023	15.065	13.275	11.832	79	4.39345	4.210	4.040	3.883
28	17.15412	14.942	13.182	11.759	80	4.18289	4.015	3.858	3.713
29	16.99683	14.827	13.096	11.693	81	3.95309	3.799	3.656	3.523
30	16.85215	14.723	13.020	11.636	82	3.74634	3.606	3.474	3.352
31	16.70511	14.617	12.942	11.578	83	3.53409	3.406	3.286	3.174
32	16.55246	14.506	12.860	11.516	84	3.32856	3.211	3.102	2.999
33	16.39072	14.387	12.771	11.448	85	3.11515	3.009	2.909	2.815
34	16.21943	14.260	12.675	11.374	86	2.92831	2.830	2.739	2.652
35	16.04123	14.127	12.573	11.295	87	2.77593	2.685	2.599	2.519
36	15.85577	13.987	12.465	11.211	88	2.68337	2.597	2.515	2.439
37	15.66586	13.843	12.354	11.124	89	2.57704	2.495	2.417	2.344
38	15.47129	13.695	12.239	11.033	90	2.41621	2.339	2.266	2.198
39	15.27184	13.542	12.120	10.939	91	2.39835	2.321	2.248	2.180
40	15.07363	13.390	12.002	10.845	92	2.49199	2.412	2.337	2.266
41	14.88314	13.245	11.890	10.757	93	2.59955	2.518	2.440	2.367
42	14.69466	13.101	11.779	10.671	94	2.64976	2.569	2.492	2.419
43	14.50529	12.957	11.668	10.585	95	2.67433	2.596	2.522	2.451
44	14.30874	12.806	11.551	10.494	96	2.62779	2.555	2.486	2.420
45	14.10460	12.648	11.428	10.397	97	2.49204	2.428	2.368	2.309
46	13.88928	12.480	11.296	10.292	98	2.33222	2.278	2.227	2.177
47	13.66208	12.301	11.154	10.178	99	2.08700	2.045	2.004	1.964
48	13.41914	12.107	10.998	10.052	100	1.65282	1.624	1.596	1.569
49	13.15312	11.892	10.823	9.908	101	1.21005	1.192	1.175	1.159
50	12.86902	11.660	10.631	9.749	102	0.76183	0.753	0.744	0.735
51	12.56581	11.410	10.422	9.573	103	0.32051	0.317	0.314	0.312

CALCULATIONS BY TABLE.

350. The preceding table of life annuities shows the sum to be paid by a person of any age, to secure an annuity of $1 during the life of the annuitant.

CASE I.

351. To find the value of a given annuity on the life of a person whose age is known.

Rule.—*Find from the table the value of a life annuity of* $1, *for the given age and rate of interest, and multiply it by the payment of the given annuity.*

REMARKS.—1. To find the value of a life-estate or widow's dower (which is a life-estate in *one third* of her husband's real estate): *Estimate the value of the property in which the life-estate is held; the yearly interest of this sum, at an agreed rate, will be a life-annuity, whose value for the given age and rate will be the value of the life-estate.*

2. The reversion of a life-annuity, life-estate, or dower is found *by deducting its value from the value of the property.*

EXAMPLES FOR PRACTICE.

1. What must be paid for a life-annuity of $650 a year, by a person aged 72 yr., int. 7%?

2. What is the life-estate and reversion in $25000, age 55 yr., int. 6%?

3. The dower and reversion in $46250, age 21 yr., int. 6%?

CASE II.

352. To find how large a life-annuity can be purchased for a given sum, by a person whose age is known.

Rule.—*Assume $1 a year for the annuity; find from the table its value for the given age and rate of interest, and divide the given cost by it; the quotient will be the payment required.*

Examples for Practice.

How large an annuity can be purchased:

1. For $500, age 26 yr., int. 6%?
2. For $1200, age 43, int. 5%?
3. For $840, age 58, int. 7%?

CASE III.

353. To find the present value of the reversion of a given annuity; that is, what remains of it, after the death of its possessor, whose age is known.

Rule.—*Find the present value of the annuity during its whole continuance; find its value during the given life; their difference will be the value of the reversion.*

Note.—It will save work, to *consider the annuity as* $1 *a year, then apply the rule, using the tables in* Art. **344** *and* Art. **350**, *and multiply the result by the given payment.*

Examples for Practice.

1. Find the present value of the reversion of a perpetuity of $500 a year, after the death of a person aged 47, int. 5%.

2. Of the reversion of a perpetuity of $165 a year after the death of a person 38 yr. old, int. 6%.

3. Of the reversion of a perpetual lease of $1600 a year after the death of A, aged 62, int. 7%.

PERSONAL INSURANCE.

DEFINITIONS.

354. 1. **Personal Insurance** is of two kinds: (1.) *Life Insurance;* (2.) *Accident Insurance.*

2. **Life Insurance** is a contract in which a company agrees, in consideration of certain premiums received, to pay a certain sum to the heirs or assigns of the insured at his death, or to himself if he attains a certain age.

3. **Accident Insurance** is indemnity against loss by accidents.

4. The **Policies** issued by life insurance companies are: (1.) *Term Policies;* (2.) *Ordinary Life Policies;* (3.) *Joint Life Policies;* (4.) *Endowment Policies;* (5.) *Reserved Endowment Policies;* (6.) *Tontine Savings Fund Policies.*

5. The chief policies are, however, the Ordinary Life and the Endowment.

6. The **Ordinary Life Policy** secures a certain sum of money at the death of the insured. Premiums may be paid annually for life, semi-annually, quarterly, or in one payment in advance; or the premiums may be paid in 5, 10, 15, or 20 annual payments.

7. An **Endowment Policy** secures to the person insured a certain sum of money at a specified time, or to his heirs or assigns if he die before that time.

REMARK.—It will be advantageous for the student to examine an "application" and a "policy" taken from some case of actual insurance; by a short study of such papers, the nature of the insurance contract will be learned more easily than by any mere verbal description; additional light may be had from the reports published by various companies.

355. 1. The following is a condensed table of one of the leading companies:

Table.

Annual Premium Rates for an Insurance of $1000.

LIFE POLICIES. Payable at death only.					ENDOWMENT POLICIES. Payable as indicated, or at death, if prior.			
Age.	Annual Payments.			Single Payment	Age.	In 10 years	In 15 years	In 20 years
	For life	20 years	10 years					
20 to 25	$19 89	$27 39	$42 56	$326 58	20 to 25	$103 91	$66 02	$47 68
26	20 40	27 93	43 37	332 58	26	104 03	66 15	47 82
27	20 93	28 50	44 22	338 83	27	104 16	66 29	47 98
28	21 48	29 09	45 10	345 31	28	104 29	66 44	48 15
29	22 07	29 71	46 02	352 05	29	104 43	66 60	48 33
30	22 70	30 36	46 97	359 05	30	104 58	66 77	48 53
31	23 35	31 03	47 98	366 33	31	104 75	66 96	48 74
32	24 05	31 74	49 02	373 89	32	104 92	67 16	48 97
33	24 78	32 48	50 10	381 73	33	105 11	67 36	49 22
34	25 56	33 26	51 22	389 88	34	105 31	67 60	49 49
35	26 38	34 08	52 40	398 34	35	105 53	67 85	49 79
36	27 25	34 93	53 63	407 11	36	105 75	68 12	50 11
37	28 17	35 83	54 91	416 21	37	106 00	68 41	50 47
38	29 15	36 78	56 24	425 64	38	106 28	68 73	50 86
39	30 19	37 78	57 63	435 42	39	106 58	69 09	51 30
40	31 30	38 83	59 09	445 55	40	106 90	69 49	51 78
41	32 47	39 93	60 60	456 04	41	107 26	69 92	52 31
42	33 72	41 10	62 19	466 89	42	107 65	70 40	52 89
43	35 05	42 34	63 84	478 11	43	108 08	70 92	53 54
44	36 46	43 64	65 57	489 71	44	108 55	71 50	54 25
45	37 97	45 03	67 37	501 69	45	109 07	72 14	55 04
46	39 58	46 50	69 26	514 04	46	109 65	72 86	55 91
47	41 30	48 07	71 25	526 78	47	110 30	73 66	56 89
48	43 13	49 73	73 32	539 88	48	111 01	74 54	57 96
49	45 09	51 50	75 49	553 33	49	111 81	75 51	59 15
50	47 18	53 38	77 77	567 13	50	112 68	76 59	60 45
51	49 40	55 38	80 14	581 24	51	113 64	77 77	61 90
52	51 78	57 51	82 63	595 66	52	114 70	79 07	63 48
53	54 31	59 79	85 22	610 36	53	115 86	80 51	65 22
54	57 02	62 22	87 94	625 33	54	117 14	82 09	67 14
55	59 91	64 82	90 79	640 54	55	118 54	83 82	69 24
56	63 00	67 60	93 78	655 99	56	120 09	85 73	
57	66 29	70 59	96 91	671 64	57	121 78	87 84	
58	69 82	73 78	100 21	687 48	58	123 64	90 15	
59	73 60	77 22	103 68	703 49	59	125 70	92 70	
60	77 63	80 91	107 35	719 65	60	127 96	95 50	
61	81 96	84 88	111 23	735 92	61	130 45		
62	86 58	89 16	115 32	752 26	62	133 19		
63	91 54	93 76	119 66	768 67	63	136 20		
64	96 86	98 73	124 28	785 10	64	139 52		
65	102 55	104 10	129 18	801 52	65	143 16		

2. Quantities considered in Life Insurance are:

1. Premium on $1000.
2. The Gain or Loss.
3. Amount of the Policy.
4. Age of the Insured.
5. The Term of years of Insurance.

These quantities give rise to **five classes** of problems, but they involve no new principles, and by the aid of the preceding tables they are easily solved. *Simple* interest is intended where interest is mentioned in the following problems:

Examples for Practice.

1. W. R. Hamilton, aged 40 years, took a life policy for $5000. Required the annual premium?

2. Conditions as above, how much would he have paid out in premiums, his death having occurred after he was 53?

3. Conditions the same, what did the premiums amount to, interest 6%?

4. James Bragg, aged 50 years, took out an endowment policy for $20000, payable in 10 years, and died after making 6 payments: how much less would he have paid out by taking a life policy for the same amount, the premium payable annually?

5. Thomas Winn, 28 years of age, took out an endowment policy for $10000, payable in 10 years; he died in 18 months: what was the gain, interest on the premiums at 6%, and how much greater would the profit have been had he taken a life policy, premiums payable annually?

6. P. Darling took out a life policy at the age of 40 years, and died just after making the tenth payment; his premiums amounted to $3975.10, interest 6%; required the amount of his policy?

7. R. C. Storey took out an endowment policy for \$10000 for 15 years; he lived to pay all of the premiums; but had he put them, instead, at 6% interest as they fell due, they would have amounted to \$15426.78: what was his age?

8. Allen Wentworth had his life insured at the age of twenty, on the life plan, for \$8000, premium payable annually in advance: how old must he be, that the sum of the premiums may exceed the policy?

9. T. B. Bullene, aged 40 years, took out a life policy for \$30000, payments to cease in 5 years, the rate being \$9.919 on the \$100; his death occurred two months after he had made the third payment: what was gained over and above the premiums, interest 6%?

10. F. M. Harrington took out an endowment policy for \$11000 when he was 42; at its maturity he had paid in premiums \$635.80 more than the face of the policy: what was the period of the endowment?

Topical Outline.

Applications of Percentage.

(With Time.)

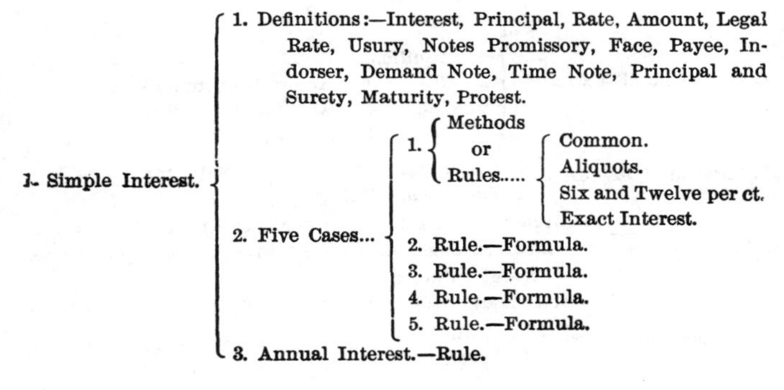

1. Simple Interest.
   1. Definitions:—Interest, Principal, Rate, Amount, Legal Rate, Usury, Notes Promissory, Face, Payee, Indorser, Demand Note, Time Note, Principal and Surety, Maturity, Protest.
   2. Five Cases...
      1. Methods or Rules.....
         - Common.
         - Aliquots.
         - Six and Twelve per ct.
         - Exact Interest.
      2. Rule.—Formula.
      3. Rule.—Formula.
      4. Rule.—Formula.
      5. Rule.—Formula.
   3. Annual Interest.—Rule.

Topical Outline.—(*Continued.*)

APPLICATIONS OF PERCENTAGE.

(*With Time.*)

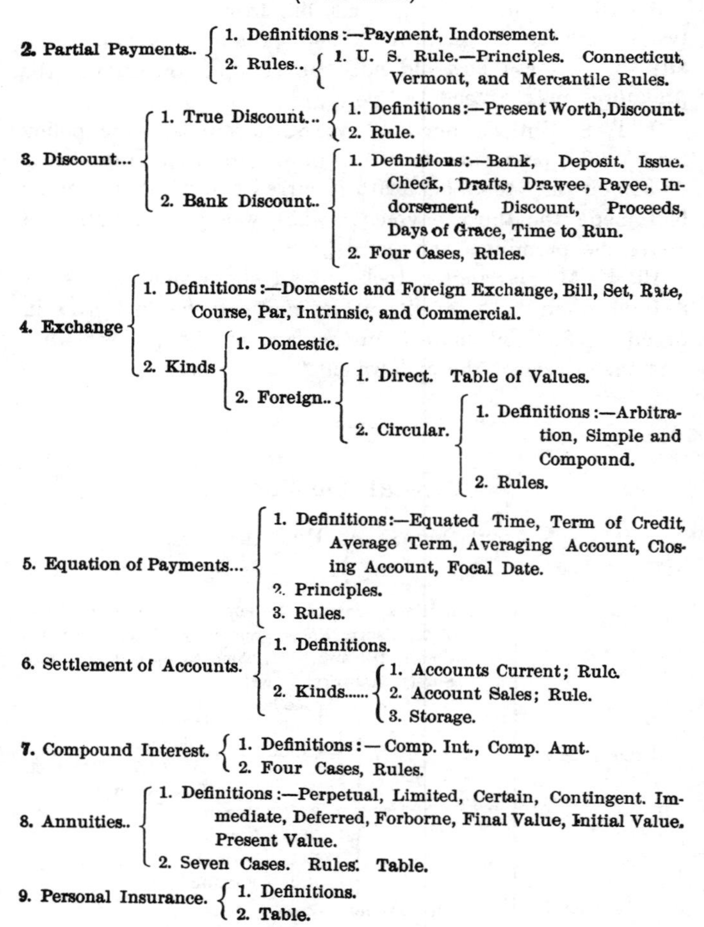

- **2. Partial Payments..**
  - 1. Definitions:—Payment, Indorsement.
  - 2. Rules..
    - 1. U. S. Rule.—Principles. Connecticut, Vermont, and Mercantile Rules.
- **3. Discount...**
  - 1. True Discount...
    - 1. Definitions:—Present Worth, Discount.
    - 2. Rule.
  - 2. Bank Discount..
    - 1. Definitions:—Bank, Deposit. Issue. Check, Drafts, Drawee, Payee, Indorsement, Discount, Proceeds, Days of Grace, Time to Run.
    - 2. Four Cases, Rules.
- **4. Exchange**
  - 1. Definitions:—Domestic and Foreign Exchange, Bill, Set, Rate, Course, Par, Intrinsic, and Commercial.
  - 2. Kinds
    - 1. Domestic.
    - 2. Foreign..
      - 1. Direct. Table of Values.
      - 2. Circular.
        - 1. Definitions:—Arbitration, Simple and Compound.
        - 2. Rules.
- **5. Equation of Payments...**
  - 1. Definitions:—Equated Time, Term of Credit, Average Term, Averaging Account, Closing Account, Focal Date.
  - 2. Principles.
  - 3. Rules.
- **6. Settlement of Accounts.**
  - 1. Definitions.
  - 2. Kinds......
    - 1. Accounts Current; Rule.
    - 2. Account Sales; Rule.
    - 3. Storage.
- **7. Compound Interest.**
  - 1. Definitions:—Comp. Int., Comp. Amt.
  - 2. Four Cases, Rules.
- **8. Annuities..**
  - 1. Definitions:—Perpetual, Limited, Certain, Contingent. Immediate, Deferred, Forborne, Final Value, Initial Value. Present Value.
  - 2. Seven Cases. Rules. Table.
- **9. Personal Insurance.**
  - 1. Definitions.
  - 2. Table.

XVII. PARTNERSHIP.

DEFINITIONS.

356. 1. **Partnership** is the association of two or more persons in a business of which they are to share the profits and the losses. The persons associated are called *partners;* together they constitute a *Firm, Company,* or *House.*

2. The **Capital** is the money employed in the business; the **Assets** or **Resources** of a firm are its property, and opposed to these are its **Liabilities** or **Debts.**

3. Partnership has **two cases:** (1.) When all the shares of the capital are continued through the same time; (2.) When the full shares are not continued through the same time. The first is called *Simple Partnership;* the second, *Compound Partnership.*

PRINCIPLE.—*Gains and losses are shared in proportion to the sums invested and the periods of investment.*

CASE I.

357. To apportion the gain or loss, when all of each partner's stock is employed through the same time.

PROBLEM.—A, B, and C are partners, with \$3000, \$4000, and \$5000 stock, respectively; if they gain \$5400, what is each one's share?

OPERATION.

3	$\frac{3}{12}$ of \$5400 =	\$1350,	A's share.
4	$\frac{4}{12}$ " 5400 =	1800,	B's "
5	$\frac{5}{12}$ " 5400 =	2250,	C's "
12		\$5400,	whole.

SOLUTION.—The whole stock is \$12000, of which A owns $\frac{3}{12}$, B $\frac{4}{12}$, C $\frac{5}{12}$; hence, by the principle stated, A should have $\frac{3}{12}$ of the gain, or \$1350; in like manner, B, \$1800; C, \$2250.

Rule.—*Divide the gain or loss among the partners in proportion to their shares of the stock.*

REMARK.—The division may be made by analysis or by simple proportion.

EXAMPLES FOR PRACTICE.

1. A and B gain in one year $3600; their store expenses are $1500. If A's stock is $2500 and B's $1875, how much does each gain?

2. A, B, and C are partners; A puts in $5000, B 6400, C $1600. C is allowed $1000 a year for personal attention to the business; their store expenses for one year are $800, and their gain, $7000. Find A's and B's gain, and C's income.

3. A, B, and C form a partnership; A puts in $24000, B $28000, C $32000; they lose $\frac{1}{6}$ of their stock by a fire, but sell the remainder at $\frac{3}{5}$ more than cost: if all expenses are $8000, what is the gain of each?

4. A, B, and C are partners; A's stock is $5760, B's, $7200; their gain is $3920, of which C has $1120: what is C's stock, and A's and B's gain?

5. A, B, and C are partners; A's stock, $8000; B's, $12800; C's, $15200; A and B together gain $1638 more than C: what is the gain of each?

6. A, B, and C have a joint capital of $27000; none of them draw from the firm, and when they quit A has $20000; B, $16000; C, $12000: how much did each contribute?

7. A, B, C, and D gain 30% on the stock; A, B, and C gain $1150; A, B, D, $1650; B, C, D, $1000; A, C, D, $1600: what was each man's stock?

CASE II.

358. To apportion the gain or loss when the full shares are not continued through the same period.

PROBLEM.—A, B, and C are partners; A puts in \$2500 for 8 mo.; B, \$4000 for 6 mo.; C, \$3200 for 10 mo.; their net gain is \$4750: divide the gain.

OPERATION.

$$\$2500 \times 8 = \$20000, \text{ A's equivalent.}$$
$$\$4000 \times 6 = 24000, \text{ B's } \text{"}$$
$$\$3200 \times 10 = 32000, \text{ C's } \text{"}$$
$$\$76000$$

$$\tfrac{20}{76} \text{ of } \$4750 = \$1250, \text{ A's share.}$$
$$\tfrac{24}{76} \text{ of } \$4750 = \$1500, \text{ B's } \text{"}$$
$$\tfrac{32}{76} \text{ of } \$4750 = \$2000, \text{ C's } \text{"}$$

SOLUTION.—A's capital (\$2500), used 8 months, is equivalent to 8 × \$2500, or \$20000, used 1 month; B's capital (\$4000), used 6 months, is equivalent to 6 × \$4000, or \$24000, used 1 month; C's capital (\$3200), used 10 months, is equivalent to 10 × \$3200, or \$32000 used 1 month. Dividing the gain (\$4750) in proportion to the stock equivalents, \$20000, \$24000, \$32000, used for the same time (1 month), the results will be the gain of each; A's \$1250, B's \$1500, C's \$2000.

Rule.—*Multiply each partner's stock by the time it is used; and divide the gain or loss in proportion to the products so obtained.*

EXAMPLES FOR PRACTICE.

1. A begins business with \$6000; at the end of 6 mo. he takes in B, with \$10000; 6 mo. after, their gain is \$3300: what is each share?

2. A and B are partners; A's stock is to B's, as 4 to 5; after 3 mo., A withdraws $\frac{2}{3}$ of his, and B $\frac{3}{4}$ of his: divide their year's gain, \$1675.

3. A, B, and C join capitals, which are as $\frac{1}{2}$, $\frac{1}{3}$, $\frac{1}{4}$; after 4 mo., A takes out $\frac{1}{2}$ of his; after 9 mo. more, their gain is \$1988: divide it.

4. A and B are partners; A puts in $2500; B, $1500; after 9 mo., they take in C with $5000; 9 mo. after, their gain is $3250: what is each one's gain?

5. A and B are partners, each contributing $1000; after 3 months, A withdraws $400, which B advances; the same is done after 3 months more; their year's gain is $800: what should each get?

6. A, B, and C are employed to empty a cistern by two pumps of the same bore; A and B go to work first, making 37 and 40 strokes respectively a minute; after 5 minutes, each makes 5 strokes less a minute; after 10 minutes, A gives way to C, who makes 30 strokes a minute until the cistern is emptied, which was in 22 minutes from the start: divide their pay, $2.

7. A and B are partners; A's capital is $4200; B's, $5600: after 4 months, how much must A put in, to entitle him to $\frac{1}{2}$ the year's gain?

8. A and B go into partnership, each with $4500. A draws out $1500, and B $500, at the end of 3 mo., and each the same sum at the end of 6 and 9 mo.; at the end of 1 yr. they quit with $2200: how must they settle?

9. A, B, C, and D go in partnership; A owns 12 shares of the stock; B, 8 shares; C, 7 shares; D, 3 shares. After 3 mo., A sells 2 shares to B, 1 to C, and 4 to D; 2 mo. afterward, B sells 1 share to C, and 2 to D; 4 mo. afterward, A buys 2 from C and 2 from D. Divide the year's gain ($18000).

10. A, with $400; B, with $500; and C, with $300, joined in business; at the end of 3 mo. A took out $200; at the end of 4 mo. B drew out $300, and after 4 mo. more, he drew out $150; at the end of 6 mo. C drew out $100; at the end of the year they close; A's gain was $225: what was the whole gain?

BANKRUPTCY.

DEFINITIONS.

359. 1. **Bankruptcy** is the inability of a person or a firm to pay indebtedness.

2. A **Bankrupt** is a person unable to pay his debts.

3. The assets of a bankrupt are usually placed in the hands of an **Assignee,** whose duty it is to convert them into cash, and divide the net proceeds among the creditors in proportion to their claims.

REMARKS.—1. This act on the part of a debtor is called *making an assignment*, and he is said to be able to pay so much *on the dollar.*

2. All necessary expenses, including assignee's fee (which is generally a certain rate per cent on the whole amount of property), must be deducted, before dividing.

4. The **amount paid on a dollar** can be found by taking such a part of one dollar as the whole property is of the whole amount of the debts; each creditor's proportion may be then found by multiplying his claim by the amount paid on the dollar.

NOTE.—Laws in regard to bankruptcy differ in the various states; usually a bankrupt who makes an honest assignment is freed by law of his remaining indebtedness, and is allowed to retain a homestead of from $500 to $5000 in value, and a small amount of personal property.

EXAMPLES FOR PRACTICE.

1. A has a lot worth $8000, good notes $2500, and cash $1500; his debts are $20000: what can he pay on $1, and what will A receive, whose claim is $4500?

SOLUTION.—$\$8000 + \$2500 + \$1500 = \$12000$, the amount of property which is $\frac{12000}{20000}$ or $\frac{3}{5}$ of the whole debts. Hence, $\frac{3}{5}$ of $\$1 = 60$ ct., the amount paid on $1, and $\$4500 \times .60 = \$2700$, the sum paid to A.

2. My assets are $2520; I owe A $1200; B, $720; C, $600; D, $1080: what does each get, and what is paid on each dollar?

3. A bankrupt's estate is worth $16000; his debts, $47500; the assignee charges 5%: what is paid on $1, and what does A get, whose claim is $3650?

Topical Outline.

Partnership.

- 1. Partnership........
  - 1. Definitions.
  - 2.
    - Case I.—Applications.
    - Case II.—Applications.
- 2. Bankruptcy........
  - 1. Definitions.
  - 2. Applications.

XVIII. ALLIGATION.

DEFINITIONS.

360. Alligation is the process of taking quantities of different values in a combination of average value. It is of two kinds, *Medial* and *Alternate.*

ALLIGATION MEDIAL.

361. Alligation Medial is the process of finding the mean or average value of two or more things of different given values.

PROBLEM.—If 3 lb. of sugar, at 5 ct. a lb, and 2 lb., at $4\frac{1}{2}$ ct. a lb., be mixed with 9 lb., at 6 ct. a lb., what per lb, is the mixture worth?

OPERATION.

Price.		Quantity.		Cost.
5 ct.	×	3	=	15 ct.
$4\frac{1}{2}$	×	2	=	9
6	×	9	=	54
		14		)78($5\frac{4}{7}$ ct

SOLUTION.—The 3 lb. at 5 ct. a lb. = 15 ct.; the 2 lb. at $4\frac{1}{2}$ ct. per lb. = 9 ct.; the 9 lb. at 6 ct. per lb. = 54 ct.; therefore, the whole 14 lb. are worth 78 ct.; and $78 \div 14 = 5\frac{4}{7}$ ct. per lb., *Ans.*

Rule.—*Find the values of the definite parts, and divide the sum of the values by the sum of the parts.*

EXAMPLES FOR PRACTICE.

1. Find the average price of 6 lb. tea, at 80 ct.; 15 lb., at 50 ct.; 5 lb., at 60 ct.; 9 lb., at 40 ct.

2. The average price of 40 hogs, at $8 each; 30, at $10 each; 16, at $12.50 each; 54, at $11.75 each.

3. How fine is a mixture of 5 pwt. of gold, 16 carats fine; 2 pwt., 18 carats fine; 6 pwt., 20 carats fine; and 1 pwt. pure gold?

4. Find the specific gravity of a compound of 15 lb. of copper, specific gravity, $7\frac{3}{4}$; 8 lb. of zinc, specific gravity, $6\frac{7}{8}$; and $\frac{1}{4}$ lb. of silver, specific gravity, $10\frac{1}{2}$.

REMARKS.—1. By the *specific gravity* of a body is usually understood, *its weight compared with the weight of an equal bulk of water;* it may be numerically expressed as the quotient of the former by the latter. Thus, a cubic inch of silver weighing $10\frac{1}{2}$ times as much as a cubic inch of water, its specific gravity $= 10\frac{1}{2}$.

2. To find the specific gravity of a body heavier than water: (1.) Find its weight in air, (2.) Suspending it by a light thread, find its weight in water and note the difference; (3.) Divide the first weight by this difference. For example: if a piece of metal weighs $1\frac{3}{4}$ oz. in air, but in water only $1\frac{1}{2}$ oz., its specific gravity $= 1\frac{3}{4} \div (1\frac{3}{4} - 1\frac{1}{2}) = 7$. (See *Norton's Natural Philosophy*, p. 152.)

5. What per cent of alcohol in a mixture of 9 gal., 86% strong; 12 gal., 92% strong; 10 gal., 95% strong; and 11 gal., 98% strong?

6. At a teacher's examination, where the lowest passable average grade was 50, an applicant received the following grades: In Orthography, 50; Reading, 25; Writing, 50; Arithmetic, 60; Grammar, 55; Geography, 55: did he succeed, or did he fail?

ALLIGATION ALTERNATE.

362. Alligation Alternate is the process of finding the proportional quantities at given particular prices or values in a required combination of given average value.

CASE I.

363. To proportion the parts, none of the quantities being limited.

PROBLEM.—1. What relative quantities of sugar, at 9 ct. a lb. and 5 ct. a lb., must be used for a compound, at 6 ct. a lb.?

SOLUTION.—If you put 1 lb. at 9 ct. in the mixture to be sold for 6 ct., you lose 3 ct.; if you put 1 lb. at 5 ct. in the mixture to be sold at 6 ct., you gain 1 ct.; 3 such lb. gain 3 ct.; the gain and loss would then be equal if 3 lb. at 5 ct. are mixed with 1 lb. at 9 ct.

OPERATION.

6	5	. . 3 lb. at 5 ct. = 15 ct.
	9	. . 1 lb. at 9 ct. = 9 ct.
		4 lb. worth 24 ct.

which is $\frac{24}{4} = 6$ ct. a lb.

PROBLEM.—2. What relative numbers of hogs, at $3, $5, $10 per head, can be bought at an average value of $7 per head?

EXPLANATION.—Writing the average price 7, and the particular values 3, 5, 10, as in the margin, we say: 3 sold for 7, is a gain of 4, which we write opposite; 5 sold for 7 is a gain of 2; 10 sold for 7 is a loss of 3. We wish to make the gains and losses *equal;* hence, each losing sale must be balanced by one which gains. To *lose* 3 *fours*, will be balanced by *gaining* 4 *threes*, and a gain of 3 *twos* will balance a loss of 2 *threes*. To indicate this in the operation, we write the *deficiency* 3, against the *excess* 4; then the excess 4 against the deficiency 3; and in another column, in the same manner, pair the 3 and 2, writing each opposite the position which the other has in the column of differences. The answer might be given in two statements of balance, thus: for each 3 of the first kind take 4 of the third, and for each 3 of the second kind take 2 of the third. Since each balance column shows only proportional parts, *we may multiply both quantities in any balance column by any number, fractional or integral*, and thus the final answers be varied indefinitely. For example, had the second balancing column been multiplied by 4, the answer would have read, 1, 4, 4, instead of 1, 1, 2. The principle just stated is of great value in removing fractions from the balance columns, when integral terms are desired.

OPERATION.

		Diff.	Balance.			*Ans.*
	3	4	3		3	1
7	5	2		3	3	1
	10	3	4	2	6	2

Rule.—1. *Write the particular values or prices in order, in a column, having the smallest at the head; write the average value in a middle position at the left and separated from the others by a vertical line.*

2. *In another column to the right, and opposite the respective values, place in order the differences between them and the average value.*

3. *Then prepare balance columns, giving to each of them two numbers, one an excess and the other a deficiency taken from the difference column; write each of these opposite the position which the other has in the difference column; so proceed until each number in the difference column has been balanced with another; then,*

The proportional quantity to be taken of each kind, will be the sum of the numbers in a horizontal line to the right of its excess or deficiency.

NOTE.—The proof of Alligation Alternate is the process of Alligation Medial.

EXAMPLES FOR PRACTICE.

1. What relative quantities of tea, worth 25, 27, 30, 32, and 45 ct. per lb., must be taken for a mixture worth 28 ct. per lb.?

REMARK.—It is evident that other results may by obtained by making the connections differently; as, 6, 17, 3, 3, 1 lb.; or, 17, 6, 1, 1, 3 lb.

2. What of sugar, at 5, $5\frac{1}{2}$, 6, 7, and 8 ct. per lb., must be taken for a mixture worth $6\frac{3}{4}$ ct. per lb.?

3. What relative quantities of alcohol, 84, 86, 88, 94, and 96% strong, must be taken for a mixture 87% strong?

4. What of gold and silver, whose specific gravities are

$19\frac{1}{4}$ and $10\frac{1}{2}$, will make a compound whose specific gravity shall be 16.84?

5. What of silver $\frac{3}{4}$ pure, and $\frac{9}{10}$ pure, will make a mixture $\frac{7}{8}$ pure?

6. What of pure gold, and 18 carats, and 20 carats fine, must be taken to make 22 carat gold?

CASE II.

364. To proportion the parts, one or more of the quantities, but not the amount of the combination, being given.

PROBLEM.—How many whole bushels of each of two kinds of wheat, worth respectively \$1.20 and \$1.40, per bushel, will, with 14 bushels, at \$1.90 per bushel, make a combination whose average value is \$1.60 per bushel?

OPERATION.

		Diff.	Bal.		Answers.									
	1.20	.40	3		1	2	3	4	5	6	7	8	9	10
1.60	1.40	.20		3	19	17	15	13	11	9	7	5	3	1
	1.90	.30	4	2	14	14	14	14	14	14	14	14	14	14

SOLUTION.—We find, by Case I, that to have that average value the parts may, in one balancing, stand 3 of the first to 4 of the third, and in another, 3 of the second to 2 of the third. By directly combining, we obtain the proportions 1, 1, and 2, and as the third must be 14, we have for one answer 7, 7, 14. But we find the other answers in the following manner.

The proportion will not be altered if in any balancing column we multiply both quantities by the same number, hence the answer can be varied as often as we can multiply or divide the columns, observing the other conditions, which are that the answers shall be *integral*, and that the number of 4's and 2's taken shall make 14. As there are more fractions than there are integers between any two limits, we try fractional multipliers in order to obtain the greatest number of answers. Observe that the number of 4's taken will not stand alone in any answer for the third kind of wheat, but will be added to some number of 2's; the number of 3's taken as any one answer for the first or second kind *will not* be increased by any other product; hence, if we use a *fractional* multiplier, it must be such that its *denominator* will *disappear* in multiplying by 3; and this shows that a fractional multiplier will not be convenient unless it can be expressed as *thirds*. Therefore, the remaining question is, How many thirds of 4 with thirds of 2 will make 14? Since $14 = \frac{42}{3}$, the question is the same as to ask, How many *whole* 4's with *whole* 2's will make 42? It is plain that there can not be more than *ten* 4's. We can take $\frac{1}{3}$ of

1 four and 19 twos, or $\frac{1}{3}$ of first column and $\frac{19}{3}$ of second.
2 fours " 17 " " $\frac{2}{3}$ " " " " $\frac{17}{3}$ " "
3 " " 15 " " $\frac{3}{3}$ " " " " $\frac{15}{3}$ " "

The answers are now obvious: write 1, 2, 3, etc., parallel with 19, 17, 15, etc., and the 14's in the third row.

Rule.—*Find the proportional parts, as in Case I, and observe the term or terms in the balance columns, standing opposite the value or price corresponding to the limited quantity; then find what multipliers will produce the given limited quantity in the required place, and of those multipliers use only such as will agree with the remaining conditions of the problem.*

Examples for Practice.

1. How many railroad shares, at 50%, must A buy, who has 80 shares that cost him 72%, in order to reduce his average to 60%?

2. How many bushels of corn, worth respectively 50, 60, and 75 ct. per bu., with 100 bu., at 40 ct. per bu., will make a mixture worth 65 ct. a bu.?

3. How much water (0 per cent) will dilute 3 gal. 2 qt. 1 pt. of acid 91% strong, to 56%?

4. A jeweller has 3 pwt. 9 gr. of old gold, 16 carats fine: how much U. S. gold, $21\frac{3}{5}$ carats fine, must he mix with it, to make it 18 carats fine?

5. How much water, with 3 pt. of alcohol, 96% strong, and 8 pt., 78%, will make a mixture 60% strong?

6. I mixed 1 gal. 2 qt. ½ pt. of water with 3 qt. 1½ pt. of pure acid; the mixture has 15% more acid than desired: how much water will reduce it to the required strength?

7. How much lead, specific gravity 11, with ½ oz. copper, sp. gr. 9, can be put on 12 oz. of cork, assuming the sp. gr. ¼, so that the three will just float, that is, have a sp. gr. (1) the same as water?

8. How many shares of stock, at 40%, must A buy, who has bought 120 shares, at 74%, 150 shares, at 68%, and 130 shares, at 54%, so that he may sell the whole at 60%, and gain 20%?

9. A buys 400 bbl. of flour, at $7.50 each, 640 bbl., at $7.25, and 960 bbl., at $6.75: how many must he buy at $5.50, to reduce his average to $6.50 per bbl.?

CASE III.

365. To proportion the parts, the amount of the whole combination being given.

PROBLEM.—If a man pay $16 for each cow, $3 for each hog, and $2 for each sheep, how many of each kind may he purchase so as to have 100 animals for $600?

OPERATION.

		Diff.	Bal.		Answers.				
	2	4	5		12	25	38	51	64
$6	3	3		10	64	50	36	22	8
	16	10	2	3	24	25	26	27	28
			7	13					

SOLUTION.—Proceeding as in Case I, we find that the given average requires 5 of the first kind with 2 of the third, and 10 of the second kind with 3 of the third; taken in two parts, 7 are required in *one* balancing, and 13 in another; these being together 20, which is contained five times in the required number, 100, if we multiply all of the terms in the balance columns by 5, we have for one answer 25 sheep, 50 hogs, and 25 cows.

As there are other multipliers affording results within the conditions, we leave the student to find the remaining answers by a process similar to that shown under Case II.

REMARK.—Suppose that we had to determine how many 7's and 11's would make 400. By trial, we find that *two* 11's taken away leave an exact number of 7's. It is now unnecessary to take single 11's or proceed by *trial* any farther; for, as 378 *is* an exact number of 7's, if we take away 11's and *leave* 7's, we must take *seven* 11's; and thus the law of continuation is obvious: 400 is composed of

```
       400
  2     22
       ---
       378...54
        77
       ---
  9    301...43
        77
       ---
 16    224...32
```

	11's......2,	9,	16,	23,	30;
With	7's......54,	43,	32,	21,	10.

Rule.—*Proportion the parts as in Case I; then, noting the sums of the balancing columns, find, by trial or by direct division, what multipliers will make those columns together equal to the given amount of the combination, and of those multipliers use only such as will agree with the remaining conditions.*

EXAMPLES FOR PRACTICE.

1. What quantities of sugar, at 3 ct. per lb. and 7 ct. per lb., with 2 lb. at 8 ct., and 5 lb. at 4 ct. per lb., will make 16 lb., worth 6 ct. per lb.?

2. How many bbl. flour, at $8 and $8.50, with 300 bbl. at $7.50, and 800 at $7.80, and 400 at $7.65, will make 2000 bbl. at $7.85 a bbl.?

3. What quantities of tea, at 25 ct. and 35 ct. a lb., with 14 lb. at 30 ct., and 20 lb. at 50 ct., and 6 lb. at 60 ct., will make 56 lb. at 40 ct. a lb.?

4. How much copper, specific gravity $7\frac{3}{4}$, with silver, specific gravity $10\frac{1}{2}$, will make 1 lb. troy, of specific gravity $8\frac{3}{5}$?

5. How much gold 15 carats fine, 20 carats fine, and pure, will make a ring 18 carats fine, weighing 4 pwt. 16 gr.?

6. A dealer in stock can buy 100 animals for $400, at the following rates,—calves, $9; hogs, $2; lambs, $1: how many may he take of each kind?

7. Hiero's crown, sp. gr. $14\frac{5}{8}$, was of gold, sp. gr. $19\frac{1}{4}$, and silver, sp. gr. $10\frac{1}{2}$; it weighed $17\frac{1}{2}$ lb.: how much gold was in it?

Topical Outline.

ALLIGATION.

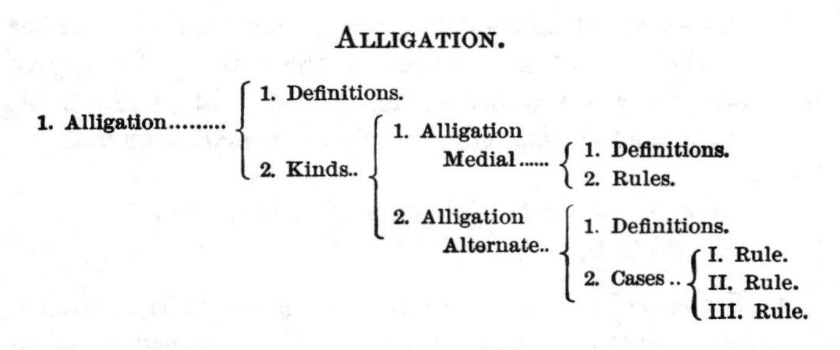

XIX. INVOLUTION.

DEFINITIONS.

366. 1. A **power** of a quantity is either that quantity itself, or the product of a number of factors each equal to that quantity.

REMARK.—Regarding unity as a base, we may say, the power of a quantity is the product arising from taking unity once as a multiplicand, with only the given quantity a certain number of times as a factor. The power takes its name from the number of times the quantity is used as factor. Unity is no power of any other positive number.

2. The **root** of a power is one of the equal factors which produce the power.

3 Powers are of different **degrees**, named from the number of times the root is taken to produce the power. The degree is indicated by a number written to the right of the root, and a little above; this index number is called an **exponent.** Thus,

5×5, or the 2*d power of* 5, is written 5^2.
$5 \times 5 \times 5$, " 3*d* " " " " 5^3.

4. The *second* power of any number is called the **square,** because the area of a square is numerically obtained by forming a second power.

5. In like manner the *third* power of any number is called its **cube,** because the solidity of a cube is numerically obtained by forming a third power.

367. To find any power of a number, higher than the first.

Rule.—*Multiply the number by itself, and continue the multiplication till that number has been used as factor as many times as are indicated by the exponent.*

NOTES.—1. The number of multiplications will be *one less* than the exponent, because the root is used *twice* in the first multiplica tion, once as multiplicand and once as multiplier.

2. When the power to be obtained is of a high degree, multiply by some of the powers instead of by the root continually; thus, to obtain the 9th power of 2, multiply its 6th power (64) by its 3d power (8); or, its 5th power (32) by its 4th power (16): the rule being, that *the product of any two or more powers of a number is that power whose degree is equal to the sum of their degrees.*

3. Any power of 1 is 1; any power of a number *greater* than 1 is *greater* than the number itself: any power of a number *less* than 1, is *less* than the number itself.

368. From Note 2, $4^3 \times 4^3 \times 4^3 \times 4^3 \times 4^3 = 4^{15}$; but the expression on the left is the 5th power of 4^3; hence, $(4^3)^5 = 4^{15}$; that is, *when the exponent of the power required is a composite number* (15), *raise the root to a power whose exponent is one of its factors* (3), *and this result to a power whose exponent is the other factor* (5).

NOTE.—Let the student carefully note the difference between *raising a power to a power*, and *multiplying* together *different* powers of the same root; thus,

$$2^3 \times 2^2 = 2^5.$$

Here we have multiplied the cube by the square and obtained the 5th power; but the 5th power is *not* the *square of the cube;* this is the *sixth* power, and we write

$$(2^2)^3 = 2^6, \text{ or } (2^3)^2 = 2^{3\times 2} = 2^6.$$

369. Any power of a fraction is equal to that power of the numerator divided by that power of the denominator.

370. The *square* of a decimal must contain *twice*, and its *cube*, *three times* as many decimal places as the root, etc.;

hence, to obtain any power of a decimal, we have the following rule:

Rule.—*Proceed as if the decimal were a whole number, and point off in the result a number of decimal places equal to the number in the root multiplied by the exponent of the power.*

Examples for Practice.

Show by involution, that:

1. $(5)^2$	equals	25.	8. $(\frac{7}{8})^5$	equals	$\frac{16807}{32768}$.
2. 14^3	"	2744.	9. $(.02)^3$	"	.000008
3. 6^5	"	7776.	10. $(5^4)^2$	"	390625.
4. 192^2	"	36864.	11. $(.046)^3$	"	.000097336
5. 1^{10}	"	1.	12. $(\frac{1}{9})^7$	"	$\frac{1}{4782969}$.
6. $(\frac{3}{5})^4$	"	$\frac{81}{625}$.	13. 2056^2	"	4227136.
7. $(2\frac{1}{4})^3$	"	$11\frac{25}{64}$.	14. $(7.62\frac{1}{2})^2$	"	$58.1406\frac{1}{4}$

371. Special processes for squaring and cubing numbers.

Problem.—Find the square of 64.

PARALLEL OPERATIONS.

$$\begin{array}{rl} 64 = & 60 + 4 = 6 \text{ tens} + 4 \text{ units.} \\ 64 & 60 + 4 \\ \hline 256 = & 60 \times 4 + 4^2 \\ 3840 = & 60^2 + 60 \times 4 \\ \hline 4096 = & 60^2 + 2(60 \times 4) + 4^2 \\ & = 3600 + 480 + 16. \end{array}$$

The operations illustrate the following principle:

Principle.—*The square of the sum of two numbers is equal to the square of the first, plus twice the product of the first by the second, plus the square of the second.* Thus:

$$25^2 = (22 + 3)^2 = 484 + 2\,(22 \times 3) + \ \ 9 = 625.$$
$$25^2 = (21 + 4)^2 = 441 + 2\,(21 \times 4) + 16 = 625.$$
$$25^2 = (20 + 5)^2 = 400 + 2\,(20 \times 5) + 25 = 625.$$

REMARK.—The usual application of this principle in *Arithmetic* is, in squaring a number as composed of *tens* and *units*. The third statement above illustrates this; and, if we represent the *tens* by t, the *units* by u, we have the following statement:

$(t + u)^2 = t^2 + 2tu + u^2$; or, in common language:

The square of any number composed of tens and units is equal to the square of the tens, + twice the product of the tens by the units, + the square of the units.

EXAMPLES FOR PRACTICE.

Square the following numbers, considering each as the sum of two quantities, and applying the principle announced in Art. **371,** on page preceding:

1. 19.
2. 29.
3. 4.
4. 40.
5. 125.
6. 59.

ILLUSTRATION.—Draw a square. From points in the sides, at equal distances from one of the corners, draw two straight lines across the figure, each parallel to two sides of the figure. These two lines will divide the square into four parts, two of them being squares and two of them rectangles. The base being composed of two lines, and the square of four parts, we see that

372. The square described on the sum of two lines is equal to the sum of the squares described on the lines, plus twice the rectangle of the lines.

REMARK.—Both the principle employed above and the illustration are frequently used in explaining the method for square root.

Problem.—Find the cube of 64.

PARALLEL OPERATIONS.

$$
\begin{array}{rll}
64^2 = & 4096 = & 60^2+2(60\times4)+4^2 \\
 & 64 = & 60+4 \\
\hline
 & 16384 = & 60^2\times4+2(60\times4^2)+4^3 \\
 & 24576 = & 60^3+2(60^2\times4)+3(60\times4^2) \\
\hline
64^3 = & 262144 = & 60^3+3(60^2\times4)+3(60\times4^2)+4^3.
\end{array}
$$

The operation illustrates the following principle:

Principle.—*The cube of any number composed of two parts, is equal to the cube of the first part, plus three times the square of the first by the second, plus three times the first by the square of the second, plus the cube of the second.* Thus:

$$25^3 = (22+3)^3 = 22^3 + 3\,(22^2 \times 3) + 3\,(22 \times 3^2) + 3^3$$
$$= 10648 + 4356 + 594 + 27 = 15625.$$

Remark.—The usual application of this principle in *Arithmetic* is, in the cubing of a number as composed of *tens* and *units*. Representing the *tens* by t, and the *units* by u, we have the following statement, which the student will express in common language, similar to that of the principle used in squaring numbers:

$$(t+u)^3 = t^3 + 3t^2u + 3tu^2 + u^3.$$

Examples for Practice.

Considering the following numbers as made, each, of two parts, cube them by the principle just stated:

(1.) 19.
(2.) 29.
(3.) 4.
(4.) 40.
(5.) 125.
(6.) 216.

XX. EVOLUTION.

DEFINITIONS.

373. 1. **Evolution** is the process of finding roots of numbers.

2. A **root** of a number is either the number itself or one of the equal factors which, without any other factor, produce the number.

Since a number is the *first root*, as also the *first power* of itself, no operation is necessary to find either of these; hence, in evolution, we seek *only* one of the *equal* factors which produce a power.

Evolution is the reverse of Involution, and is sometimes called the *Extraction of Roots.*

3. *Roots*, like *powers*, are of different degrees, 2d, 3d, 4th, etc.; the degree of a root is always the same as the degree of the power to which that root must be raised to produce the given number.

Thus, the 3d root of 343 is 7, since 7 must be raised to the 3d power, to produce 343; the 5th root of 1024 is 4, since 4 must be raised to the 5th power, to produce 1024.

Since the 2d and 3d powers are called the *square* and *cube*, so the 2d and 3d roots are called the *square* root and *cube* root.

4. To indicate the root of a number, we use the **Radical Sign** ($\sqrt{}$), or a *fractional exponent.*

The radical sign is placed before the number; the degree of the root is shown by the small figure between the branches of the radical sign, called the **Index** of the root.

Thus, $\sqrt[3]{18}$ signifies the cube root of 18; $\sqrt[5]{9}$ signifies the 5th root of 9. The square root is usually indicated without the index 2; thus, $\sqrt{10}$ is the same as $\sqrt[2]{10}$.

5. The root of a number may be indicated by a *fractional exponent whose numerator is* 1, *and whose denominator is the index of the root to be expressed.*

Thus, $\sqrt{7}=7^{\frac{1}{2}}$, and $\sqrt[3]{5}=5^{\frac{1}{3}}$; similarly, $4^{\frac{2}{3}}=16^{\frac{1}{3}}$, the numerator indicating a power and the denominator a root.

6. A **perfect power** is a number whose root can be exactly expressed in the ordinary notation; as 32, whose fifth root is 2.

7. An **imperfect power** is a number whose root can not be exactly expressed in the ordinary notation; as 10, whose square root is 3.1622+

8. The squares and cubes of the first nine numbers are as follows:

Numbers,	1	2	3	4	5	6	7	8	9
Squares,	1	4	9	16	25	36	49	64	81
Cubes,	1	8	27	64	125	216	343	512	729

9. The **Square Root** of a number is one of the two equal factors which, without any other factor, produce that number; thus, $7\times7=49$, and $\sqrt{49}=7$.

10. The **Cube Root** of a number is one of the three equal factors which, without any other, produce that number; thus, $3\times3\times3=27$, and $\sqrt[3]{27}=3$.

374. Concerning powers and roots in the ordinary decimal notation, we state the following principles:

PRINCIPLES.—1. *The square of any number has twice as many, or one less than twice as many, figures as the number itself has.*

2. *There will be as many figures in the square root of a perfect power as there are periods of two figures each in the power, beginning with units, and also a figure in the root corresponding to a part of such period at the left in the power.*

3. *The cube of a number has three times as many figures, or one or two less than three times as many, as the number itself has.*

4. *There will be as many figures in the cube root of a perfect power, as there are periods of three figures each in the power, beginning with units, and also a figure corresponding to any part of such a period at the left hand.*

Exercises.

1. Prove that there will be six figures in the cube of the greatest integer of two figures.

2. Prove that there will be twelve figures in the fourth power of the greatest integer of three figures.

EXTRACTION OF THE SQUARE ROOT.

First Explanation.

Problem.—What is the length of the side of a square containing 576 sq. in.?

OPERATION.

```
        5̇7 6̇ ( 2 0
                4
 4 0 | 4 0 0   2 4 in., Ans.
   4 | 1 7 6
 4 4 | 1 7 6
```

Solution.—The length required will be expressed by the square root of 576; by Principle 2, we know that the root can have no less than two places of figures; and since the square of 3 tens is greater, and that of 2 tens less than 576, the root must be less than 30 and greater than 20; hence, 2 is the first figure of the root, and 400 the greatest square of tens contained in 576. Let the first of the accompanying figures represent the square whose side is to be found. We see that the side must be greater than 20, and that the given area exceeds by 176 sq. in. the square whose side is 20 in. long. It is also evident from the figure that the 176 sq. in. may be regarded as made of three parts, two of them being rectangles and one a small square; these parts are of the same width, and, if that width be ascertained and added to 20 in., the required side will be found. The two 20-inch rectangles, with the small square, may be considered as making one long rectangle of the required width, as shown in the figure on the right; and, as the exact area of that rectangle is 176 sq. in., if we knew its length, its true width would be found by *dividing* the area *by* the length (Art. 197, 7); but we do know that the *length is greater than* 40 *in.*, and hence, that the width is, in inches, *less* than the quotient of 176 by 40; and since, in 176, 40 is contained *more than four* times, but *not five* times, 4 is the highest number we need try for the width. Now, as the *true length* of that rectangle is 40 in. *increased by the true width*, the proper *way* to try 4, is to add it to 40 and multiply the sum by 4; thus, 40 in. + 4 in. = 44 in.; and 44 × 4 = 176. This shows that 4 in. is the width of the rectangle, and hence the required side is 20 in. + 4 in. = 24 in., *Ans.*

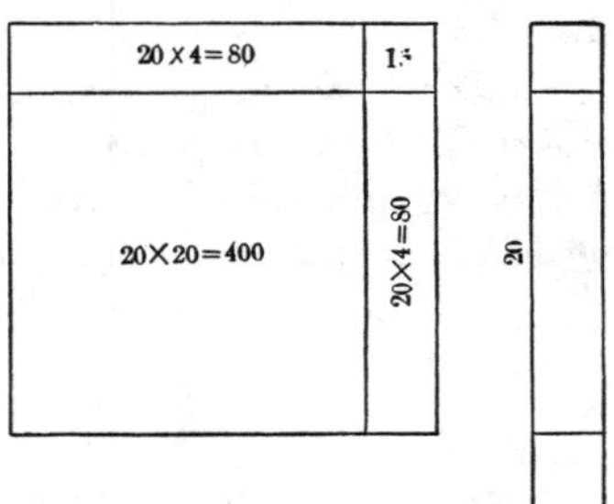

REMARKS.—1. Since 17 contains 4 as often *integrally*, as 176 contains 40, it is convenient to use simply the 17 as dividend with 4 as divisor, and then annex the quotient to the divisor and to the first figure of the root.

2. At the first step we ascertained that the whole root was greater than 2 tens and less than 3 tens; at the next step we learned that the units were not equal to 5, and by trial they were found to be 4. The whole process was a gradual approach to the exact root,—one figure at a time. It is important for the student to note that in the processes of evolution there must be steps of trial. Even the higher branches will not exempt from all trial work. The most valuable rules pertaining to such numerical operations, simply *narrow* the trial by making the limits obvious. Thus our device above showed the second part of the root less than 5; an actual trial showed it to be exactly 4.

```
    |5 8 0 8 1 ( 2 4 1
    |4
 44 |1 8 0
    |1 7 6
         |4 8 1
     481 |4 8 1
```

3. If the power had been 58081, we should have found there were three figures in the root; here, as in the former case, 4 is found the greatest figure

which can stand in ten's place, and we may treat the 24 tens exactly as we treated the 2 tens in the first illustration.

SECOND EXPLANATION.

We learned in Art. 371 that the square of a number composed of tens and units is equal to the square of the tens, plus twice the product of the tens by the units, plus the square of the units.

The square of $(20 + 4)$, or 24^2, is $20^2 + 2 \times (20 \times 4) + 4^2$. Now, if the square of the tens be taken away, there will remain $2 \times (20 \times 4) + 4^2 = 40$ times 4, and 4 times 4, or, simply $(40 + 4) \times 4$. We see then that if the square of the tens be taken away, the *remainder* is a product whose larger factor is the *double of the tens, increased by the units,* the smaller factor being simply the units.

Suppose, then, in seeking the square root of 1764, we have found the tens of the root to be 4; the remainder 164 must be the product of the units by a factor which is equal to the sum of twice the tens and once the units. If we knew the units, that larger factor could be found by doubling the tens and adding the units; if, on the other hand, we knew the larger factor, the units could be found by direct division; we *do* know that larger factor to be more than 80, and hence that the units factor is *less* than the *exact number* of times 164 contains 80. Therefore, the units figure can not be so great as 3, and the largest we need try is 2. The proper *way* to try 2, is to add it to 80, and then multiply by 2; this being done, we see that, the product being equal to 164, 2 *is* the exact number of units, 80 + 2 the larger factor exactly, and 42 the exact root.

OPERATION.

```
      1764(40
      1600  2
 80 |  164 42, Ans.
  2 |  164
 82 |
```

NOTE.—These successive steps showed, that the first figure *was* the root as nearly as tens could express it; with the second figure we found the root exactly. Had the power been 1781, the 42 would still have been the true root *as far as tens and units* could express it; and at the next step, seeking a figure in *tenth's* place, we would have found the true root, 42.2, as far as expressible by *tens, units, and tenths.* Continuing this operation, we find 42.201895 to be the root, *true* as far as *millionths* can express it; so, in any case, when *a figure* is correctly found, the *true* root can not differ from the whole root obtained, by so much as a *unit in the place of that figure.*

375. To extract the square root of a number written in the decimal notation, as integer, fraction, or mixed number.

Rule.—1. *Point off the number into periods of two figures each, commencing with units.*

2. *Find the greatest square in the first period on the left; place its root on the right, like a quotient in division; subtract the square from the period, and to the remainder bring down the next period for a dividend.*

3. *Double the root already found, as if it were units, and write it on the left for a trial divisor; find how often this is contained in the dividend, exclusive of the right-hand figure, annexing the quotient to the root and to the divisor; then multiply the complete divisor by the quotient, subtract the product from the dividend, and to the remainder bring down the next period as before.*

4. *Double the root as before, place it on the left as a trial divisor, proceeding as with the former divisor and quotient figure; continue the operation until the remainder is nothing, or until the lowest required decimal order of the root has been obtained.*

Notes.—1. If any product be found too large, the last figure of the root is too large.

2. The number of decimal places in the power must be *even;* hence the number of decimal periods can be increased only by annexing ciphers in *pairs.* Contracted division may be used to find the lower orders of an imperfect root.

3. When a remainder is greater than the previous divisor it does not follow that the last figure of the root is too small, unless that remainder be large enough to contain *twice the part of the root already found and 1 more;* for this would be the complete divisor, and would be contained in the remainder if the root were increased by 1. Hence, *the square of any number must be increased by a unit more than twice the number itself, to make the square of the next higher.*

Thus, $125^2=15625$; simply add $250+1$, and find $15876=126^2$.

EXAMPLES FOR PRACTICE.

1. $\sqrt{2809.}$
2. $\sqrt{1444.}$
3. $\sqrt{11881.}$
4. $\sqrt{185640625.}$
5. $\sqrt{80012304.}$
6. $\sqrt{6203794.}$
7. $\sqrt{3444736.}$
8. $\sqrt{57600.}$
9. $\sqrt{16499844.}$
10. $\sqrt{49098049.}$
11. $\sqrt{73005.}$
12. $\sqrt{386^3.}$

13. Find the square root of 3, true to the 7th decimal place.

14. Find the square root of 9.869604401089358, true to the 7th decimal place.

15. $\sqrt{.030625} \times \sqrt{40.96} \times \sqrt{.00000625} =$ what?

16. $\sqrt{(126)^2 \times (58)^2 \times (604)^2} =$

17. $\sqrt{12.96 \times \text{sq. rt. of } \frac{75}{108}} =$

REMARK.—The remainder, at any point, is equal to the square of an *unknown* part, plus twice the product of that part by a *known* part. The remainder may also be considered as the *difference of two squares*, which is always equal to the *product* of the *sum* of the roots by their *difference*.

376. The square root of the product of any number of quantities is equal to the product of their square roots; thus, $\sqrt{16 \times .49} = 4 \times .7 = 2.8$

377. The square root of a common fraction is equal to the square root of the numerator, divided by the square root of the denominator.

REMARK.—It is advantageous to multiply both terms by what will render the denominator a square.

EXAMPLES FOR PRACTICE.

1. $\sqrt{\frac{6}{7}} =$
2. $\sqrt{34\frac{5}{8}} =$
3. $\sqrt{\frac{4}{7}} =$
4. $\sqrt{272\frac{3}{12}} =$

5. $\sqrt{6\frac{2}{5}} =$
6. $\sqrt{\frac{28}{57} \times \frac{392}{2527} \times \frac{35}{38}} \times \sqrt{3} =$
7. $\sqrt{123.454321 \times .81} =$
8. $\sqrt{1.728 \times 4.8 \times \frac{3}{7}} =$

EXTRACTION OF THE CUBE ROOT.

First Explanation.

Problem.—What is the edge of a cube whose solid inches number 13824?

OPERATION.

```
                      1̇3824̇(20
                        8000   4
20²×3    =1200   5824  24, Ans.
20 ×3×4=240
        4²=  16
           1456   5824
```

Solution.—The length required will be expressed by the cube root of 13824. By Prin. 4, we know the root can have no less than two places of figures; also, since the cube of 3 tens is greater and that of 2 tens is less than 13824, the root must be less than 30 and more than 20; consequently, 2 is the first figure of the root, and 8000 is the greatest cube of tens contained in 13824. Let the first of the accompanying figures represent the cube whose edge is to be found. We see that it must be greater than 20 inches, and that the given solidity exceeds by 5824 cu. in. the cube whose edge is 20 inches, and which for convenience we will call *A*. (See Fig. 2.) It is also evident from the second figure, where the separate parts are shown, that the 5824 cu. in. may be regarded as made up of seven parts, three of them being square blocks (*B*) 20 in. long, three of them being rectangular blocks (*O*) of the same length, and one a small cube (*C*). These parts are of the same thickness, and if that thickness be ascertained and added to 20 in., the required edge will be found. The square blocks and the oblong blocks, with the small cube, may be considered as standing in line

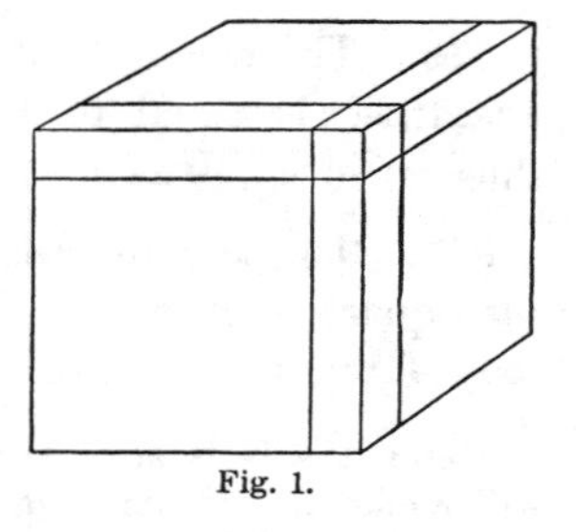

Fig. 1.

(Fig. 3) and forming one oblong solid of uniform thickness. Now, as the exact solidity of that solid is 5824 cu. in., if we knew its side surface, its true thickness would be found by *dividing* the number expressing the solidity *by* the number expressing the surface. But we do know that side surface to be greater than 3 times 400 sq. in., and hence the thickness must be *less* than the quotient of 5824 by 1200; and since in 5824, 1200 is contained 4 times but not 5 times, 4 is the *highest* number we need try for the width; as the exact surface of one side is equal to 1200 sq. in., increased by 3 rectangles 20 in. long, and a small square also, each of a width equal to the required thickness, the proper *way* to try 4 for that thickness is, to multiply it by 3 times 20, then by itself, and, adding the products to 1200, multiply the sum by 4; thus, 1200 sq. in. + 80 sq. in. × 3 + 16 sq. in. = 1456 sq. in., and 1456 × 4 = 5824, which

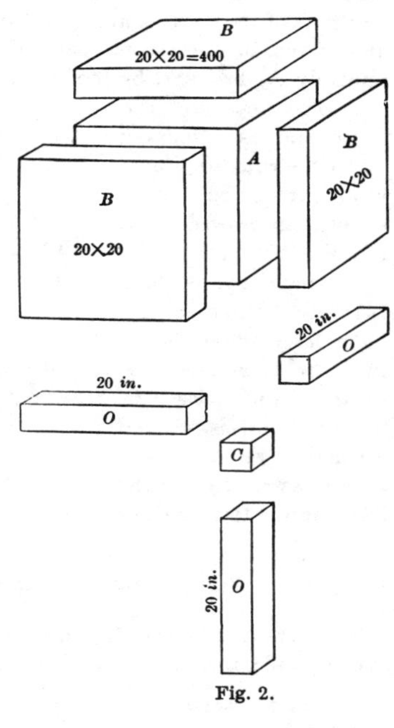

Fig. 2.

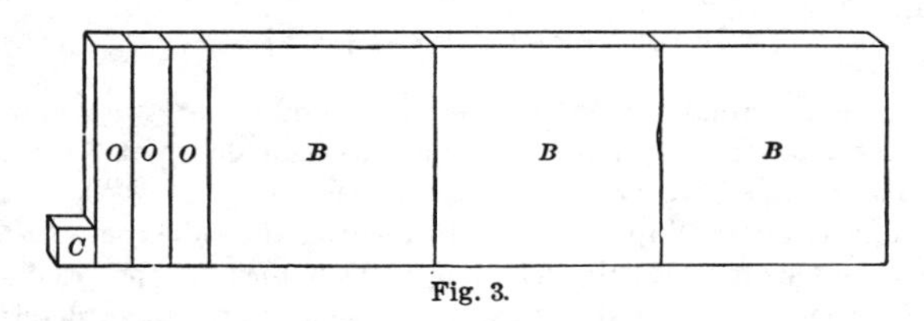

Fig. 3.

shows that 4 in. *is* the true thickness, and 20 in. + 4 in. = the required edge, 24 in., *Ans.*

NOTE.—Fig. 1 shows also the complete cube with the section lines marked.

REMARKS.—1. Since 3 times the square of 2 tens is equal to 300 times the square of 2, it is allowable to use simply the square of the first part of the root as units, multiplying by 300 for a trial divisor; and so, too, in the second part of the trial work, it will answer to multiply the first part by the last found figure and by 30.

2. The steps are trial steps, and as remarked under the rule for square root, our artifices have simply narrowed the range of the trial.

3. Had the power been 14172488, there would have been three figures in the root; and here, as in the former case, the second figure is 4; and we may treat the 24 tens exactly as we treated the 2 tens in the former illustration.

OPERATION.

```
                        14172488(242
                         8
 2²×300=1200           | 6172
 2×4×30=  240          |
     4²=   16          |
         1456          | 5824
                         348488
24²×300=172800         |
24×2×30=  1440         |
     2²=     4         |
        174244         | 348488
```

SECOND EXPLANATION.

We have seen (page 346) how the cube of a number, of two figures, is composed; that, for example,

$$24^3=8000+3(20^2\times4)+3(20\times4^2)+4^3.$$

Here we see that if the cube of the tens be taken away, there will remain

$$(20^2\times3+20\times4\times3+4^2)\times4;$$

that is, the remainder may be taken as a product of two factors, of which the smaller is the units, and the larger made up of 3 times the square of the tens, with 3 times the tens by the units, with also the *square* of the units. Suppose, then, in seeking the cube root of 74088, we find the tens to be 4; the remainder 10088 must be the product of units by a factor composed of three parts, such as we have described. If we knew the larger factor, the units could be obtained by direct division; but we *do* know that larger factor to be greater than 3 times the square of 40; hence, we know the units must be *less* than the quotient of 10088 by 4800, and consequently 2 is the largest figure we need try for units. The *way* to try 2, is to compose a larger factor after the manner just described; hence, multiply 2 by 3 times 40, add 2^2 or 4, add the sum to the 4800, and multiply by 2, which, being done, shows that 2 is the units figure, and that 42 is the root sought.

OPERATION.

	74088(42
	64
$4^2 \times 300 = 4800$	10088
$4 \times 2 \times 30 = 240$	
$2^2 = 4$	
5044	10088

REMARK.—Three times the square of the tens is the convenient trial divisor. This is in most instances a greater part of the complete divisor; for example, the least number of tens above *one* ten is 2, and the greatest figure in unit's place can not exceed 9; the cube of 29 is 24389, the first complete divisor is 1821, the first trial divisor being 1200, a greater part of it.

378. To extract the cube root of a number written in the decimal notation as whole number, fraction, or mixed number.

Rule.—1. *Beginning with units, separate the number into periods of three figures each; the extreme left period may have but one or two figures, but the extreme right, whether of units or decimal orders, must have three places, by the annexing of ciphers if necessary.*

2. *Find the greatest cube in the highest period, place its root on the right as a quotient in division, and then subtract the cube from the period, bringing down to the right of the remainder the next period to complete a dividend.*

3. *Square the root found as if it were units, multiply it by* 300, *and place the product on the left as a trial divisor; find how often it is contained in the dividend, and place the quotient figure to the right of the root; multiply the quotient by* 30 *times the preceding part of the root as units, square also the quotient, and add the two results to the trial divisor; then multiply the sum by the quotient, and subtract the product from the dividend, annexing to the remainder another period as before.*

4. *Square the whole root as before, multiply by* 300, *pro-*

ceeding as with the former trial divisor, quotient, and additions; continue the operation until the remainder is nothing, or until the lowest required decimal order of the root has been obtained.

NOTES.—1. Should any product exceed the dividend, the quotient figure is too large.

2. If any remainder is larger than the previous divisor, it does not follow that the last quotient figure is too small, *unless the remainder is large enough to contain* 3 *times the square of that part of the root already found, with* 3 *times that part of the root, and* 1 *more;* for this is the proper divisor if the root is increased by 1.

3. Should decimal periods be required beyond those with which the operation begins, the operator may annex three ciphers to each new remainder.

4. When the operator has obtained one more than half the required decimal figures of the root, the last complete divisor and the last remainder may be used in the manner of contracted division.

379. The cube root of any product is equal to the product of the cube roots of the factors. Thus,

$$\sqrt[3]{250 \times 4 \times 648 \times 9} = \sqrt[3]{125 \times 8 \times 216 \times 27} = 5 \times 2 \times 6 \times 3.$$

380. The cube root of a common fraction is equal to the quotient of the cube root of the numerator by the cube root of the denominator.

REMARKS.—1. When the terms are not both perfect cubes, multiply both by the square of the denominator, or by some smaller factor which will make the denominator a cube.

2. Reduce mixed numbers to improper fractions, or the fractional parts of such numbers to decimals.

EXAMPLES FOR PRACTICE.

1. $\sqrt[3]{512.}$
2. $\sqrt[3]{19683.}$
3. $\sqrt[3]{7301384.}$
4. $\sqrt[3]{94818816.}$

5. $\sqrt[3]{1067462648}$.
6. $\sqrt[3]{5.088448}$
7. $\sqrt[3]{22188.041}$
8. $\sqrt[3]{32.65}$
9. $\sqrt[3]{.0079}$
10. $\sqrt[3]{3.0092}$
11. $\sqrt[3]{\frac{23}{729}}$.
12. $\sqrt[3]{25}$.
13. $\sqrt[3]{11}$.
14. $\sqrt[3]{\frac{2}{3}}$.
15. $\sqrt[3]{\frac{4}{15}}$.
16. $\sqrt[3]{171.416328875}$
17. $\sqrt[3]{7011}$.
18. $\sqrt[3]{\frac{48}{4394}}$.
19. $\sqrt[3]{\frac{2}{3} \text{ of } \frac{4}{11}}$.

EXTRACTION OF ANY ROOT.

381. We have seen in the chapter on Involution, that if a *power* be *raised* to a power, the new exponent is the product of the given exponent by the number of times the power is taken as a factor; that, for example, 2^3 raised to the 4th power, is $2^{3\times4} = 2^{12}$. Consequently, reversing that process, a power may be separated into equal factors, if the given exponent be a composite number; thus, $2^{12} = 2^{4\times3}$, and consequently $\sqrt[4]{2^{12}} = 2^3$, or $\sqrt[3]{2^{12}} = 2^4$; for 2^{12} equals either $2^4 \times 2^4 \times 2^4$, or $2^3 \times 2^3 \times 2^3 \times 2^3$.

It is important to note the distinction between separating a power into factors which are powers, and separating that power into *equal* factors, having the same roots and exponents. The latter separation would be extracting a root. Thus,

$$2^5 = 2^3 \times 2^2 \text{ because equal to } 2^{3+2}.$$

But the *square root* of 2^5 is *not* 2^3, nor is the *cube* root of 2^5 equal to 2^2. If 2^3 be taken *twice as a factor* we have 2^6, and $\sqrt{2^6} = 2^3$; similarly, $\sqrt[3]{2^6} = 2^2$.

382. *A root of any required degree may be extracted, by separating the number denoting the degree, into its factors, and extracting successively the roots denoted by those factors.* Thus, the 9th root is the cube root of the cube root, and the 6th root the square root of the cube root.

HORNER'S METHOD.

383. Horner's Method, named from its inventor, **Mr.** W. G. Horner, of Bath, England, may be advantageously applied in extracting any root, especially if the degree of the root be not a composite number.

Rule for Extracting any Root.—1. *Make as many columns as there are units in the index of the root to be extracted; place the given number at the head of the right-hand column, and ciphers at the head of the others.*

2. *Commencing at the right, separate the given number into periods of as many figures as there are columns; extract the required root to within unity, of the left-hand period, for the* 1*st figure of the root.*

3. *Write this figure in the* 1*st column, multiply it then by itself, and set it in the* 2*d column; multiply this again by the same figure, and set it in the* 3*d column, and so on, placing the last product in the right-hand column, under that part of the given number from which the figure was derived, and subtracting it from the figures above it.*

4. *Add the same figure to the* 1*st column again, multiply the result by the figure again, adding the product to the* 2*d column, and so on, stopping at the next to the last column.*

5. *Repeat this process, leaving off one column at the right every time, until all the columns have been thus dropped; then annex one cipher to the number in the* 1*st column, two to the number in the* 2*d column, and so on, to the number in the last column, to which the next period of figures from the given number must be brought down.*

6. *Divide the number in the last column by the number in the previous column as a trial divisor (making allowance for completing the divisor); this will give the* 2*d figure of the root, which must be used precisely as the* 1*st figure of the root has been; and so on, till all the periods have been brought down.*

PROBLEM.—Extract the fourth root of 68719476736.

```
 0          0            0
 5          25           125          68719476736(512 Ans.
 5          50           375          625
 10         75          *500000     * 621947
  5         75            15201       515201
 15        *15000        515201     * 1067466736
  5           201         15403       1067466736
*200        15201       *530604000
   1          202          3129368
 201        15403        533733368
   1          203
 202       *1560600
   1           4084
 203        1564684
   1
*2040
    2
 2042
```

NOTE.—It is convenient to denote by * the place where a column is dropped; *i. e.*, reached for the last time by the use of the root figure in hand.

384. The process may often be shortened by this

Contracted Method.—*Obtain one less than half of the figures required in the root as the rule directs; then, instead of annexing ciphers and bringing down a period to the last numbers in the columns, leave the remainder in the right-hand column for a dividend; cut off the right-hand figure from the last number of the previous column, two right-hand figures from the last number in the column before that, and so on, always cutting off one more figure for every column to the left.*

With the number in the right-hand column and the one in the previous column, determine the next figure of the root, and use it as directed in the rule, recollecting that the figures cut off are not used except in carrying the tens they produce.

This process is continued until the required number of figures is obtained, observing that when all the figures in the last number of any column are cut off, that column will be no longer used.

REMARK.—Add to the 1st column mentally; multiply and add to the next column in one operation: multiply and subtract from the right-hand column in like manner.

PROBLEM.—Extract the cube root of 44.6 to six decimals.

```
0        0          44.600(3.546323
3        9           17600
6        2700          1725000
90       3175           238136
95       367500          12182
100      371716            865
1050     375948            111
1054     37659
1058     37723
1062
```

REMARK.—The trial divisors may be known by ending in two ciphers; the complete divisors stand just beneath them. After getting 3 figures of the root, contract the operation by last rule.

EXAMPLES FOR PRACTICE.

1. Extract the square root of 15625.
2. Extract the cube root of 68719476736.
3. Extract the fifth root of 14348907.
4. The cube root of 151.
5. $\sqrt[4]{97.41}$
6. $\sqrt[4]{1.08}$
7. $\sqrt[5]{\frac{5}{12}}$.
8. $\sqrt[5]{35.2}$
9. $\sqrt[12]{782757789696.}$
10. $\sqrt[9]{1367631.}$

APPLICATIONS OF SQUARE ROOT AND CUBE ROOT.

DEFINITIONS.

385. 1. A **Triangle** is a figure which has three sides and three angles; as, ABC, MNP.

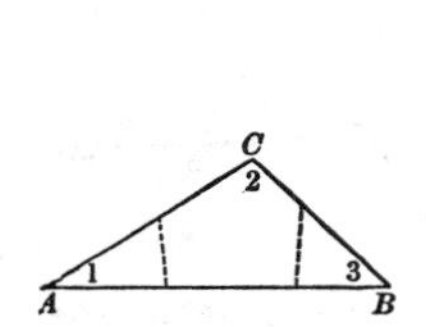

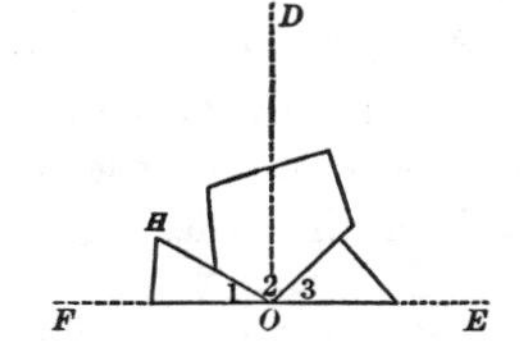

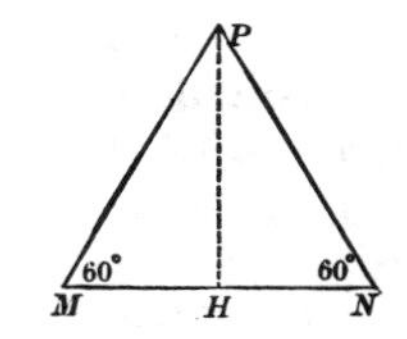

2. The **Base** of a triangle is that side upon which it is supposed to stand; as, AB, MN.

3. The **Altitude** of a triangle is the perpendicular distance from the base to the vertex of the angle opposite; as, HP.

REMARK.—The three angles of a triangle are together equal to 180°, or two right angles. The proof of this belongs to Geometry, but a fair *illustration* may be made in the manner indicated above. Mark the angles of a card or paper triangle, 1, 2, 3; and by two cuts divide it into three parts. Place the marked angles with their vertices as at O, and it will be seen that the pieces fit a straight edge through O, while the angles cover just twice 90°, or EOD + FOD. Any angle less than 90°, as HOF, is an *acute angle;* any angle greater than 90°, as HOE, is an *obtuse angle.*

4. An **Equilateral Triangle** is a triangle having three equal sides; as, MNP.

REMARK.—The angles of an equilateral triangle are 60° each; hence, six equilateral triangles can be formed about the same point as a vertex, each angle at the vertex being measured by the sixth of a circumference. (Art. **204.**)

RIGHT-ANGLED TRIANGLES.

386. A **Right-angled Triangle** is a triangle having one right angle; as IGK, where G = 90°.

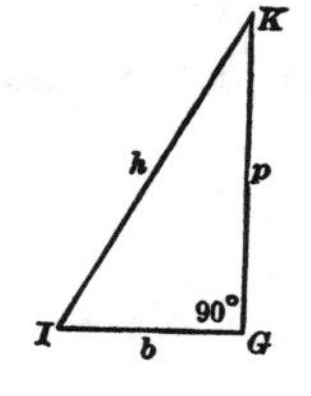

The side opposite the right angle is called the *hypothenuse;* the other two sides are called the *base* and *perpendicular.*

It is demonstrated in Geometry that the square described upon the hypothenuse of a right-angled triangle is equal to the sum of the squares described on the other two sides.

ILLUSTRATION.—A practical proof of this may be made in the following manner, especially valuable when the triangle has no equal sides. It will be a useful and entertaining exercise for the pupil.

Let the triangle be described upon a card, and let it stand upon the hypothenuse, as AEB does. Make three straight cuts;—one, perpendicular from A, through the smaller square; one, perpendicular from B, through the larger square, and one at right angles from the end of the second cut. The two squares are thus divided

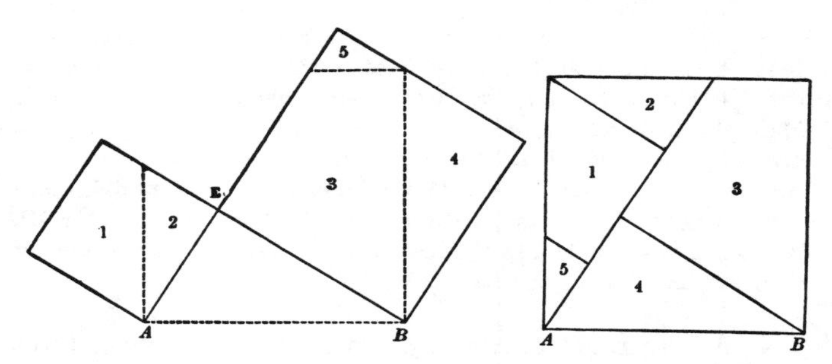

into five parts, which may be marked, and arranged, as here shown, in a square equal to one described on the hypothenuse.

REMARK.—The perpendicular in an equilateral triangle divides the base into two equal parts, and also divides the opposite angle into two which are 30° each. From this it follows that if, in a right-angled triangle, one angle is 30°, the side opposite that angle is half the hypothenuse; and, conversely, if one side be half the hypothenuse, the angle opposite will be 30°.

387. To find the hypothenuse when the other two sides are given.

Rule.—*Add together the squares of the base and perpendicular, and extract the square root of the sum.*

388. To find one side when the hypothenuse and the other side are given.

Rule.—*Subtract the square of the given side from the square of the hypothenuse, and extract the square root of the difference*; or,

Multiply the square root of the sum of the hypothenuse and side by the square root of their difference.

Representing the three sides by the initial letters h, p, b, we have the following

FORMULAS.—1. $h = \sqrt{p^2 + b^2}$.
2. $p = \sqrt{h^2 - b^2}$; or, $p = \sqrt{h + b} \times \sqrt{h - b}$.
3. $b = \sqrt{h^2 - p^2}$; or, $b = \sqrt{h + p} \times \sqrt{h - p}$.

EXAMPLES FOR PRACTICE.

1. Find the length of a ladder reaching 12 ft. into the street, from a window 30 ft. high.
2. What is the *diagonal*, or line joining the opposite corners, of a square whose side is 10 ft.?
3. What is saved by following the diagonal instead of the sides, 69 rd. and 92 rd., of a rectangle?
4. A boat in crossing a river 500 yd. wide, drifted with the current 360 yd.; how far did it go?

REMARK.—Integers expressing the sides of right-angled triangles may be found to any extent in the following manner: Take any two unequal numbers; the *sum* of their *squares* may represent a hypothenuse; the *difference* of their squares will then stand for one side, and *double their product* for the remaining side. Thus, from 3 and 2, form 13, 12, 5; from 4 and 1, form 17, 15, 8; from 5 and 2, form 29, 21, 20.

PARALLEL LINES AND SIMILAR FIGURES.

DEFINITIONS.

389. 1. **Parallel Lines** are lines which have the same direction. The shortest distance between two straight parallels is, at all points, equal to the same perpendicular line.

2. **Similar Figures** are figures having the same number of sides, and their like dimensions proportional.

REMARKS.—1. Similar figures have their corresponding angles equal.

2. If a line be drawn through any triangle parallel to one of the sides, the other two sides are divided proportionally, and the triangle marked off, is similar to the whole triangle. An illustration of this has already been furnished in solving Ex. 8, Art. 231.

3. All equilateral triangles are similar; the same is true of all squares, all circles, all spheres.

3. The areas of similar figures are to each other as the squares of their like dimensions.

4. The solidities of similar solids are to each other as the cubes of their like dimensions.

GENERAL EXERCISES IN EVOLUTION AND ITS APPLICATIONS.

1. One square is $12\frac{1}{4}$ times another: how many times does the side of the 1st contain the side of the 2d?

2. The diagonals of two similar rectangles are as 5 to 12. how many times does the larger contain the smaller?

3. The lengths of two similar solids are 4 in. and 50 in.; the 1st contains 16 cu. in.: what does the 2d contain?

4. The solidities of two balls are 189 cu. in. and 875 cu. in.; the diameter of the 2d is $17\frac{1}{2}$ in.; find the diameter of the 1st.

5. In extracting the square root of a perfect power, the last complete dividend was found 4725: what was the power?

6. What number multiplied by $\frac{2}{7}$ of itself makes 504?

7. Separate $91252\frac{1}{2}$ into three factors which are as the numbers 1, $2\frac{1}{2}$, and 3.

8. What integer multiplied by the next greater, makes 1332?

9. The length and breadth of a ceiling are as 6 and 5; if each dimension were one foot longer, the area would be 304 sq. ft.: what are the dimensions?

10. In extracting the cube root of a perfect integral power, the operator found the last complete dividend 241984: what was the power?

11. If we cut from a cubical block enough to make each dimension one inch shorter, it will lose 1657 cubic inches: what is the solidity?

12. A hall standing east and west, is 46 ft. by 22 ft., and $12\frac{1}{2}$ ft. high: what is the length of the shortest path a fly can travel, by walls and floor, from a southeast lower corner to a northwest upper corner?

13. How many stakes can be driven down upon a space 15 ft. square, allowing no two to be nearer each other than $1\frac{1}{2}$ ft., and how many allowing no two to be nearer than $1\frac{1}{4}$ ft.?

14. What integer is that whose square root is 5 times its cube root?

15. If the true annual rate of interest be 10%, what would be the true rate for each 73 days, if the interest be

compounded through the year? Prove the result by contracted multiplication. (Art. **334**, Rem. 2.)

16. If a field be in the form of an equilateral triangle whose altitude is 4 rods, what would be the cost of fencing it in, at 75 ct. a rod?

Topical Outline.

Powers and Roots.

- 1. Involution
  - 1. Definitions.
  - 2. Terms
    - 1. Powers.
    - 2. Root.
    - 3. Degree.
    - 4. Exponent.
  - 3. Squaring and Cubing
    - 1. Algebraic Statements.
    - 2. Numerical Illustrations.
    - 3. Geometrical Illustration.
    - 4. Principles.
- 2. Evolution
  - 1. Definitions.
  - 2. Terms
    - 1. Root.
    - 2. Powers
      - 1. Perfect.
      - 2. Imperfect.
    - 3. Radical.
    - 4. Index.
  - 3. Square Root
    - 1. First explanation (Geometrical).
    - 2. Second explanation (Algebraic).
    - 3. Rule.
  - 4. Cube Root
    - 1. First explanation (Geometrical).
    - 2. Second explanation (Algebraic).
    - 3. Rule.
  - 5. Roots in General—Horner's Method.
  - 6. Applications.

XXI. SERIES.

DEFINITIONS.

390. 1. A **Series** is any number of quantities having a fixed order, and related to each other in value according to a fixed law. These quantities are called *Terms;* the first and last are called *Extremes*, and the others *Means.*

The *Law* of a series is a statement by which, from some necessary number of the terms the others may be computed.

2. There are many different kinds of series. Those usually treated in Arithmetic are distinguished as *Arithmetical* and *Geometrical;* these series are commonly called *Progressions.*

ARITHMETICAL PROGRESSION.

391. 1. An **Arithmetical Progression** is a series in which any term differs from the preceding or following by a fixed number. That fixed number is called the **common difference;** and the series is *Ascending* or *Descending*, accordingly as the *first term* is the *least* or the *greatest.*

Thus, 1, 3, 5, 7, 9, is an ascending series, whose common difference is 2; but if it were written in a reverse order (or, if we treated 9 as the first term), the series would be descending.

2. Every Arithmetical Progression may be considered under the relations of **five quantities**, such that any *three of them being given*, the others may be found. These five are conveniently represented as follows:

First term, ... a.
Last term, ... l.
Number of terms, ... n.
Common difference, . . . d.
Sum of all the terms, . . s.

3. These give rise to twenty different cases, but all the calculations may be made from the principles stated in the two following cases.

NOTE.—Some of the problems arising under this subject are, properly, Algebraic exercises. Nothing will be presented here, however, which is beyond analysis by means of principles and processes exhibited in this book. The formulas given are easily understood, and the student will find it a very simple operation to write the numbers in place of their corresponding letters, and work according to the signs. The formulas are presented as a *convenience.*

CASE I.

392. One extreme, the common difference, and the number of terms being given, to find the other extreme.

PROBLEM.—Find the 20th term of the arithmetical series 1, 4, 7, 10, etc.

SOLUTION.—Here the series may be considered as made of 20 terms, and we seek the last. The com. diff. is 3, and the terms are composed thus: 1, $1+3$, $1+6$, $1+9$, etc.; and it is obvious that, as the *addition* of the com. diff. *commences* in forming the second term, it is taken *twice* in the third term, *three times* in the fourth, and so on; similarly therefore it must be taken 19 times in forming the 20th, and the simple operation is, $1+(20-1)\times 3=58$, *Ans.*

FORMULA.—$l=a+(n-1)d$; or $l=a-(n-1)d$.

Rule.—*Multiply the number of terms less one by the common difference, add the product to the given extreme when the larger is sought, subtract it from the given extreme when the smaller is sought.*

EXAMPLES FOR PRACTICE.

1. Find the 12th term of the series 3, 7, 11, etc.
2. Find the 18th term of the series 100, 96, etc.
3. Find the 64th term of the series $3\frac{1}{2}$, $5\frac{3}{4}$, etc.

4. Find the 10th term of the series .025, .037, etc.

5. Find the 1st term of the series 68, 71, 74, having 19 terms.

6. Find the 1st term of the series 117, $123\frac{1}{2}$, 130, having 6 terms.

7. Find the first term of the series $18\frac{3}{4}$, $12\frac{1}{2}$, $6\frac{1}{4}$, having 365 terms.

CASE II.

393. The extremes and the number of terms being given, to find the sum of the series.

PROBLEM.—What is the sum of 9 terms of the series 1, 4, 7, 10, etc.?

OPERATION.

$$1 + (9 - 1) \times 3 = 25.$$

$$\frac{1 + 25}{2} \times 9 = 117, \textit{ Ans.}$$

EXPLANATION.—Writing the series in full, in the common order, and also in a reverse order, we have

$$\text{Sum} = 1 + 4 + 7 + 10 + 13 + 16 + 19 + 22 + 25.$$
$$\text{Sum} = 25 + 22 + 19 + 16 + 13 + 10 + 7 + 4 + 1.$$

Twice the sum $= 26 + 26 + 26 + 26 + 26 + 26 + 26 + 26 + 26 = 9$ times the sum of the extremes; $\therefore$ the sum $= \frac{1}{2}$ of $9 \times 26 = 117$, *Ans.* If we add a term whose place is a certain distance beyond the first, to another whose place is equally distant from the last, the sum will be the same as that of the extremes, and hence, as the above illustrates, the double of any such series is equal to the product of the number of terms by the sum of the extremes.

FORMULA.—$s = \frac{(a + l)\,n}{2}$.

Rule.—*Multiply the sum of the extremes by the number of terms, and divide by* 2.

EXAMPLES FOR PRACTICE.

1. Find the sum of the arithmetical series whose extremes are 850 and 0, and number of terms, 57.

2. Extremes, 100 and .0001: number of terms, 12345.

3. What is the sum of the arithmetical series 1, 2, 3, etc., having 10000 terms?

4. Of 1, 3, 5, etc., having 1000 terms?

5. Of 999, 888, 777, etc., having 9 terms?

6. Of 4.12, 17.25, 30.38, etc., having 250 terms?

7. Whose 5th term is 21; 20th term, 60; number of terms, 46?

Examples for Practice.

Remark.—It is not deemed necessary to formulate a special rule for each class of examples here introduced. The following are presented as exercises in analysis, each depending on one or more of the principles above stated.

1. Find the common difference of a series whose extremes are 8 and 28, and number of terms, 6.

2. Extremes are $4\frac{1}{2}$ and $20\frac{3}{4}$, and number of terms, 14.

3. Insert one arithmetical mean between 8 and 54.

4. Insert five arithmetical means between 6 and 30.

5. Insert two arithmetical means between 4 and 40.

6. Insert four arithmetical means between 2 and 3.

7. What is the number of terms in a series whose extremes are 9 and 42, and common difference, 3?

8. Whose extremes are 3 and $10\frac{1}{2}$, and common difference, $\frac{3}{8}$?

9. In the series 10, 15 . . . 500?

10. What principal, on annual interest at 10%, will, in 50 yr., amount to $4927.50?

GEOMETRICAL PROGRESSION.

394. 1. A **Geometrical Progression** is a series in which any term after the first is the product of the preceding term by a fixed number. That fixed number is called the **ratio;** and the series is *ascending* or *descending* accordingly as the *first term* is the *least* or the *greatest.*

Thus, 1, 3, 9, 27, is an ascending progression, and the common multiplier is the ratio of 3 to 1 (Art. **228**, 2), or of any term to the preceding; considering 27 as the first term, the same series may be called descending.

2. Any Geometrical Progression may be considered under the relations of **five quantities,** of which *three must be known* in order to find the others. These five are thus represented:

First term, ...	a.
Last term, ...	l.
Ratio, ...	r.
Number of terms, ...	n.
Sum of all the terms, . .	s.

3. These quantities give rise to 20 different classes of problems, but all of the necessary calculations depend upon principles set forth in the following cases.

NOTE.—Some of the problems arising from these quantities require such an application of the formulas as can not be understood without a knowledge of Algebra.

CASE I.

395. From one extreme, the common ratio and the number of terms, to find the other extreme.

PROBLEM.—Find the 8th term of the series 1, 2, 4, 8, etc.

STATEMENT.

$2^7 \times 1 = 128$, *Ans.*

EXPLANATION.—Here, 1 being the first term, and 2 the ratio, we see that the series may be formed thus: 1, 1×2, 1×2^2, 1×2^3, etc., the ratio being raised to its *second* power in forming the 3d term, to its *third* power in forming the 4th; and so, similarly, the 8th term $=1\times 2^7=128$, *Ans.*

FORMULA.—$l=ar^{n-1}$.

Rule.—*Consider the given extreme as the first term, and multiply it by that power of the ratio whose degree is denoted by the number of terms less one.*

EXAMPLES FOR PRACTICE.

1. Find the last term in the series 64, 32, etc., of 12 terms.
2. In 2, 5, $12\frac{1}{2}$, etc., having 6 terms.
3. In 100, 20, 4, etc., having 9 terms.
4. 1st term, 4; common ratio, 3; find the 10th term.

5. 3d term, 16; common ratio, 6: find the 9th term.

6. 33d term, 1024; common ratio, $\frac{3}{4}$: find the 40th term.

7. Find the 1st term of the series 90, 180, of 6 terms.
8. Of $\frac{729}{2048}$, $\frac{2187}{4096}$, having 11 terms.

CASE II.

396. From one extreme, the ratio, and the number of terms, to find the sum of the terms.

PROBLEM.—The first term is 3, the ratio 4, the number of terms 5; required the sum of the series.

OPERATION.

$$\frac{4\times 3\times 4^4-3}{4-1}=1023, \textit{ Ans.}$$

EXPLANATION.—Writing the whole series, we have:

$$S = 3 + 12 + 48 + 192 + 768.$$
$$\text{Also, } 4\,S = \quad 12 + 48 + 192 + 768 + 3072.$$

It is evident that the lower line exceeds the upper by the difference between 3072 and 3; this difference may be written $4 \times 768 - 3$, or $4 \times 3 \times 4^4 - 3$, and as this is 3 *times the series,* we have, *once* the series =

$$\frac{4 \times 3 \times 4^4 - 3}{4 - 1} = 1023, \ \textit{Ans.}$$

Now, observing the form, note that we have multiplied the last term by the ratio, then subtracted the first term, and then divided by the ratio less one. If the series had stood with 768 for first term, and the multiplier $\frac{1}{4}$, we should have had

$$768 + 192 + 48 + 12 + 3 \qquad = S,$$
$$192 + 48 + 12 + 3 + \tfrac{3}{4} = \tfrac{1}{4}\,S;$$

and thus $768 - \frac{3}{4} = \frac{3}{4}$ of the series; hence, we can write,

$$\frac{768 - \frac{1}{4} \times 3}{1 - \frac{1}{4}} = 1023, \text{ as before.}$$

Here we have taken the product of the last term by the ratio *from* the first term, and have divided by the excess of unity above the ratio. In either case, therefore, we have illustrated

Rule I.—*Find the last term and multiply it by the ratio; then find the difference between this product and the first term, and divide by the difference between the ratio and unity.*

FORMULA.—$S = \dfrac{rl - a}{r - 1}$; or, $S = \dfrac{a - rl}{1 - r}$.

The first answer, above given, may take another form, thus:

$\dfrac{4 \times 4^4 \times 3 - 3}{4 - 1}$ is the same as $\dfrac{3(4^5 - 1)}{4 - 1}$, where appear the ratio 4, the first term 3, and the number of terms 5. This form, often used when the series is ascending, has the following general statement

$$S = \frac{a(r^n - 1)}{r - 1},$$

and corresponds to the following rule:

Rule II.—*Raise the ratio to a power denoted by the number of terms, subtract 1, divide the remainder by the ratio less 1, and multiply the quotient by the first term.*

NOTES.—1. The amount of a debt at compound interest for a number of complete intervals, is the last term of a geometrical progression, whose ratio is 1 + the rate per cent. The table (Art. **335**) shows the powers of the ratio. For example, the period being 4 yr., the *number of terms* is *five;* the first term is the principal, and the power of the ratio required by Case I, is the *fourth.*

2. The amount of an annuity at compound interest is conveniently found by the Formula corresponding to Rule II. The table (Art. **335**) is available, and the work very simple. Thus, if the annuity be \$200, the time 40 yr., and the rate 6%, we have $a = 200$, $n = 40$, $r = 1.06$ Then, writing these values in the Formula, we have:

$$\text{Amount} = \frac{\$200(1.06^{40}-1)}{1.06-1} = \frac{\$200 \times 9.2857179}{.06} = \$30952.39, \textit{ Ans.}$$

3. If the series be a descending one having an infinite number of terms, the last term is 0, and the product required by Rule I is 0.

EXAMPLES FOR PRACTICE.

1. Find the sum of 6, 12, 24, etc., to 10 terms.
2. Of 16384, 8192, etc., to 20 terms.
3. Of $\frac{2}{3}$, $\frac{4}{9}$, $\frac{8}{27}$, etc., to 7 terms.

Find the sum of the following infinite geometrical series:

4. Of 1, $\frac{1}{2}$, $\frac{1}{4}$, etc.
5. Of $\frac{3}{5}$, $\frac{9}{25}$, $\frac{27}{125}$, etc.
6. Of $\frac{1}{2}$, $\frac{3}{8}$, $\frac{9}{32}$, etc.
7. Of $\frac{7}{6}$, 1, $\frac{6}{7}$, etc.
8. Of $.\dot{3}\dot{6} = .3636$, etc. $= \frac{36}{100} + \frac{36}{10000}$, etc.
9. Of $.\dot{3}4920\dot{6}$, of $.\dot{4}8\dot{0}$, of $.\dot{6}$.
10. Find the amount of an annuity of \$50, the time being 53 yr., the rate per cent 10.
11. Applying the formula used in the last example, to any case of the same kind, prove the truth of the rule, in Case IV, of Annuities.

12. Calculate a table of amounts of an annuity of $1, for any number of years from 1 to 6, at 8%.

REMARK.—It is not considered necessary to give special rules for finding the ratio, and the number of terms when these are unknown; so far as these are admissible here, they involve no principles beyond what are presented in the matter already given.

EXAMPLES FOR PRACTICE.

1. Find the common ratio: first term, 8; fourth term, 512.
2. First term, $4\frac{15}{16}$; eleventh term, 49375000000.
3. Sixteenth term, 729; twenty-second term, 1000000.
4. Insert 1 geometric mean between 63 and 112.
5. Four geometric means between 6 and 192.
6. Three geometric means between $\frac{1}{36864}$, and $\frac{1}{9}$.
7. Two geometric means between 14.08 and 3041.28

Topical Outline.

SERIES.

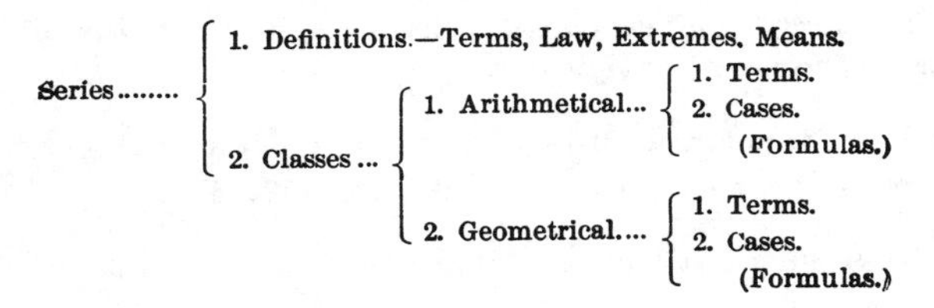

XXII. MENSURATION.

DEFINITIONS.

397. 1. **Geometry** is that branch of mathematics which treats of quantity having extension and form. When a quantity is so considered, it is called *Magnitude.*

2. There are four kinds of Magnitude known to Geometry: *Lines*, *Angles*, *Surfaces*, and *Solids.* A *point* has position, but not magnitude.

3. **Mensuration** is the application of Arithmetic to Geometry; it may be defined also as the art of computing *lengths*, *areas*, and *volumes.*

LINES.

398. 1. A **line** is that which has length only.

2. A **straight line** is the shortest distance between two points.

3. A **broken line** is a line made of connected straight lines of different directions.

4. A **curve**, or **curved line**, is a line having no part straight.

The word "line," used without the qualifying word "curve," is understood to mean a *straight* line.

5. A **horizontal line** is a line parallel with the horizon, or with the water level. (See Art. **389**; 1.)

6. A **vertical line** is a line perpendicular to a horizontal plane.

ANGLES.

399. An **angle** is the opening, or inclination, of two lines which meet at a point. (Art. **204.**)

REMARK.—Angles differing from right angles are called *oblique angles.* (See Art. **385,** 3, Rem., and Art. **386.**)

SURFACES.

POLYGONS.

400. 1. A **surface** is that which has length and breadth without thickness.

A *solid* has length, breadth, and thickness.

A line is meant when we speak of the *side* of a limited surface, or the *edge* of a *solid;* a surface is meant when we speak of the *side* or the *base* of a solid.

2. A **Plane** is a surface such that any two points in it can be joined by a straight line which lies wholly in the surface. The application of a straight-edge is the test of a plane.

3. A **plane figure** is any portion of a plane bounded by lines.

4. A **polygon** is a portion of a plane inclosed by straight lines; the **perimeter** of a polygon is the whole boundary.

5. **Area** is surface defined in amount. For the numerical expression of area, a square is the measuring unit. (Art. **197.**)

6. A polygon is **regular** when it has all its sides equal, and all its angles equal.

7. A polygon having three sides is a triangle; having four sides, a quadrilateral; five sides, a pentagon; six sides, a hexagon, etc.

Quadrilateral.

The six diagrams following represent regular polygons.

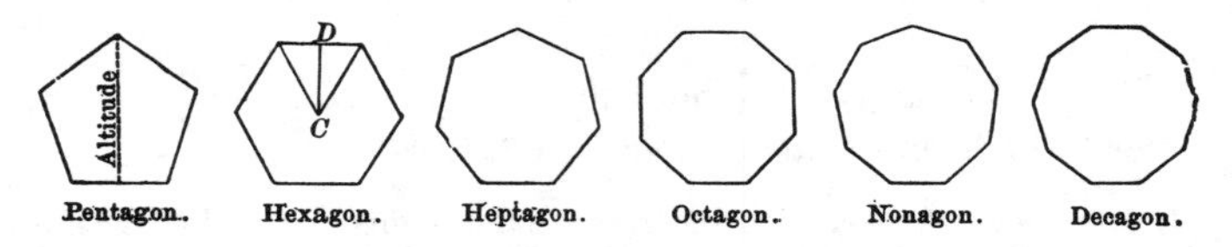

8. The **diagonal** of a polygon is the straight line joining two angles not adjacent; as, PN, on the preceding page.

9. The **base** is the side on which a figure is supposed to stand.

10. The **altitude** of a polygon is the perpendicular distance from the highest point, or one of the highest points, to the line of the base.

11. The **center** of a regular polygon is the point within, equally distant from the middle points of the sides; the **apothem** of such a polygon is the perpendicular line drawn from the center to the middle of a side; as, C a center, and CD an apothem.

Triangles.

401. Triangles are classified with respect to their angles, and also with respect to their sides.

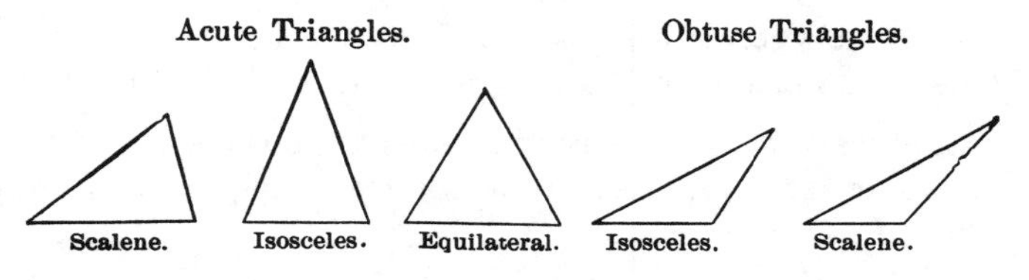

1. A triangle is **right-angled** when it has one right angle; it is **acute-angled** when each angle is acute; it is **obtuse-angled** when one angle is obtuse. These three classes may be named right triangles, acute triangles, obtuse triangles; the last two classes are sometimes called **oblique** triangles.

2. A triangle is **scalene** when it has no equal sides; **isosceles,** when it has two equal sides; and **equilateral,** when its three sides are equal.

A right triangle can be scalene, as when the sides are 3, 4, 5; or, isosceles, as when it is one of the halves into which a diagonal divides a square. An obtuse triangle can be scalene or isosceles; an acute triangle can be scalene, isosceles, or equilateral.

QUADRILATERALS.

402. Quadrilaterals are of three classes:

1. A **Trapezium** is a quadrilateral having no two sides parallel.

2. A **Trapezoid** is a quadrilateral having two and only two sides parallel.

3. A **Parallelogram** is a quadrilateral having two pairs of parallel sides.

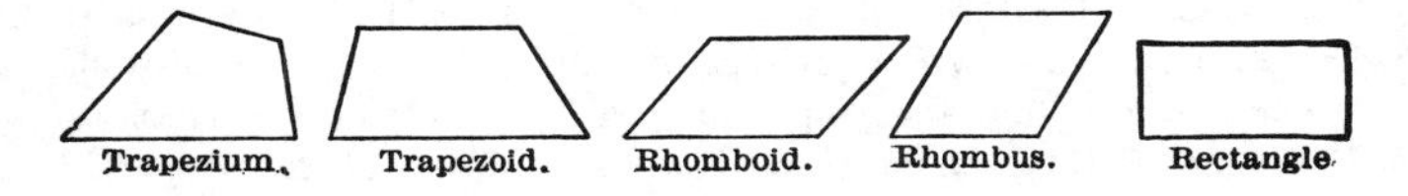

403. Parallelograms are of three classes:

1. A **Rhomboid** is a parallelogram having one pair of parallel sides greater than the other, and no right angle.

2. A **Rhombus** is a parallelogram whose four sides are equal.

3. A **Rectangle** is a parallelogram whose angles are all right angles; when the rectangle has four equal sides, it is a **square.**

A square is a rhombus whose angles are 90°; it is also the form of the unit for surface measure. It may properly be defined, an *equilateral rectangle.*

Square.

AREAS.

TRIANGLES AND QUADRILATERALS.

404. The **general rules** depend on the principles stated in the following remarks:

REMARKS.—1. The area of a rectangle is equal to the product of its length by its breadth. (Art. 197, Ex.)

2. The diagonal of a rectangle divides it into two equal triangles. The accompanying figure illustrates this; EHI = HEF. Observe also that the perpendicular GL divides the whole into two rectangles; EGL is half of one of them, LGI the half of the other, and *both* these smaller triangles make EGI, which must therefore be half of the whole; EGI and EHI have the same base and equal altitudes.

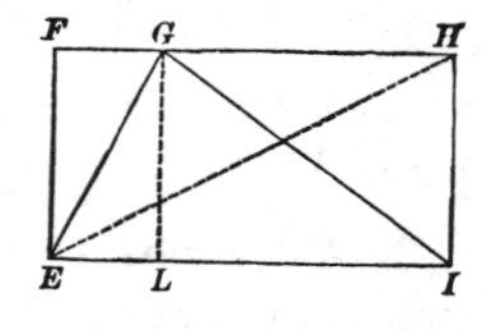

If the triangle EGH be supposed to stand on GH as a *base*, its altitude is EF; the perpendicular which represents the height is, in such a case, said to fall on the base *produced; i. e.*, extended. The triangle EGH is equal to the half of GHIL. Any triangle has an area equal to the product of half the base by the altitude.

Observe also that the trapezoid EFGI is made of two triangles, EGF and EGI; each of these has the altitude of the trapezoid, and each has one of the parallel sides for a base. Hence, the area of each being $\frac{1}{2}$ its base $\times$ the common altitude, the two areas, or the whole trapezoid, must equal the *half of both* bases $\times$ the altitude.

3. If the piece EGF were taken off and put on the right of EGHI, the line EF being placed on HI, the *whole area* would be the *same*, but the *perimeter* would be *increased*, and the figure would be a rhomboid. Different quadrilaterals may have *equal areas* and *unequal boundaries;* also, they may have the sides *in the same order and equal*, with *unequal areas.* To find accurately the area of a quadrilateral, more must be known than merely the four sides in order. A regular polygon has a greater area than any other figure of the same perimeter.

4. When triangles have equal bases their areas are to each other as their altitudes; the altitudes being equal, their areas are as their bases. The area of any triangle is equal to half the product of the perimeter by the radius of the inscribed circle.

General Rules.

I. To find the area of a parallelogram.

Rule.—*Multiply one of two parallel sides by the perpendicular distance between them.*

II. To find the area of a triangle.

Rule.—*Take half the product of the base by the altitude.*

III. To find the area of a trapezoid.

Rule.—*Multiply half the sum of the parallel sides by the altitude.*

Note.—The following is demonstrated in Geometry:

IV. To find the area of a triangle when the sides are given.

Rule.—*Add the three sides together and take half the sum; from the half sum take the sides separately; multiply the half sum and the three remainders together, and extract the square root of the product.*

Notes.—1. The area of a trapezium may be found by applying this rule to the parts when the sides are known and the diagonal is given in length and in special position as between the sides. The area of *any* polygon may be found by dividing it into triangles and measuring their bases and altitudes.

2. The area of a rhombus is equal to half the product of its diagonals; these are at right angles.

Examples for Practice.

1. Find the area of a parallelogram whose base is 9 ft. 4 in. and altitude 2 ft. 5 in.
2. Of an oil cloth 42 ft. by 5 ft. 8 in.

3. How many tiles 8 in. square in a floor 48 ft. by 10 ft.?

4. Find the area of a triangle whose base is 72 rd. and altitude 16 rd.

5. Base 13 ft. 3 in.; altitude 9 ft. 6 in.

6. Sides 1 ft. 10 in.; 2 ft.; 3 ft. 2 in.

7. Sides 15 rd.; 18 rd.; 25 rd.

8. What is the area of a trapezoid whose bases are 9 ft. and 21 ft., and altitude 16 ft.?

9. Bases 43 rd. and 65 rd.; altitude 27 rd.?

10. What is the area of a figure made up of 3 triangles whose bases are 10, 12, 16 rd. and altitudes 9, 15, $10\frac{1}{2}$ rd.?

11. Whose sides are 10, 12, 14, 16 rd. in order, and distance from the starting point to the opposite corner, 18 rd.?

12. How much wainscoting in a room 25 ft. long, 18 ft. wide, and 14 ft. 3 in. high, allowing a door 7 ft. 2 in. by 3 ft. 4 in., and two windows, each 5 ft. 8 in. by 3 ft. 6 in., and a chimney 6 ft. 4 in. by 5 ft. 6 in.; charging for the door and windows half-work?

13. What is the perimeter of a rhombus, one diagonal being 10 rd., and the area $86.60\frac{1}{4}$ sq. rd.?

14. Find the cost of flooring and joisting a house of 3 floors, each 48 ft. by 27 ft., deducting from each floor for a stairway 12 ft. by 8 ft. 3 in., allowing 9 in. rests for the joists; estimating the flooring and joisting *between* the walls at $1.46 a sq. yd., and the joisting *in* the walls at 76 ct. a sq. yd.; each row of rests being measured 48 ft. long by 9 in. wide.

15. What is the area of a square farm whose diagonal is 20.71 ch. longer than a side?

16. How many sq. yd. of plastering in a room 30 ft.

long, 25 ft. wide, and 12 ft. high, deducting 3 windows, each 8 ft. 2 in. by 5 ft.; 2 doors each 7 ft. by 3 ft. 6 in.; and a fire-place 4 ft. 6 in. by 4 ft. 10 in.; the sides of the windows being plastered 15 in. deep? And what will it cost, at 25 ct. a sq. yd.?

17. From a point in the side and 8 ch. from the corner of a square field containing 40 A., a line is run, cutting off $19\frac{1}{2}$ A.: how long is the line?

18. How much painting on the sides of a room 20 ft. long, 14 ft. 6 in. wide, and 10 ft. 4 in. high, deducting a fire-place 4 ft. 4 in. by 4 ft., and 2 windows each 6 ft. by 3 ft. 2 in.?

19. Find the cost of glazing the windows of a house of 3 stories, at 20 ct. a sq. ft. Each story has 4 windows, 3 ft. 10 in. wide; those in the 1st story are 7 ft. 8 in. high; those in the 2nd, 6 ft. 10 in. high; in the 3d, 5 ft. 3 in. high.

Regular Polygons and the Circle.

405. Any **regular polygon** may be divided into equal isosceles triangles, by lines from the center to the vertices; the apothem is their common altitude, and the perimeter the sum of their bases.

406. To find the area of a regular polygon.

Rule.—*Multiply the perimeter by half the apothem.*

All regular polygons of the same number of sides are similar figures. (Art. **389,** Rem. 1.)

407. 1. The **circle,** as already defined (Art. **204**), is a figure bounded by a uniform curve.

2. Any line drawn in a circle, having its ends in the curve, is called a **chord;** as AB, BD.

3. The portion of the curve which is cut off by such a line is called an **arc,** and the space between the chord and the arc is called a **segment.**

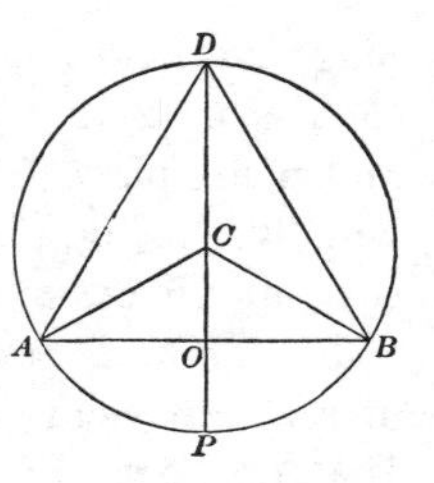

Thus, the curve APB is an arc, AB is the chord of that arc, and these inclose a segment whose base is AB, and whose height is OP.

4. If a line be drawn from the middle of a chord to the center, it will be perpendicular to the chord; so also, a line perpendicular at the middle of a chord, will, if extended, pass through the center, and bisect either of the arcs standing on that chord. Thus, AB is bisected by the perpendicular CO, and the arc AP = PB; so the arc AD = BD.

5. A **tangent** to a circle is a straight line having only one point in common with the curve; it simply touches the circle; a **secant** enters the figure from without.

If with C as a center, and CO as a radius, a circle were drawn in the equilateral triangle ABD, the sides would be tangent to the circle; the circle would be *inscribed* in the triangle. The circle of which CB is radius, is *circumscribed* about ABD.

6. The space inclosed by two radii and an arc, is called a **sector**; as, ACP.

The arc of that sector is the same fraction of the whole circumference that the area of the sector is of the whole circle.

Calculations Pertaining to the Circle.

408. The accompanying diagrams present (Fig. 1) a regular polygon of six sides, (Fig. 2) one of twelve sides, and (Fig. 3) a circle divided into twenty-four sectors.

Remarks.—1. The hexagon is composed of six equilateral triangles, and hence if OB be 1, the side AB = 1, and it is easy to compute the apothem, $\sqrt{1-\frac{1}{4}} = .86602540378$

2. If the distance from center to vertices be unchanged, and a regular polygon of twelve sides be formed about the same center, it will differ less from a circle whose radius is OB, than the hexagon differs from such a circle. This is evident from the second figure; and if the polygon be made of twenty-four sides (the distance from center to vertices remaining the same), it will be still nearer the circle in shape and size; in the space of the diagram, one of the twenty-four triangles forming such a polygon would differ very little from one of the twenty-four sectors here shown. The circle is regarded as composed of an infinite number of triangles whose common altitude is the radius and the sum of whose bases is the circumference. Hence, the area $= \frac{1}{2}$ the sum of bases $\times$ altitude; or,

Area of circle $= \frac{1}{2}$ circumference $\times$ radius.
Area of circle $= \frac{1}{4}$ circumference $\times$ diameter.

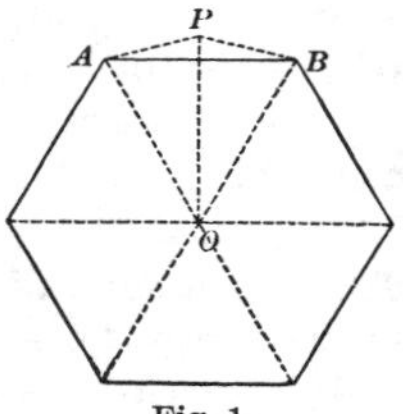

Fig. 1.

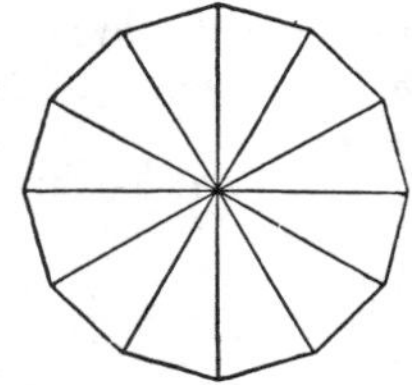
Fig. 2.

3. Since the perimeter of the hexagon is 6, it is easy to compute the next perimeter shown, which is 12 times AP or BP. The apothem being found above, subtract it from OP or 1, and obtain .13397459622 the perpendicular of a right-angled triangle; then, the base of that triangle being .5, the half of AB, find the hypothenuse .517638090205, = PB. Now, if we treat PB as we treated AB we can find the apothem of the second figure, and then find one of the 24 sides of another polygon, still more nearly equal to the circle. If these operations be continued, we shall find results 8th and 9th as follows:

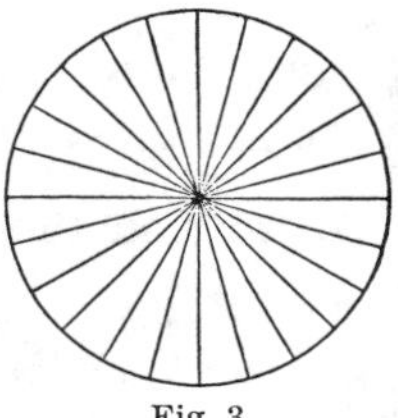
Fig. 3.

Perimeter of polygon of 1536 sides = 6.28318092
Perimeter of polygon of 3072 sides = 6.28318420

If the distance from center to vertices be taken $\frac{1}{2}$ instead of 1, the results will be

3.141590 + and 3.141592 +

Hence, if the circle of diameter 1, be taken as a polygon of 1536 sides, and then as a polygon of 3072 sides, the expressions for perimeter do not differ at the fourth decimal place. The number 3.1416 is usually given, although by more expeditious methods than that above illustrated, the calculation has been carried to a great number of decimal places, of which the following correctly shows eighteen:

3.141592653589793238

This important ratio, of circumference to diameter, is represented by the Greek letter π (*pi.*).

4. Since circumference = diameter $\times \pi$, and area = $\frac{1}{4}$ circumference $\times$ diameter, we have area $= \frac{\pi}{4} \times$ square of diameter. Representing the circumference by c, area by C, diameter by d, radius by R, we have the following formulas:

$$c = d\pi; \qquad C = d^2\frac{\pi}{4}; \qquad C = R^2\pi.$$

$$d = \frac{c}{\pi}; \qquad d = \sqrt{\frac{4C}{\pi}}; \qquad R = \sqrt{\frac{C}{\pi}}.$$

General Rules.

409. Pertaining to the circle we have the following general rules:

I. To find the circumference:

1. *Multiply the diameter by* 3.1415926; or,
2. *Divide the area by* $\frac{1}{4}$ *of the diameter;* or,
3. *Extract the square root of* 12.56637 *times the area.*

II. To find the diameter:

1. *Divide the circumference by* 3.1415926; or,
2. *Divide the area by* .785398, *and extract the square root.*

III. To find the area:

1. *Multiply the diameter by* $\frac{1}{4}$ *of the circumference;* or,
2. *Multiply the square of the diameter by* .785398; or,
3. *Multiply the square of the radius by* 3.1415926

IV. To find the area of a sector of a circle:

1. *Multiply the arc by one half the radius;* or,

2. *Take such a fraction of the whole area as the arc is of the whole circumference.*

V. To find the area of a segment less than a semicircle:

1. *Subtract from the area of the sector having the same arc, the area of the triangle whose base is the base of the segment, and whose vertex is the center of the circle;* or,

2. *Divide the cube of the height by twice the base, and increase the quotient by two thirds of the product of height and base.*

REMARK.—Add the triangle to the sector, if the segment be greater than a semicircle. The second rule gives an approximate result.

NOTES.—1. The side of a square inscribed in a circle is to radius as $\sqrt{2}$ is to 1.

2. The side of an inscribed equilateral triangle is to radius as $\sqrt{3}$ is to 1.

3. If radius be 1, the side of the inscribed regular pentagon is 1.1755; heptagon, .8677; nonagon, .6840; undecagon, .5634

EXAMPLES FOR PRACTICE.

1. What are the circumferences whose diameters are 16, $22\frac{1}{4}$, 72.16, and 452 yd.?

2. What are the diameters whose circumferences are 56, $182\frac{1}{2}$, 316.24, and 639 ft.?

3. Find the areas of the circles with diameters 10 ft.; 2 ft. 5 in.; 13 yd. 1 ft.

4. Whose circumferences are 46 ft.; 7 ft. 3 in.; 6 yd. 1 ft. 4 in.

5. Circum. 47.124 ft., diameter 15 ft.

6. If we saw down through $\frac{1}{3}$ of the diameter of a round log uniformly thick, what portion of the log is cut in two?

7. What fraction of a round log of uniform thickness is the largest squared stick which can be cut out of it?

SOLIDS.

DEFINITIONS.

410. 1. A **Solid** is that which has length, breadth, and thickness.

A solid may have plane surfaces, curved surfaces, or both. A *curved surface* is one no part of which is a plane.

2. The **faces** of a solid are the polygons formed by the intersections of its bounding planes; the lines of those intersections are called **edges.**

3. A **Prism** is a solid having two bases which are parallel polygons, and faces which are parallelograms.

A prism is triangular, quadrangular, etc., according to the shape of its base. The first of the figures here given represents a *quadrangular* prism, the second a *pentagonal* prism.

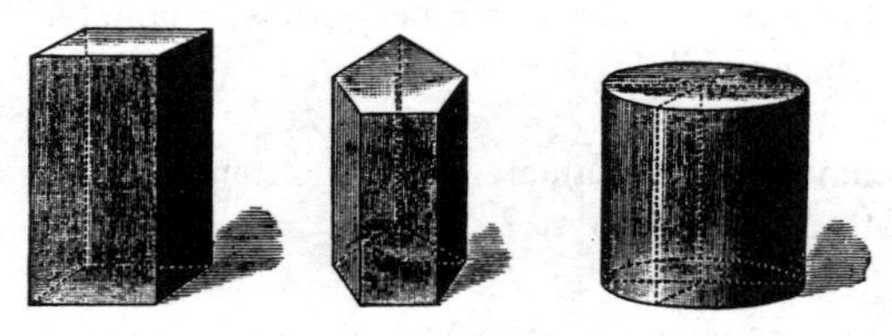

4. A **right prism** is a prism whose faces are rectangles.

5. A **Parallelopiped** is a prism whose faces are parallelograms. Its bounding surfaces are six parallelograms.

The first figure above represents a parallelopiped whose faces are rectangles.

6. A **Cube** is a parallelopiped whose faces are squares.

7. A **Cylinder** is a solid having two bases which are equal parallel circles, and having an equal diameter in any parallel plane between them.

8. A **Pyramid** is a solid with only one base, and whose faces are triangles with a common vertex.

9. A **Cone** is a solid whose base is a circle, and whose other surface is convex, terminating above in a point called the vertex.

10. A **frustum** of a pyramid or cone is the solid which remains when a portion having the vertex is cut off by a plane parallel to the base.

11. A **Sphere** is a solid bounded by a curved surface, every point of which is at the same distance from a point within, called the center. The *diameter* of a sphere is a straight line passing through the center and having its ends in the surface; the *radius* is the distance from the center to the surface.

A *segment* of a sphere is a portion cut off by one plane, or between two planes; its *bases* are circles, and its *height* is the portion of the diameter which is cut off with it.

12. The **slant height** of a *pyramid* is the perpendicular distance from the vertex to one of the sides of a base; the **slant height** of a *cone* is the straight line drawn from the vertex to the circumference of the base.

13. The **altitude** of any solid is the perpendicular distance between the planes of its bases, or the perpendicular distance from its highest point to the plane of the base.

14. The **Volume** of a solid is the number of solid units it contains; the assumed unit of measure is a cube. (Art. **199.**)

15. Solids are **similar** when their like lines are proportional, and their corresponding angles equal.

General Rules.

I. To find the convex surface of a prism or cylinder:

Rule.—*Multiply the perimeter of the base by the altitude.*

II. To find the volume of a prism or cylinder:

Rule.—*Multiply the base by the altitude.*

III. To find the convex surface of a pyramid or cone:

Rule.—*Multiply the perimeter of the base by one half the slant height.*

IV. To find the volume of a pyramid or cone:

Rule.—*Multiply the base by one third of the altitude.*

V. To find the convex surface of a frustum of a pyramid or cone:

Rule.—*Multiply half the sum of the perimeters of the bases by the slant height.*

VI. To find the solidity of a frustum of a pyramid or cone:

Rule.—*To the sum of the two bases add the square root of their product, and multiply the amount by one third of the altitude.*

VII. To find the surface of a sphere:

Rule.—*Multiply the circumference by the diameter.*

VIII. To find the volume of a sphere:

Rule.—*Multiply the cube of the diameter by* .5235987

NOTES.—1. Similar solids are to each other as the cubes of their like dimensions.

2. The sphere is regarded as composed of an infinite number of cones whose common altitude is the radius, and the sum of whose bases is the whole surface of the sphere.

3. The cone is regarded as a pyramid of an infinite number of faces, and the cylinder as a prism of an infinite number of faces.

EXAMPLES FOR PRACTICE.

1. Find the convex surface of a right prism with altitude $11\frac{1}{4}$ in., and sides of base $5\frac{1}{4}$, $6\frac{1}{2}$, $8\frac{3}{4}$, $10\frac{1}{2}$, 9 in.

2. Of a right cylinder whose altitude is $1\frac{3}{4}$ ft., and the diameter of whose base is 1 ft. $2\frac{1}{2}$ in.

6 sq

3. Find the whole surface of a right triangular prism, the sides of the base 60, 80, and 100 ft.; altitude 90 ft.

4. The whole surface of a cylinder; altitude 28 ft.; circumference of the base 19 ft.

5. Find the convex surface and whole surface of a right pyramid whose slant height is 391 ft.; the base 640 ft. square.

6. Of a right cone whose slant height is 66 ft. 8 in.; radius of the base 4 ft. 2 in.

7. Find the solidity of a pyramid whose altitude is 1 ft. 2 in., and whose base is a square $4\frac{1}{2}$ in. to a side.

8. Whose altitude is 15.24 in., and whose base is a triangle having each side 1 ft.

9. What is the solidity of a prism whose bases are squares 9 in. on a side, and whose altitude is 1 ft. 7 in.?

10. Whose altitude is $6\frac{1}{2}$ ft., and whose bases are parallelograms 2 ft. 10 in. long by 1 ft. 8 in. wide?

11. Whose altitude is 7 in., and whose base is a triangle with a base of 8 in. and an altitude of 1 ft.?

12. Whose altitude is 4 ft. 4 in., and whose base is a triangle with sides of 2, $2\frac{1}{2}$, and 3 ft.?

13. What is the solidity of a cylinder whose altitude is $10\frac{1}{2}$ in., and the diameter of whose base is 5 in.?

14. Find the convex surface of a frustum of a pyramid with slant height $3\frac{1}{3}$ in., lower base 4 in. square, upper base $2\frac{1}{2}$ in. square.

15. The convex surface and whole surface of the frustum of a cone, the diameters of the bases being 7 in. and 3 in., and the slant height 5 in.

16. Find the solidity of a frustum of a pyramid whose altitude is 1 ft. $4\frac{1}{2}$ in.; lower base, $10\frac{2}{3}$ in. square; upper, $4\frac{1}{4}$ in. square.

17. Of a frustum of a cone, the diameters of the bases being 18 in. and 10 in., and the altitude 16 in.

18. What are the surfaces of two spheres whose diameters are 27 ft. and 10 in.?

19. Find the solidity of a sphere whose diameter is 6 mi., and surface 113.097335 sq. mi.

20, Of a sphere whose diameter is 4 ft.

21. Of a sphere whose surface is 40115 sq. mi.

22. By what must the diameter of a sphere be multiplied to make the edge of the largest cube which can be cut out of it?

MISCELLANEOUS MEASUREMENTS.

Masons' and Bricklayers' Work.

411. Masons' work is sometimes measured by the cubic foot, and sometimes by the *perch*. The latter is $16\frac{1}{2}$ ft. long, $1\frac{1}{2}$ ft. wide, and 1 ft. deep, and contains $16\frac{1}{2} \times 1\frac{1}{2} \times 1 = 24\frac{3}{4}$ cu. ft., or 25 cu. ft. nearly.

412. To find the number of perches in a piece of masonry.

Rule.—*Find the solidity of the wall in cubic feet by the rules given for mensuration of solids, and divide it by* $24\frac{3}{4}$.

Note.—Brick work is generally estimated by the thousand bricks; the usual size being 8 in. long, 4 in. wide, and 2 in. thick. When bricks are laid in mortar, an allowance of $\frac{1}{10}$ is made for the mortar.

Examples for Practice.

1. How many perches of 25 cu. ft. in a pile of building-stone 18 ft. long, $8\frac{1}{2}$ ft. wide, and 6 ft. 2 in. high?

2. Find the cost of laying a wall 20 ft. long, 7 ft. 9 in. high, and with a mean breadth of 2 ft., at 75 ct. a perch.

3. The cost of a foundation wall 1 ft. 10 in. thick, and 9 ft. 4 in. high, for a building 36 ft. long, 22 ft. 5 in. wide outside, at $2.75 a perch, allowing for 2 doors 4 ft. wide.

4. The cost of a brick wall 150 ft. long, 8 ft. 6 in. high, 1 ft. 4 in. thick, at $7 a thousand, allowing $\frac{1}{10}$ for mortar?

5. How many bricks of ordinary size will build a square chimney 86 ft. high, 10 ft. wide at the bottom, and 4 ft. at the top outside, and 3 ft. wide inside all the way up?

SUGGESTION.—Find the solidity of the whole chimney, then of the hollow part; the difference will be the solid part of the chimney.

GAUGING.

413. Gauging is finding the contents of vessels, in bushels, gallons, or barrels.

414. To gauge any vessel in the form of a rectangular solid, cylinder, cone, frustum of a cone, etc.

Rule.—*Find the solidity of the vessel in cubic inches by the rules already given; this divided by* 2150.42, *will give the contents in bushels; by* 231, *will give it in wine gallons, which may be reduced to barrels by dividing the uumber by* $31\frac{1}{2}$.

NOTE.—In applying the rule to cylinders, cones, and frustums ot cones, instead of multiplying the square of half the diameter by 3.14159265, and dividing it by 231, *multiply the square of the diameter by* .0034, which amounts to the same, and is shorter.

EXAMPLES FOR PRACTICE.

1. How many bushels in a bin 8 ft. 3 in. long, 3 ft. 5 in. high, and 2 ft. 10 in. wide?

2. How many wine gallons in a bucket in the form of a frustum of a cone, the diameters at the top and bottom being 13 in and 10 in., and depth 12 in.?

3. How many barrels in a cylindrical cistern 11 ft. 6 in. deep and 7 ft. 8 in wide?

4. In a vat in the form of a frustum of a pyramid, 5 ft. deep, 10 ft. square at top, 9 ft. square at bottom?

415. To find the contents in gallons of a cask or barrel.

REMARK.—When the staves are straight from the bung to each end, consider the cask as two frustums of a cone, and calculate its contents by the last rule; but when the staves are curved, use this rule:

Rule.—*Add to the head diameter (inside) two thirds of the difference between the head and bung diameters; but if the staves are only slightly curved, add six tenths of this difference; this gives the mean diameter; express it in inches, square it, multiply it by the length in inches, and this product by* .0034: *the product will be the contents in wine gallons.*

NOTE.—After finding the mean diameter, the contents are found as if the cask were a cylinder.

EXAMPLES FOR PRACTICE.

1. Find the number of gallons in a cask of cider whose staves are straight from bung to head, the length being 26 in., the bung diameter 16 inches, and head diameter 13 in.

2. In a barrel of vinegar, with staves slightly curved, length 2 ft. 10 in., bung diameter 1 ft. 9 in., head 1 ft. 6 in.

3. In a cask of milk with curved staves, length 5 ft. 4 in., bung diameter 3 ft. 6 in., head diameter 3 ft.

LUMBER MEASURE.

416. To find the amount of square-edged inch boards that can be sawed from a round log.

REMARK.—The following is much used by lumber-men, and is sufficiently accurate for practical purposes. It is known as Doyle's Rule.

Rule.—*From the diameter in inches subtract* 4; *the square of the remainder will be the number of square feet of inch boards yielded by a log* 16 *feet in length.*

EXAMPLES FOR PRACTICE.

1. How much square-edged inch lumber can be cut from a log 32 inches in diameter, and 20 feet long?

OPERATION.

$32-4=28;\ 28\times28\times\frac{20}{16}=980$ feet.

Or, $\frac{5}{4}\times28\times28=980$ feet.

2. In a log 24 in. in diameter, and 12 ft. long?
3. In a log 25 in. in diameter, and 24 ft. long?

4. In a log 50 in. in diameter, and 12 ft. long?

TO MEASURE GRAIN AND HAY.

417. Grain is usually estimated by the bushel, and sold by weight; **Hay,** by the ton.

REMARKS.—1. The standard bushel contains 2150.4 cubic inches. A cubic foot is nearly .8 of a bushel.

2. Hay well settled in a mow may be estimated (approximately) at 550 cubic feet for clover, and 450 cubic feet for timothy, per ton.

418. To find the quantity of grain in a wagon or in a bin:

Rule.—*Multiply the contents in cubic feet by* .8

REMARKS.—1. If it be corn on the cob, deduct one half.
2. For corn not "shucked," deduct two thirds for cob and shuck.

419. To find the quantity of hay in a stack, rick, or mow:

Rule.—*Divide the cubical contents in feet by* 550 *for clover, or by* 450 *for timothy; the quotient will be the number of tons.*

EXAMPLES FOR PRACTICE.

1. How many bushels of shelled corn, or corn on the cob, or corn not shucked, will a wagon-bed hold that is $10\frac{1}{2}$ feet long, $3\frac{1}{2}$ feet wide, and 2 feet deep?

2. In a bin 40 feet long, 16 wide, and 10 feet high?

3. A hay-mow contains 48000 cubic feet: how many tons of well settled clover or timothy will it hold?

Topical Outline.

MENSURATION.

1. General Definitions:—Geometry, Magnitude, Mensuration.

2. Lines
- Straight.
  - Parallel.
  - Perpendicular.
  - Horizontal.
  - Vertical.
  - Diagonal.
- Broken.
- Curved.

MENSURATION.—(*Continued.*)

- **3. Angles**
  - Right.
  - Oblique
    - Acute.
    - Obtuse.
- **4. Surfaces.**
  - 1. General Definitions:—Plane, Plane figure, Area, Polygons, Regular, Perimeter, Similar, Center, Altitude, Base, Apothem.
  - 2. Triangles (Rules.)
    - 1. Right
      - Isosceles.
      - Scalene.
    - 2. Oblique
      - Acute
        - Scalene.
        - Isosceles.
        - Equilateral.
      - Obtuse
        - Scalene.
        - Isosceles.
  - 3. Quadrilateral (Rules.)
    - 1. Trapezium.
    - 2. Trapezoid.
    - 3. Parallelogram.
      - 1. Rhomboid.
      - 2. Rhombus — Square.
      - 3. Rectangle — Square.
  - 4. Circle
    - 1. Terms:—Circumference, Radius, Chord, Diameter, Segment, Sector, Tangent, Secant.
    - 2. Calculations, Value of π.
    - 3. Formulas, Rules.
- **5. Solids**
  - 1. General Definitions:—Solid, Base, Face, Edge, Similar.
  - 2. Prism (Rules.)
    - Triangular. (Right.)
    - Quadrangular
      - Parallelopiped. (Right.) — Cube.
    - Pentagonal, etc.
  - 3. Pyramid (Rules.) (Frustum.)
    - Altitude.
    - Slant Height.
    - Convex Surface.
  - 4. Cone (Rules.) (Frustum.)
    - Altitude.
    - Slant Height.
    - Convex Surface.
  - 5. Cylinder (Rules.)
    - Surface.
    - Solidity.
  - 6. Sphere
    - 1. Terms:—Radius, Diameter, Segment.
    - 2. Convex Surface, Rule.
    - 3. Solidity, Rule.
  - 7. General Formulas.
  - 8. Miscellaneous Applications
    - Masons' and Bricklayers' Work.
    - Gauging.
    - Lumber Measure.
    - Measuring Grain and Hay.

XXIII. MISCELLANEOUS EXERCISES.

NOTE.—Nos. 1 to 50 are to be solved mentally.

1. If I gain $\frac{1}{4}$ ct. apiece by selling eggs at 7 ct. a dozen, how much apiece will I gain by selling them at 9 ct. a dozen?

2. If I gain $\frac{1}{2}$ ct. apiece by selling apples at 3 for a dime, how much apiece would I lose by selling them 4 for a dime?

3. If I sell potatoes at $37\frac{1}{2}$ ct. per bu., my gain is only $\frac{3}{5}$ of what it would be, if I charged 45 ct. per bu.: what did they cost me?

4. If I sell my oranges for 65 ct., I gain $\frac{3}{8}$ ct. apiece more than if I sold them for 50 ct.: how many oranges have I?

5. If I sell my pears at 5 ct. a dozen, I lose 16 ct.; if I sell them at 8 ct. a dozen, I gain 11 ct.: how many pears have I, and what did they cost me?

6. If I sell eggs at 6 ct. per dozen, I lose $\frac{2}{3}$ ct. apiece; how much per dozen must I charge to gain $\frac{2}{3}$ ct. apiece?

7. One eighth of a dime is what part of 3 ct.?

8. If I lose $\frac{3}{8}$ of my money, and spend $\frac{4}{7}$ of the remainder, what part have I left?

9. A's land is $\frac{1}{7}$ less in quantity than B's, but $\frac{1}{20}$ better in quality: how do their farms compare in value?

10. If $\frac{2}{3}$ of A's money equals $\frac{4}{5}$ of B's, what part of B's equals $\frac{5}{8}$ of A's?

11. I gave A $\frac{5}{14}$ of my money, and B $\frac{7}{12}$ of the remainder: who got the most, and what part?

12. A is $\frac{2}{3}$ older than B, and B $\frac{2}{3}$ older than C: how many times C's age is A's?

13. Two thirds of my money equals $\frac{4}{5}$ of yours; if we put our money together, what part of the whole will I own?

14. How many thirds in $\frac{1}{2}$?

15. Reduce $\frac{4}{5}$ to thirds; $\frac{5}{6}$ to ninths; and $\frac{3}{5}$ to a fraction, whose numerator shall be 8.

16. What fraction is as much larger than $\frac{4}{5}$ as $\frac{2}{3}$ is less than $\frac{4}{5}$?

17. After paying out $\frac{1}{4}$ and $\frac{1}{5}$ of my money, I had left $8 more than I had spent: what had I at first?

18. In 12 yr. I shall be $\frac{7}{5}$ of my present age: how long since was I $\frac{5}{7}$ of my present age?

19. Four times $\frac{2}{9}$ of a number is 12 less than the number what is the number?

20. A man left $\frac{5}{11}$ of his property to his wife, $\frac{2}{3}$ of the remainder to his son, and the balance, $4000, to his daughter: what was the estate?

21. I sold an article for $\frac{1}{4}$ more than it cost me, to A, who sold it for $6, which was $\frac{2}{5}$ less than it cost him: what did it cost me?

22. A is $\frac{2}{3}$ older than B; their father, who is as old as both of them, is 50 yr. of age: how old are A and B?

23. A pole was $\frac{3}{7}$ under water; the water rose 8 ft., and then there was as much under water as had been above water before: how long is the pole?

24. A is $\frac{3}{4}$ as old as B; if he were 4 yr. older, he would be $\frac{9}{10}$ as old as B; how old is each?

25. A's money is $4 more than $\frac{2}{3}$ of B's, and $5 less than $\frac{3}{4}$ of B's: how much has each?

26. Two thirds of A's age is $\frac{3}{4}$ of B's, and A is $3\frac{1}{2}$ yr. the older: how old is each?

27. If 3 boys do a work in 7 hr., how long will it take a man who works $4\frac{1}{2}$ times as fast as a boy?

28. If 6 men can do a work in $5\frac{1}{2}$ days, how much time would be saved by employing 4 more men?

29. A man and 2 boys do a work in 4 hr.: how long would it take the man alone if he worked equal to 3 boys?

30. A man and a boy can mow a certain field in 8 hr.; if the boy rests $3\frac{3}{4}$ hr., it takes them $9\frac{1}{2}$ hr.; in what time can each do it?

31. Five men were employed to do a work; two of them failed to come, by which the work was protracted $4\frac{1}{2}$ days: in what time could the 5 have done it?

32. Three men can do a work in 5 days; in what time can 2 men and 3 boys do it, allowing 4 men to work equal to 9 boys?

33. A man and a boy mow a 10-acre field; how much more does the man mow than the boy, if 2 men work equal to 5 boys?

34. Six men can do a work in $4\frac{1}{3}$ days; after working 2 days, how many must join them so as to complete it in $3\frac{2}{5}$ da.?

35. Eight men can do a certain amount of work in $6\frac{3}{4}$ days; after beginning, how soon must they be joined by 2 more so as to complete it in $5\frac{7}{8}$ days?

36. Seven men can build a wall in $5\frac{1}{2}$ days; if 10 men are employed, what part of his time can each rest, and the work be done in the same time?

37. Nine men can do a work in $8\frac{1}{3}$ days; how many days may 3 remain away, and yet finish the work in the same time by bringing 5 more with them?

38. Ten men can dig a trench in $7\frac{1}{2}$ days; if 4 of them are absent the first $2\frac{1}{2}$ days, how many other men must they then bring with them to complete the work in the same time?

39. At what times between 6 and 7 o'clock are the hour-hand and minute-hand 20 min. apart? $10\frac{10}{11}$ min. after 6, and $54\frac{6}{11}$ min. after 6.

40. At what times between 4 and 5 o'clock is the minute-hand as far from 8 as the hour-hand is from 3?

41. At what time between 5 and 6 o'clock is the minute-hand midway between 12 and the hour-hand? when is the hour-hand midway between 4 and the minute-hand?

42. A, B, and C dine on 8 loaves of bread; A furnishes 5 loaves; B, 3 loaves; C pays the others 8d. for his share: how must A and B divide the money?

43. A boat makes 15 mi. an hour down stream, and 10 mi. an hour up stream: how far can she go and return in 9 hr.?

44. I can pasture 10 horses or 15 cows on my ground; if I have 9 cows, how many horses can I keep?

45. A's money is 12 % of B's, and 16 % of C's; B has $100 more than C: how much has A?

46. Eight men hire a coach; by getting 6 more passengers, the expense to each is diminished $$1\frac{3}{4}$: what do they pay for the coach?

47. A company engage a supper; being joined by $\frac{2}{5}$ as many more, the bill of each is 60 ct. less: what would each have paid if none had joined them?

48. By mixing 5 lb. of good sugar with 3 lb. worth 4 ct. a lb. less, the mixture is worth $8\frac{1}{2}$ ct. a lb.: find the prices of the ingredients.

49. By mixing 10 lb. of good sugar with 6 lb. worth only $\frac{2}{3}$ as much, the mixture is worth 1 ct. a lb. less than the good sugar: find the prices of the ingredients and of the mixture.

50. A and B have the same amount of money; if A had $20 more, and B $10 less, A would have $2\frac{1}{3}$ times as much as B: what amount has each?

---

51. A and B pay $1.75 for a quart of varnish, and 10 ct. for the bottle; A contributes $1, B, the rest: they divide the varnish equally, and A keeps the bottle: which owes the other, and how much?

52. How far does a man walk while planting a field of corn 285 ft. square, the rows being 3 ft. apart and 3 ft. from the fences?

53. Land worth $1000 an acre, is worth how much a front foot of 90 ft. depth; reserving $\frac{1}{10}$ for streets?

54. I buy stocks at 20 % discount, and sell them at 10 % premium: what per cent. do I gain?

55. I invest, and sell at a loss of 15 %; I invest the proceeds again, and sell at a gain of 15 %: do I gain or lose on the two speculations, and how many per cent.?

56. I sell at 8 % gain, invest the proceeds, and sell at an advance of $12\frac{1}{2}$ %; invest the proceeds again, and sell at 4 % loss, and quit with $1166.40: what did I start with?

57. I can insure my house for $2500, at $\frac{8}{10}$ % premium annually, or permanently by paying down 12 annual premiums: which should I prefer, and how much will I gain by it if money is worth 6 % per annum to me?

58. A owes B $1500, due in 1 yr. 10 mon. He pays him $300 cash, and a note of 6 mon. for the balance: what is the face of the note, allowing interest at 6 %?

59. If I charge 12 % per annum compound interest, payable quarterly, what rate per annum is that?

60. How many square inches in one face of a cube which contains 2571353 cu. in.?

61. What is the side of a cube which contains as many cubic inches, as there are square inches in its surface?

62. The boundaries of a square and circle are each 20 ft.; which is the greater, and how much?

63. If I pay $1000 for a 5 yr. lease, and $200 for repairs, how much rent payable quarterly is that equal to, allowing 10% interest, payable quarterly?

64. What is the value of a widow's dower in property worth $3000, her age being 40, and interest 5%?

65. What principal must be loaned Jan. 1, at 9 %, to be re-paid by five notes of $200 each, payable on the first day of each of the five succeeding months?

66. After spending 25 % of my money, and 25 % of the remainder, I had left $675: what had I at first?

67. I had a 60-day note discounted at 1 % a month, and paid $4.80 above true interest: what was the face of the note?

68. Invested $10000; sold out at a loss of 20%: how much must I borrow at 4%, so that, by investing all I have at 18%, I may retrieve my loss?

69. If ¼ of an inch of rain fall, how many bbl. will be caught by a cistern which drains a roof 52 ft. by 38 ft.?

70. A father left $20000 to be divided among his 4 sons, aged 6 years, 8 years, 10 years, and 12 years respectively, so that each share, placed at 4½% compound interest, should amount to the same when its possessor became of age (21 yr.): what were the shares?

71. $30000 of bonds bearing 7% interest, payable semi-annually, and due in 20 yr. are bought so as to yield 8% payable semi-annually: what is the price?

72. A man wishes to know how many hogs at $9, sheep at $2, lambs at $1, and calves at $9 per head, can be bought for $400, having, of the four kinds, 100 animals in all. How many different answers can be given?

73. The stocks of 3 partners, A, B, and C, are $350, $220, and $250, and their gains $112, $88, and $120 respectively; find the time each stock was in trade, B's time being 2 mon. longer than A's.

74. By discounting a note at 20% per annum, I get 22½% per annum interest: how long does the note run?

75. A receives $57.90, and B $29.70, from a joint speculation: if A invested $7.83⅓ more than B, what did each invest?

76. A borrows a sum of money at 6%, payable semi-annually, and lends it at 12%, payable quarterly, and clears $2450.85 a year: what is the sum?

77. Find the sum whose true discount by simple interest for 4 yr. is $25 more at 6% than at 4% per annum.

78. I invested $2700 in stock at 25% discount, which pays 8% annual dividends: how much must I invest in stock at 4% discount and paying 10% annual dividends, to secure an equal income?

79. Exchanged $5200 of stock bearing 5% interest at 69%, for stock bearing 7% interest at 92%, the interest on each stock having been just paid: what is my cash gain, if money is worth 6% to me?

80. Bought goods on 4 mon. credit; after 7 mon. I sell them for $1500, 2½% off for cash; my gain is 15%, money being worth 6%: what did I pay for the goods?

81. The 9th term of a geometric series is 137781, and the 13th term 11160261: what is the 4th term?

82. My capital increases every year by the same per cent.; at the end of the 3d year it was $13310; at the end of the 7th year it was $19487.171: what was my original capital, and the rate of gain?

83. Find the length of a minute-hand, whose extreme point moves 4 inches in 3 min. 28 sec.

84. Three men own a grindstone, 2 ft. 8 in. in diameter: how many inches must each grind off to get an equal share, allowing 6 in. waste for the aperture?

85. I sold an article at 20% gain; had it cost me $300 more, I would have lost 20%: find the cost.

86. A boat goes 16¼ miles an hour down stream, and 10 mi. an hour up stream: if it is 22½ hr. longer in coming up than in going down, how far down did it go?

87. Had an article cost 10% less, the number of % gain would have been 15 more: what was the % gain?

88. Bought a check on a suspended bank at 55%; exchanged it for railroad bonds at 60%, which bear 7% interest: what rate of interest do I receive on the amount of money invested?

89. Bought sugar for refinery; 6% is wasted in the process; 30% becomes molasses, which is sold at 40 per cent. less than the same weight of sugar cost; at what per cent advance on the first cost must the clarified sugar be sold, so as to yield a profit of 14% on the investment?

90. There is coal now on the dock, and coal is running on also, from a shoot, at a uniform rate. Six men can clear the dock in one hour, but 11 men can clear it in 20 minutes: how long would it take 4 men?

91. A merchant sold his apples, losing 4%; keeping $18 of the proceeds, he gave the remainder to an agent to buy rye, 8% commission; he lost in all $32: what were the apples worth?

92. A clock gaining 3½ min. a day was started right at noon of the 22d of February, 1804: what was the true time when that clock

showed noon a week afterward; and, if kept going, when did it next show true time?

93. The number of square inches in one side of a right-angled triangular board is 144, and the base is half the height; required the areas of the different triangles which can be marked off by lines parallel to the base, at 12, 13, 14, $14\frac{1}{2}$ inches from the smaller end.

94. Suppose a body falls 16 ft. the first second, 48 ft. the next, 80 the next, and so on, *constantly* increasing, how far will it have fallen in 4 sec.; in $4\frac{1}{2}$ sec.; in 5 sec?

95. A man traveling at a *constant* increase, is observed to have gone 1 mile the first hour, 3 miles the next, 5 the next, and so on: how far will he have gone in $6\frac{1}{2}$ hr.?

96. The number of men in a side rank of a solid body of militia, is to the number in front as 2 to 3; if the length and breadth be increased so as to number each 4 men more, the whole body will contain 2320 men: how many does it now contain?

97. A grocer at one straight cut took off a segment of a cheese which had $\frac{1}{4}$ of the circumference, and weighed 3 lb.: what did the whole cheese weigh?

98. A wooden wheel of uniform thickness, 4 ft. in diameter, stands in mud 1 ft. deep: what fraction of the wheel is out of the mud?

99. My lot contains 135 sq. rd., and the breadth to length is as 3 to 5: what is the width of a road which shall extend from one corner half round the lot, and occupy $\frac{1}{4}$ of the ground?

100. A circular lot 15 rd. in diameter is to have three circular grass beds just touching each other and the large boundary: what must be the distance between their centers, and how much ground is left in the triangular space about the main center?

101. I have an inch board 5 ft. long, 17 in. wide at one end, and 7 in. at the other: how far from the larger end must it be cut straight across, so that the solidities of the two parts shall be equal?

102. Four equal circular pieces of uniform thickness, the largest possible, are to be cut from a circular plate of the same thickness, and worth $67: supposing there is no waste, what is the worth of each of the four, and what is the worth of the outer portion which is left?

103. A 12-inch ball is in the corner where walls and floor are at right angles: what must be the diameter of another ball which can touch that ball while both touch the same floor and the same walls?

104. A workman had a squared log twice as long as wide or deep; he made out of it a water-trough, of sides, ends, and bottom each 3 inches thick, and having 11772 solid inches: what is the capacity of it in gallons?

105. A tin vessel, having a circular mouth 9 in. in diameter, a bottom $4\frac{1}{2}$ in. in diameter, and a depth of 10 in., is $\frac{1}{4}$ part full of water: what is the diameter of a ball which can be put in and just be covered by the water?

Who is Joseph Ray?

Joseph Ray lived as a contemporary of Abraham Lincoln. Youth in that generation finished their schoolbooks and then read the Bible, sang from the hymnbooks of Lowell Mason, and read Roman and Greek classics in the original languages. It was not unusual for a blacksmith to carry a Greek New Testament under his cap for reading during his lunch break. The literacy rate, even on the frontier, was higher than today's rate.

Ray was Professor of Mathematics for twenty-five years at a preparatory school in Ohio. He had no use for indolence and sham. He was always delighted to join his students in sports. He knew how to use balls, marbles, and tops as concrete illustrations to help young children make the transfer from solid objects to abstract figures.

From the Presidency of Abraham Lincoln to that of Teddy Roosevelt, few Americans went to school or were taught at home without considerable exposure to either Ray's Arithmetics or McGuffey's Readers—usually both. Ray and McGuffey challenged students to excellent accomplishment. Their influence on our country has certainly eclipsed Mann's and rivaled Dewey's, yet education histories, edited by humanists, seldon mention these men.

Ray's classic Arithmetics are now brought to a new generation which is in search of excellence.

Ray's Arithmetic Series

Primary Arithmetic. Reading, writing and understanding of numbers to 100; adding and subtracting with sums to 20; multiplication and division to 10's; and signs and vocabulary needed for this level of arithmetic.

Intellectual Arithmetic. Reading, writing and understanding of higher whole numbers, fractions, and mixed numbers; addition, subtraction, multiplication and division of higher numbers; computation of simple fractions; beginning ratio and percentage; and signs and vocabulary needed for all these operations.

Practical Arithmetic. Roman numerals; carrying in addition and borrowing in subtraction; measurement and compound numbers; factors; decimals and percentage; ratio and proportion; powers and roots; beginning geometry; advanced vocabulary.

Higher Arithmetic. Philosophical understandings; principles and properties of numbers; advanced study of common and decimal fractions, measurements, ratio, proportion, percentage, powers, and roots; series; business math; geometry.

Test Examples. A supply of problems for making tests to accompany study in the *Practical Arithmetic* and the *Higher Arithmetic.*

Key to Ray's Primary, Intellectual and Practical Arithmetics. Answers to problems in the three lower books.

Key to Ray's Higher Arithmetic. Answers to problems in the *Higher Arithmetic.*

Parent-Teacher Guide. Gives unit by unit helps for teaching; suggests grade levels for each book; provides progress chart samples for each grade and tests for each unit.

These books are available from:
Mott Media
Fenton, Michigan 48430
www.mottmedia.com

McGuffey's Reading Series

Primer. Begins with the alphabet, moves to simple one-syllable words such as *cat* and *fox*, then on to more difficult one-syllable words such as *horse* and *spring*.

Pictorial Primer. Begins with the alphabet. First lessons have simple three- and four-word sentences with no paragraphing.

First Reader. Follows the Primers at second or third grade. Words usually have the main alphabetic sounds, and few have silent letters. Helps to build fluency in using the phonics principles learned at primer levels.

Second Reader. Begins with fairly easy one- and two-syllable words and progresses to more difficult words. Reading selections on a variety of topics. Can be used to about fifth grade.

Third Reader. For students who have mastered basic reading skills and are fluent in easy reading. This Reader develops more advanced vocabulary and thinking skills. Can be used for two or three years, beginning at about sixth grade.

Fourth Reader. Develops advanced comprehension, requiring students to understand a variety of viewpoints and think about abstract ideas. Readings on themes such as life values, truth, religion, and freedom. Can be used at high school level.

Progressive Speller. Can be used at all grade levels. Begins with phonics rules. Covers spelling difficulty from one-syllable words to very complex words.

Parent-Teacher Guide. Explains how to teach reading and has specific ideas for using the lessons in this series of Readers. Also gives guidelines for helping children grow in all the language arts—spelling, writing, speaking, penmanship, grammar.

Materials for Teaching Phonics and Spelling

Phonics Made Plain by Michael S. Brunner. Flashcards for teaching sounds of the letters and letter combinations which make up the code for reading and writing. Wall chart for classroom reference and teaching of sounds. Instruction for use.

The ABC's and All Their Tricks by Margaret M. Bishop. A comprehensive book of phonics knowledge for teachers. Useful for study, reference, solving students' spelling problems, and answering any phonics question you can think of. Tells how to teach children of all ages who have difficulty with phonics.

A Measuring Scale for Ability in Spelling by Leonard P. Ayres. The classic book from 1915, describes the research which identified one thousand most commonly used words and arranged them into difficulty groups to use for testing children's spelling level and for teaching spelling. Includes complete directions for testing.

Mrs. Silver's Phonics Workbook and ***Mrs. Silver's Phonics Workbook Teacher's Edition*** by Claudine Silver. A lesson on each single letter and on some digraphs such as *ck*, *sh*, and *th*. Ideas enough to spend several days on each lesson and learn it well. Each sound is correlated with a Bible verse and with science, art, and other school subjects. Can be used at kindergarten or first grade level. With the McGuffey series, it should be used before the Primers. Both the student workbook and the teacher's edition are required for this program.

Phonics in Song by Leon V. Metcalf. Book of catchy tunes for teaching each letter of the alphabet and the digraphs *ch*, *sh*, *th*, and *wh*. A sing-along CD is also available.

These materials are available at:

www.mottmedia.com